The Nuetron

" The Partner of The Proton "

Edited by Paul F. Kisak

Contents

Chapter 1

Neutron

This article is about the subatomic particle. For other uses, see Neutron (disambiguation).

The **neutron** is a subatomic particle, symbol n or n0, with no net electric charge and a mass slightly larger than that of a proton. Protons and neutrons, each with mass approximately one atomic mass unit, constitute the nucleus of an atom, and they are collectively referred to as nucleons.[4] Their properties and interactions are described by nuclear physics.

The nucleus consists of Z protons, where Z is called the atomic number, and N neutrons, where N is the neutron number. The atomic number defines the chemical properties of the atom, and the neutron number determines the isotope or nuclide.[5] The terms isotope and nuclide are often used synonymously, but they refer to chemical and nuclear properties, respectively. The atomic mass number, symbol A, equals Z+N. For example, carbon has atomic number 6, and its abundant carbon-12 isotope has 6 neutrons, whereas its rare carbon-13 isotope has 7 neutrons. Some elements occur in nature with only one stable isotope, such as fluorine (see stable nuclide). Other elements occur as many stable isotopes, such as tin with ten stable isotopes. Even though it is not a chemical element, the neutron is included in the table of nuclides.[6]

Within the nucleus, protons and neutrons are bound together through the nuclear force, and neutrons are required for the stability of nuclei. Neutrons are produced copiously in nuclear fission and fusion. They are a primary contributor to the nucleosynthesis of chemical elements within stars through fission, fusion, and neutron capture processes.

The neutron is essential to the production of nuclear power. In the decade after the neutron was discovered in 1932,[7] neutrons were used to effect many different types of nuclear transmutations. With the discovery of nuclear fission in 1938,[8] it was quickly realized that, if a fission event produced neutrons, each of these neutrons might cause further fission events, etc., in a cascade known as a nuclear chain reaction.[5] These events and findings led to the first self-sustaining nuclear reactor (Chicago Pile-1, 1942) and the first nuclear weapon (Trinity, 1945).

Free neutrons, or individual neutrons free of the nucleus, are effectively a form of ionizing radiation, and as such, are a biological hazard, depending upon dose.[5] A small natural "neutron background" flux of free neutrons exists on Earth, caused by cosmic ray muons, and by the natural radioactivity of spontaneously fissionable elements in the Earth's crust.[9] Dedicated neutron sources like neutron generators, research reactors and spallation sources produce free neutrons for use in irradiation and in neutron scattering experiments.

1.1 Description

Neutrons and protons are both nucleons, which are attracted and bound together by the nuclear force to form atomic nuclei. The nucleus of the most common isotope of the hydrogen atom (with the chemical symbol "H") is a lone proton. The nuclei of the heavy hydrogen isotopes deuterium and tritium contain one proton bound to one and two neutrons, respectively. All other types of atomic nuclei are composed of two or more protons and various numbers of neutrons. The most common nuclide of the common chemical element lead, ^{208}Pb has 82 protons and 126 neutrons, for example.

The free neutron has a mass of about 1.675×10^{-27} kg (equivalent to 939.6 MeV/c^2, or 1.0087 u).[3] The neutron has a mean square radius of about 0.8×10^{-15} m, or 0.8 fm,[10] and it is a spin-½ fermion.[11] The neutron has a magnetic moment with a negative value, because its orientation is opposite to the neutron's spin.[12] The neutron's magnetic moment causes its motion to be influenced by magnetic fields. Although the neutron has no net electric charge, it does have a slight distribution of charge within it. With its positive electric charge, the proton is directly influenced by electric fields, whereas the response of the neutron to this force is much weaker.

A free neutron is unstable, decaying to a proton, electron and antineutrino with a mean lifetime of just under 15 minutes (881.5±1.5 s). This radioactive decay, known as beta decay,[13] is possible since the mass of the neutron is slightly greater than the proton. The free proton is stable. Neutrons or protons bound in a nucleus can be stable or unstable, however, depending on the nuclide. Beta decay, in which neutrons decay to protons, or vice versa, is governed by the weak force, and it requires the emission or absorption of electrons and neutrinos, or their antiparticles.

Protons and neutrons behave almost identically under the influence of the nuclear force within the nucleus. The concept of isospin, in which the proton and neutron are viewed as two quantum states of the same particle, is used to model the interactions of nucleons by the nuclear or weak forces. Because of the strength of the nuclear force at short distances, the binding energy of nucleons is more than seven orders of magnitude larger than the electromagnetic energy binding electrons in atoms. Nuclear reactions (such as nuclear fission) therefore have an energy density that is more than ten million times that of chemical reactions. Because of the mass–energy equivalence, nuclear binding energies add or subtract from the mass of nuclei. Ultimately, the ability of the nuclear force to store energy arising from the electromagnetic repulsion of nuclear components is the basis for most of the energy that makes nuclear reactors or bombs possible. In nuclear fission, the absorption of a neutron by a heavy nuclide (e.g., uranium-235) causes the nuclide to become unstable and break into light nuclides and additional neutrons. The positively charged light nuclides then repel, releasing electromagnetic potential energy.

The neutron is classified as a hadron, since it is composed of quarks, and as a baryon, since it is composed of three quarks.[14] The finite size of the neutron and its magnetic moment indicate the neutron is a composite, rather than elementary, particle. The neutron consists of two down quarks with charge $-\frac{1}{3}e$ and one up quark with charge $+\frac{2}{3}e$, although this simple model belies the complexities of the Standard Model for nuclei.[15] The masses of the three quarks sum to only about 12 MeV/c^2, whereas the neutron's mass is about 940 MeV/c^2, for example.[15] Like the proton, the quarks of the neutron are held together by the strong force, mediated by gluons.[16] The nuclear force results from secondary effects of the more fundamental strong force.

1.2 Discovery

Main article: Discovery of the neutron

The story of the discovery of the neutron and its properties is central to the extraordinary developments in atomic physics that occurred in the first half of the 20th century, leading ultimately to the atomic bomb in 1945. In the 1911 Rutherford model, the atom consisted of a small positively charged massive nucleus surrounded by a much larger cloud of negatively charged electrons. In 1920 Rutherford suggested the nucleus consisted of positive protons and neutrally-charged particles, suggested to be a proton and an electron bound in some way.[17] Electrons were assumed to reside within the nucleus because it was known that beta radiation consisted of electrons emitted from the nucleus.[17] Rutherford called these uncharged particles *neutrons*, by the Latin root for *neutralis* (neuter) and the Greek suffix *-on* (a suffix used in the names of subatomic particles, i.e. *electron* and *proton*).[18][19] References to the word *neutron* in connection with the atom can be found in the literature as early as 1899, however.[20]

Throughout the 1920s, physicists assumed that the atomic nucleus was composed of protons and "nuclear electrons"[21][22] but there were obvious problems. It was difficult to reconcile the proton–electron model for nuclei with the Heisenberg uncertainty relation of quantum mechanics.[23][24] The Klein paradox,[25] discovered by Oskar Klein in 1928, presented further quantum mechanical objections to the notion of an electron confined within a nucleus.[23] Observed properties of atoms and molecules were inconsistent with the nuclear spin expected from proton–electron hypothesis. Since both protons and electrons carry an intrinsic spin of ½ $\hbar$, there is no way to arrange an odd number of spins ±½ $\hbar$ to give a spin integer multiple of $\hbar$. Nuclei with integer spin are common, e.g., ^{14}N.

In 1931, Walther Bothe and Herbert Becker found that if alpha particle radiation from polonium fell on beryllium, boron, or lithium, an unusually penetrating radiation was produced. The radiation was not influenced by an electric field, so Bothe and Becker assumed it was gamma radiation.[26][27] The following year Irène Joliot-Curie and Frédéric Joliot in Paris showed that if this "gamma" radiation fell on paraffin, or any other hydrogen-containing compound, it ejected protons of very high energy.[28] Neither Rutherford nor James Chadwick at the Cavendish Laboratory in Cambridge were convinced by the gamma ray interpretation.[21] Chadwick quickly performed a series of experiments that showed that the new radiation consisted of uncharged particles with about the same mass as the proton.[7][29][30] These particles were neutrons. Chadwick won the Nobel Prize in Physics for this discovery in 1935.[2]

Models for atomic nucleus consisting of protons and neutrons were quickly developed by Werner Heisenberg[31][32][33] and others.[34][35] The proton–neutron model explained the puzzle of nuclear spins. The origins of beta radiation were explained by Enrico Fermi in 1934 by the process of beta decay, in which the neutron decays to a proton by *creating* an electron and a (as yet undiscovered) neutrino.[36] In 1935 Chadwick and his doctoral student Maurice Goldhaber, reported the first accurate measurement of the mass of the neutron.[37][38]

By 1934, Fermi had bombarded heavier elements with neutrons to induce radioactivity in elements of high atomic number. In 1938, Fermi received the Nobel Prize in Physics *"for his demonstrations of the existence of new radioactive elements produced by neutron irradiation, and for his related discovery of nuclear reactions brought about by slow neutrons"*.[39] In 1938 Otto Hahn, Lise Meitner, and Fritz Strassmann discovered nuclear fission, or the fractionation of uranium nuclei into light elements, induced by neutron bombardment.[40][41][42] In 1945 Hahn received the 1944 Nobel Prize in Chemistry *"for his discovery of the fission of heavy atomic nuclei."* [43][44][45] The discovery of nuclear fission would lead to the development of nuclear power and the atomic bomb by the end of World War II.

1.3 Beta decay and the stability of the nucleus

Under the Standard Model of particle physics, the only possible decay mode for the neutron that conserves baryon number is for one of the neutron's quarks to change flavour via the weak interaction. The decay of one of the neutron's down quarks into a lighter up quark can be achieved by the emission of a W boson. By this process, the Standard Model description of beta decay, the neutron decays into a proton (which contains one down and two up quarks), an electron, and an electron antineutrino.

Since interacting protons have a mutual electromagnetic repulsion that is stronger than their attractive nuclear interaction, neutrons are a necessary constituent of any atomic nucleus that contains more than one proton (see diproton and neutron–proton ratio).[46] Neutrons bind with protons and one another in the nucleus via the nuclear force, effectively moderating the repulsive forces between the protons and stabilizing the nucleus.

See also: Beta-decay stable isobars and Neutron emission

1.3.1 Free neutron decay

Outside the nucleus, free neutrons are unstable and have a mean lifetime of 881.5±1.5 s (about 14 minutes, 42 seconds); therefore the half-life for this process (which differs from the mean lifetime by a factor of ln(2) = 0.693) is 611.0±1.0 s (about 10 minutes, 11 seconds).[13] Beta decay of the neutron, described above, can be denoted by the radioactive decay:[47]

$$n0 \rightarrow p+ + e- + \nu$$
$$e$$

where p+, e−, and ν
e denote the proton, electron and electron antineutrino, respectively. For the free neutron the decay energy for this process (based on the masses of the neutron, proton, and electron) is 0.782343 MeV. The maximal energy of the beta decay electron (in the process wherein the neutrino receives a vanishingly small amount of kinetic energy) has been measured at 0.782 ± .013 MeV.[48] The latter number is not well-enough measured to determine the comparatively tiny rest mass

of the neutrino (which must in theory be subtracted from the maximal electron kinetic energy) as well as neutrino mass is constrained by many other methods.

A small fraction (about one in 1000) of free neutrons decay with the same products, but add an extra particle in the form of an emitted gamma ray:

n0 → p+ + e− + ν
e + γ

This gamma ray may be thought of as a sort of "internal bremsstrahlung" that arises as the emitted beta particle interacts with the charge of the proton in an electromagnetic way. Internal bremsstrahlung gamma ray production is also a minor feature of beta decays of bound neutrons (as discussed below).

A very small minority of neutron decays (about four per million) are so-called "two-body (neutron) decays", in which a proton, electron and antineutrino are produced as usual, but the electron fails to gain the 13.6 eV necessary energy to escape the proton, and therefore simply remains bound to it, as a neutral hydrogen atom (one of the "two bodies"). In this type of free neutron decay, in essence all of the neutron decay energy is carried off by the antineutrino (the other "body").

The transformation of a free proton to a neutron (plus a positron and a neutrino) is energetically impossible, since a free neutron has a greater mass than a free proton.

1.3.2   Bound neutron decay

Main article: Atomic nucleus

While a free neutron has a half life of about 10.2 min, most neutrons within nuclei are stable. According to the nuclear shell model, the protons and neutrons of a nuclide are a quantum mechanical system organized into discrete energy levels with unique quantum numbers. For a neutron to decay, the resulting proton requires an available state at lower energy than the initial neutron state. In stable nuclei the possible lower energy states are all filled, meaning they are each occupied by two protons with spin up and spin down. The Pauli exclusion principle therefore disallows the decay of a neutron to a proton within stable nuclei. The situation is similar to electrons of an atom, where electrons have distinct atomic orbitals and are prevented from decaying to lower energy states, with the emission of a photon, by the exclusion principle.

Neutrons in unstable nuclei can decay by beta decay as described above. In this case, an energetically allowed quantum state is available for the proton resulting from the decay. One example of this decay is carbon-14 (6 protons, 8 neutrons) that decays to nitrogen-14 (7 protons, 7 neutrons) with a half-life of about 5,730 years.

Inside a nucleus, a proton can transform into a neutron via inverse beta decay, if an energetically allowed quantum state is available for the neutron. This transformation occurs by emission of an antielectron (also called positron) and an electron neutrino:

p+ → n0 + e+ + ν
e

The transformation of a proton to a neutron inside of a nucleus is also possible through electron capture:

p+ + e− → n0 + ν
e

Positron capture by neutrons in nuclei that contain an excess of neutrons is also possible, but is hindered because positrons are repelled by the positive nucleus, and quickly annihilate when they encounter electrons.

1.3.3   Competition of beta decay types

Three types of beta decay in competition are illustrated by the single isotope copper-64 (29 protons, 35 neutrons), which has a half-life of about 12.7 hours. This isotope has one unpaired proton and one unpaired neutron, so either the proton or the neutron can decay. This particular nuclide (though not all nuclides in this situation) is almost equally likely to decay through proton decay by positron emission (18%) or electron capture (43%), as through neutron decay by electron emission (39%).

1.4 Intrinsic properties

1.4.1 Electric charge

The total electric charge of the neutron is $0\ e$. This zero value has been tested experimentally, and the present experimental limit for the charge of the neutron is $-2(8) \times 10^{-22}\ e$,[49] or $-3(13) \times 10^{-41}$ C. This value is consistent with zero, given the experimental uncertainties (indicated in parentheses). By comparison, the charge of the proton is, of course, $+1\ e$.

1.4.2 Electric dipole moment

Main article: Neutron electric dipole moment

The Standard Model of particle physics predicts a tiny separation of positive and negative charge within the neutron leading to a permanent electric dipole moment.[50] The predicted value is, however, well below the current sensitivity of experiments. From several unsolved puzzles in particle physics, it is clear that the Standard Model is not the final and full description of all particles and their interactions. New theories going beyond the Standard Model generally lead to much larger predictions for the electric dipole moment of the neutron. Currently, there are at least four experiments trying to measure for the first time a finite neutron electric dipole moment, including:

- Cryogenic neutron EDM experiment being set up at the Institut Laue–Langevin[51]

- nEDM experiment under construction at the new UCN source at the Paul Scherrer Institute[52]

- nEDM experiment being envisaged at the Spallation Neutron Source[53]

- nEDM experiment being built at the Institut Laue–Langevin[54]

1.4.3 Magnetic moment

Main article: Neutron magnetic moment

Even though the neutron is a neutral particle, the magnetic moment of a neutron is not zero. Since the neutron is a neutral particle, it is not affected by electric fields, but with its magnetic moment it is affected by magnetic fields. The magnetic moment of the neutron is an indication of its quark substructure and internal charge distribution.[55] The value for the neutron's magnetic moment was first directly measured by Luis Alvarez and Felix Bloch at Berkeley, California in 1940,[56] using an extension of the magnetic resonance methods developed by Rabi. Alvarez and Bloch determined the magnetic moment of the neutron to be $\mu_n = -1.93(2)\ \mu N$, where μN is the nuclear magneton.

1.4.4 Structure and geometry of charge distribution

An article published in 2007 featuring a model-independent analysis concluded that the neutron has a negatively charged exterior, a positively charged middle, and a negative core.[57] In a simplified classical view, the negative "skin" of the neutron assists it to be attracted to the protons with which it interacts in the nucleus. (However, the main attraction between neutrons and protons is via the nuclear force, which does not involve charge.)

The simplified classical view of the neutron's charge distribution also "explains" the fact that the neutron magnetic dipole points in the opposite direction from its spin angular momentum vector (as compared to the proton). This gives the neutron, in effect, a magnetic moment which resembles a negatively charged particle. This can be reconciled classically with a neutral neutron composed of a charge distribution in which the negative sub-parts of the neutron have a larger average radius of distribution, and therefore contribute more to the particle's magnetic dipole moment, than do the positive parts that are, on average, nearer the core.

1.4.5   Mass

The mass of a neutron cannot be directly determined by mass spectrometry due to lack of electric charge. However, since the mass of protons and deuterons can be measured by mass spectrometry, the mass of a neutron can be deduced by subtracting proton mass from deuteron mass, with the difference being the mass of the neutron plus the binding energy of deuterium (expressed as a positive emitted energy). The latter can be directly measured by measuring the energy (B_d) of the single 0.7822 MeV gamma photon emitted when neutrons are captured by protons (this is exothermic and happens with zero-energy neutrons), plus the small recoil kinetic energy (E_{rd}) of the deuteron (about 0.06% of the total energy).

$$m_n = m_d - m_p + B_d - E_{rd}$$

The energy of the gamma ray can be measured to high precision by X-ray diffraction techniques, as was first done by Bell and Elliot in 1948. The best modern (1986) values for neutron mass by this technique are provided by Greene, et al.[58] These give a neutron mass of:

m_neutron = 1.008644904(14) u

The value for the neutron mass in MeV is less accurately known, due to less accuracy in the known conversion of u to MeV:[59]

m_neutron = 939.56563(28) MeV/c^2.

Another method to determine the mass of a neutron starts from the beta decay of the neutron, when the momenta of the resulting proton and electron are measured.

1.4.6   Anti-neutron

Main article: Antineutron

The antineutron is the antiparticle of the neutron. It was discovered by Bruce Cork in the year 1956, a year after the antiproton was discovered. CPT-symmetry puts strong constraints on the relative properties of particles and antiparticles, so studying antineutrons yields provide stringent tests on CPT-symmetry. The fractional difference in the masses of the neutron and antineutron is $(9\pm6)\times10^{-5}$. Since the difference is only about two standard deviations away from zero, this does not give any convincing evidence of CPT-violation.[13]

1.5   Neutron compounds

1.5.1   Dineutrons and tetraneutrons

Main articles: Dineutron and Tetraneutron

The existence of stable clusters of 4 neutrons, or tetraneutrons, has been hypothesised by a team led by Francisco-Miguel Marqués at the CNRS Laboratory for Nuclear Physics based on observations of the disintegration of beryllium-14 nuclei. This is particularly interesting because current theory suggests that these clusters should not be stable.

The dineutron is another hypothetical particle. In 2012, Artemis Spyrou from Michigan State University and coworkers reported that they observed, for the first time, the dineutron emission in the decay of ^{16}Be. The dineutron character is evidenced by a small emission angle between the two neutrons. The authors measured the two-neutron separation energy to be 1.35(10) MeV, in good agreement with shell model calculations, using standard interactions for this mass region.[60]

1.5.2 Neutronium and neutron stars

Main articles: Neutronium and Neutron star

At extremely high pressures and temperatures, nucleons and electrons are believed to collapse into bulk neutronic matter, called neutronium. This is presumed to happen in neutron stars.

The extreme pressure inside a neutron star may deform the neutrons into a cubic symmetry, allowing tighter packing of neutrons.[61]

1.6 Detection

Main article: Neutron detection

The common means of detecting a charged particle by looking for a track of ionization (such as in a cloud chamber) does not work for neutrons directly. Neutrons that elastically scatter off atoms can create an ionization track that is detectable, but the experiments are not as simple to carry out; other means for detecting neutrons, consisting of allowing them to interact with atomic nuclei, are more commonly used. The commonly used methods to detect neutrons can therefore be categorized according to the nuclear processes relied upon, mainly neutron capture or elastic scattering. A good discussion on neutron detection is found in chapter 14 of the book *Radiation Detection and Measurement* by Glenn F. Knoll (John Wiley & Sons, 1979).

1.6.1 Neutron detection by neutron capture

A common method for detecting neutrons involves converting the energy released from neutron capture reactions into electrical signals. Certain nuclides have a high neutron capture cross section, which is the probability of absorbing a neutron. Upon neutron capture, the compound nucleus emits more easily detectable radiation, for example an alpha particle, which is then detected. The nuclides 3He, 6Li, 10B, 233U, 235U, 237Np and 239Pu are useful for this purpose.

1.6.2 Neutron detection by elastic scattering

Neutrons can elastically scatter off nuclei, causing the struck nucleus to recoil. Kinematically, a neutron can transfer more energy to light nuclei such as hydrogen or helium than to heavier nuclei. Detectors relying on elastic scattering are called fast neutron detectors. Recoiling nuclei can ionize and excite further atoms through collisions. Charge and/or scintillation light produced in this way can be collected to produce a detected signal. A major challenge in fast neutron detection is discerning such signals from erroneous signals produced by gamma radiation in the same detector.

Fast neutron detectors have the advantage of not requiring a moderator, and therefore being capable of measuring the neutron's energy, time of arrival, and in certain cases direction of incidence.

1.7 Sources and production

Main articles: Neutron source, neutron generator and research reactor

Free neutrons are unstable, although they have the longest half-life of any unstable sub-atomic particle by several orders of magnitude. Their half-life is still only about 10 minutes, however, so they can be obtained only from sources that produce them freshly.

Natural neutron background. A small natural background flux of free neutrons exists everywhere on Earth. In the atmosphere and deep into the ocean, the "neutron background" is caused by muons produced by cosmic ray interaction with the atmosphere. These high energy muons are capable of penetration to considerable depths in water and soil. There, in striking atomic nuclei, among other reactions they induce spallation reactions in which a neutron is liberated from the nucleus. Within the Earth's crust a second source is neutrons produced primarily by spontaneous fission of uranium and thorium present in crustal minerals. The neutron background is not strong enough to be a biological hazard, but it is of importance to very high resolution particle detectors that are looking for very rare events, such as (hypothesized) interactions that might be caused by particles of dark matter.[9] Recent research has shown that even thunderstorms can produce neutrons with energies of up to several tens of MeV.[62]

Even stronger neutron background radiation is produced at the surface of Mars, where the atmosphere is thick enough to generate neutrons from cosmic ray muon production and neutron-spallation, but not thick enough to provide significant protection from the neutrons produced. These neutrons not only produce a Martian surface neutron radiation hazard from direct downward-going neutron radiation but may also produce a significant hazard from reflection of neutrons from the Martian surface, which will produce reflected neutron radiation penetrating upward into a Martian craft or habitat from the floor.[63]

Sources of neutrons for research. These include certain types of radioactive decay (spontaneous fission and neutron emission), and from certain nuclear reactions. Convenient nuclear reactions include tabletop reactions such as natural alpha and gamma bombardment of certain nuclides, often beryllium or deuterium, and induced nuclear fission, such as occurs in nuclear reactors. In addition, high-energy nuclear reactions (such as occur in cosmic radiation showers or accelerator collisions) also produce neutrons from disintigration of target nuclei. Small (tabletop) particle accelerators optimized to produce free neutrons in this way, are called neutron generators.

In practice, the most commonly used small laboratory sources of neutrons use radioactive decay to power neutron production. One noted neutron-producing radioisotope, californium−252 decays (half-life 2.65 years) by spontaneous fission 3% of the time with production of 3.7 neutrons per fission, and is used alone as a neutron source from this process. Nuclear reaction sources (that involve two materials) powered by radioisotopes use an alpha decay source plus a beryllium target, or else a source of high-energy gamma radiation from a source that undergoes beta decay followed by gamma decay, which produces photoneutrons on interaction of the high energy gamma ray with ordinary stable beryllium, or else with the deuterium in heavy water. A popular source of the latter type is radioactive antimony-124 plus beryllium, a system with a half-life of 60.9 days, which can be constructed from natural antimony (which is 42.8% stable antimony-123) by activating it with neutrons in a nuclear reactor, then transported to where the neutron source is needed.[64]

Nuclear fission reactors naturally produce free neutrons; their role is to sustain the energy-producing chain reaction. The intense neutron radiation can also be used to produce various radioisotopes through the process of neutron activation, which is a type of neutron capture.

Experimental nuclear fusion reactors produce free neutrons as a waste product. However, it is these neutrons that possess most of the energy, and converting that energy to a useful form has proved a difficult engineering challenge. Fusion reactors that generate neutrons are likely to create radioactive waste, but the waste is composed of neutron-activated lighter isotopes, which have relatively short (50–100 years) decay periods as compared to typical half-lives of 10,000 years for fission waste, which is long primarily to the long half-life of alpha-emitting transuranic actinides.[65]

1.7.1 Neutron beams and modification of beams after production

Free neutron beams are obtained from neutron sources by neutron transport. For access to intense neutron sources, researchers must go to a specialist neutron facility that operates a research reactor or a spallation source.

The neutron's lack of total electric charge makes it difficult to steer or accelerate them. Charged particles can be accelerated, decelerated, or deflected by electric or magnetic fields. These methods have little effect on neutrons. However, some effects may be attained by use of inhomogeneous magnetic fields because of the neutron's magnetic moment. Neutrons can be controlled by methods that include moderation, reflection, and velocity selection. Thermal neutrons can be polarized by transmission through magnetic materials in a method analogous to the Faraday effect for photons. Cold neutrons of wavelengths of 6–7 angstroms can be produced in beams of a high degree of polarization, by use of magnetic mirrors and magnetized interference filters.[66]

1.8 Applications

The neutron plays an important role in many nuclear reactions. For example, neutron capture often results in neutron activation, inducing radioactivity. In particular, knowledge of neutrons and their behavior has been important in the development of nuclear reactors and nuclear weapons. The fissioning of elements like uranium-235 and plutonium-239 is caused by their absorption of neutrons.

Cold, *thermal* and *hot* neutron radiation is commonly employed in neutron scattering facilities, where the radiation is used in a similar way one uses X-rays for the analysis of condensed matter. Neutrons are complementary to the latter in terms of atomic contrasts by different scattering cross sections; sensitivity to magnetism; energy range for inelastic neutron spectroscopy; and deep penetration into matter.

The development of "neutron lenses" based on total internal reflection within hollow glass capillary tubes or by reflection from dimpled aluminum plates has driven ongoing research into neutron microscopy and neutron/gamma ray tomography.[67][68][69]

A major use of neutrons is to excite delayed and prompt gamma rays from elements in materials. This forms the basis of neutron activation analysis (NAA) and prompt gamma neutron activation analysis (PGNAA). NAA is most often used to analyze small samples of materials in a nuclear reactor whilst PGNAA is most often used to analyze subterranean rocks around bore holes and industrial bulk materials on conveyor belts.

Another use of neutron emitters is the detection of light nuclei, in particular the hydrogen found in water molecules. When a fast neutron collides with a light nucleus, it loses a large fraction of its energy. By measuring the rate at which slow neutrons return to the probe after reflecting off of hydrogen nuclei, a neutron probe may determine the water content in soil.

1.9 Medical therapies

Main articles: Fast neutron therapy and Neutron capture therapy of cancer

Because neutron radiation is both penetrating and ionizing, it can be exploited for medical treatments. Neutron radiation can have the unfortunate side-effect of leaving the affected area radioactive, however. Neutron tomography is therefore not a viable medical application.

Fast neutron therapy utilizes high energy neutrons typically greater than 20 MeV to treat cancer. Radiation therapy of cancers is based upon the biological response of cells to ionizing radiation. If radiation is delivered in small sessions to damage cancerous areas, normal tissue will have time to repair itself, while tumor cells often cannot.[70] Neutron radiation can deliver energy to a cancerous region at a rate an order of magnitude larger than gamma radiation[71]

Beams of low energy neutrons are used in boron capture therapy to treat cancer. In boron capture therapy, the patient is given a drug that contains boron and that preferentially accumulates in the tumor to be targeted. The tumor is then bombarded with very low energy neutrons (although often higher than thermal energy) which are captured by the boron-10 isotope in the boron, which produces an excited state of boron-11 that then decays to produce lithium-7 and an alpha particle that have sufficient energy to kill the malignant cell, but insufficient range to damage nearby cells. For such a therapy to be applied to the treatment of cancer, a neutron source having an intensity of the order of billion (10^9) neutrons per second per cm^2 is preferred. Such fluxes require a research nuclear reactor.

1.10   Protection

Exposure to free neutrons can be hazardous, since the interaction of neutrons with molecules in the body can cause disruption to molecules and atoms, and can also cause reactions that give rise to other forms of radiation (such as protons). The normal precautions of radiation protection apply: Avoid exposure, stay as far from the source as possible, and keep exposure time to a minimum. Some particular thought must be given to how to protect from neutron exposure, however. For other types of radiation, e.g. alpha particles, beta particles, or gamma rays, material of a high atomic number and with high density make for good shielding; frequently, lead is used. However, this approach will not work with neutrons, since the absorption of neutrons does not increase straightforwardly with atomic number, as it does with alpha, beta, and gamma radiation. Instead one needs to look at the particular interactions neutrons have with matter (see the section on detection above). For example, hydrogen-rich materials are often used to shield against neutrons, since ordinary hydrogen both scatters and slows neutrons. This often means that simple concrete blocks or even paraffin-loaded plastic blocks afford better protection from neutrons than do far more dense materials. After slowing, neutrons may then be absorbed with an isotope that has high affinity for slow neutrons without causing secondary capture radiation, such as lithium-6.

Hydrogen-rich ordinary water affects neutron absorption in nuclear fission reactors: Usually, neutrons are so strongly absorbed by normal water that fuel enrichment with fissionable isotope is required. The deuterium in heavy water has a very much lower absorption affinity for neutrons than does protium (normal light hydrogen). Deuterium is, therefore, used in CANDU-type reactors, in order to slow (moderate) neutron velocity, to increase the probability of nuclear fission compared to neutron capture.

1.11   Neutron temperature

Main article: Neutron temperature

1.11.1   Thermal neutrons

A *thermal neutron* is a free neutron that is Boltzmann distributed with kT = 0.0253 eV (4.0×10^{-21} J) at room temperature. This gives characteristic (not average, or median) speed of 2.2 km/s. The name 'thermal' comes from their energy being that of the room temperature gas or material they are permeating. (see *kinetic theory* for energies and speeds of molecules). After a number of collisions (often in the range of 10–20) with nuclei, neutrons arrive at this energy level, provided that they are not absorbed.

In many substances, thermal neutron reactions show a much larger effective cross-section than reactions involving faster neutrons, and thermal neutrons can therefore be absorbed more readily (i.e., with higher probability) by any atomic nuclei that they collide with, creating a heavier — and often unstable — isotope of the chemical element as a result.

Most fission reactors use a neutron moderator to slow down, or *thermalize* the neutrons that are emitted by nuclear fission so that they are more easily captured, causing further fission. Others, called fast breeder reactors, use fission energy neutrons directly.

1.11.2   Cold neutrons

Cold neutrons are thermal neutrons that have been equilibrated in a very cold substance such as liquid deuterium. Such a *cold source* is placed in the moderator of a research reactor or spallation source. Cold neutrons are particularly valuable for neutron scattering experiments.

1.11.3   Ultracold neutrons

Ultracold neutrons are produced by inelastically scattering cold neutrons in substances with a temperature of a few kelvins, such as solid deuterium or superfluid helium. An alternative production method is the mechanical deceleration of cold neutrons.

1.11.4   Fission energy neutrons

Main article: nuclear fission

A *fast neutron* is a free neutron with a kinetic energy level close to 1 MeV (1.6×10^{-13} J), hence a speed of ~14000 km/s (~ 5% of the speed of light). They are named *fission energy* or *fast* neutrons to distinguish them from lower-energy thermal neutrons, and high-energy neutrons produced in cosmic showers or accelerators. Fast neutrons are produced by nuclear processes such as nuclear fission. Neutrons produced in fission, as noted above, have a Maxwell–Boltzmann distribution of kinetic energies from 0 to ~14 MeV, a mean energy of 2 MeV (for U-235 fission neutrons), and a mode of only 0.75 MeV, which means that more than half of them do not qualify as fast (and thus have almost no chance of initiating fission in fertile materials, such as U-238 and Th-232).

Fast neutrons can be made into thermal neutrons via a process called moderation. This is done with a neutron moderator. In reactors, typically heavy water, light water, or graphite are used to moderate neutrons.

1.11.5   Fusion neutrons

For more details on this topic, see Nuclear fusion § Criteria and candidates for terrestrial reactions.

D–T (deuterium–tritium) fusion is the fusion reaction that produces the most energetic neutrons, with 14.1 MeV of kinetic energy and traveling at 17% of the speed of light. D–T fusion is also the easiest fusion reaction to ignite, reaching near-peak rates even when the deuterium and tritium nuclei have only a thousandth as much kinetic energy as the 14.1 MeV that will be produced.

14.1 MeV neutrons have about 10 times as much energy as fission neutrons, and are very effective at fissioning even non-fissile heavy nuclei, and these high-energy fissions produce more neutrons on average than fissions by lower-energy neutrons. This makes D–T fusion neutron sources such as proposed tokamak power reactors useful for transmutation of transuranic waste. 14.1 MeV neutrons can also produce neutrons by knocking them loose from nuclei.

On the other hand, these very high energy neutrons are less likely to simply be captured without causing fission or spallation. For these reasons, nuclear weapon design extensively utilizes D–T fusion 14.1 MeV neutrons to cause more fission. Fusion neutrons are able to cause fission in ordinarily non-fissile materials, such as depleted uranium (uranium-238), and these materials have been used in the jackets of thermonuclear weapons. Fusion neutrons also can cause fission in substances that are unsuitable or difficult to make into primary fission bombs, such as reactor grade plutonium. This physical fact thus causes ordinary non-weapons grade materials to become of concern in certain nuclear proliferation discussions and treaties.

Other fusion reactions produce much less energetic neutrons. D–D fusion produces a 2.45 MeV neutron and helium-3 half of the time, and produces tritium and a proton but no neutron the other half of the time. D–^{3}He fusion produces no neutron.

1.11.6   Intermediate-energy neutrons

A fission energy neutron that has slowed down but not yet reached thermal energies is called an epithermal neutron.

Cross sections for both capture and fission reactions often have multiple resonance peaks at specific energies in the epithermal energy range. These are of less significance in a fast neutron reactor, where most neutrons are absorbed before slowing down to this range, or in a well-moderated thermal reactor, where epithermal neutrons interact mostly with moderator nuclei, not with either fissile or fertile actinide nuclides. However, in a partially moderated reactor with more interactions of epithermal neutrons with heavy metal nuclei, there are greater possibilities for transient changes in reactivity that might make reactor control more difficult.

Ratios of capture reactions to fission reactions are also worse (more captures without fission) in most nuclear fuels such as plutonium-239, making epithermal-spectrum reactors using these fuels less desirable, as captures not only waste the one neutron captured but also usually result in a nuclide that is not fissile with thermal or epithermal neutrons, though still fissionable with fast neutrons. The exception is uranium-233 of the thorium cycle, which has good capture-fission ratios at all neutron energies.

1.11.7 High-energy neutrons

These neutrons have much more energy than fission energy neutrons and are generated as secondary particles by particle accelerators or in the atmosphere from cosmic rays. They can have energies as high as tens of joules per neutron. These neutrons are extremely efficient at ionization and far more likely to cause cell death than X-rays or protons.[72][73]

1.12 See also

- Ionizing radiation
- Isotope
- List of particles
- Neutronium
- Neutron magnetic moment
- Neutron radiation and the Sievert radiation scale
- Nuclear reaction
- Thermal reactor
- Nucleosynthesis
  - Neutron capture nucleosynthesis
  - R-process
  - S-process

1.12.1 Neutron sources

- Neutron generator
- Neutron sources

1.12.2 Processes involving neutrons

- Neutron bomb
- Neutron diffraction
- Neutron flux
- Neutron transport

1.13 References

[1] Ernest Rutherford. Chemed.chem.purdue.edu. Retrieved on 2012-08-16.

[2] 1935 Nobel Prize in Physics. Nobelprize.org. Retrieved on 2012-08-16.

[3] Mohr, P.J.; Taylor, B.N. and Newell, D.B. (2011), "The 2010 CODATA Recommended Values of the Fundamental Physical Constants" (Web Version 6.0). The database was developed by J. Baker, M. Douma, and S. Kotochigova. (2011-06-02). National Institute of Standards and Technology, Gaithersburg, Maryland 20899.

[4] Thomas, A.W.; Weise, W. (2001), *The Structure of the Nucleon*, Wiley-WCH, Berlin, ISBN 3-527-40297-7

[5] Glasstone, Samuel; Dolan, Philip J., eds. (1977), *The Effects of Nuclear Weapons, Third Edition*, U.S. Dept. of Defense and Energy Research and Development Administration, U.S. Government Printing Office, ISBN 1-60322-016-X

[6] Nudat 2. Nndc.bnl.gov. Retrieved on 2010-12-04.

[7] Chadwick, James (1932). "Possible Existence of a Neutron". *Nature* **129** (3252): 312. Bibcode:1932Natur.129Q.312C. doi:10.1038/129312a0.

[8] O. Hahn and F. Strassmann (1939). "Über den Nachweis und das Verhalten der bei der Bestrahlung des Urans mittels Neutronen entstehenden Erdalkalimetalle ("On the detection and characteristics of the alkaline earth metals formed by irradiation of uranium with neutrons")". *Naturwissenschaften* **27** (1): 11–15. Bibcode:1939NW.....27...11H. doi:10.1007/BF01488241.. The authors were identified as being at the Kaiser-Wilhelm-Institut für Chemie, Berlin-Dahlem. Received 22 December 1938.

[9] M. J. Carson et al. (2004). "Neutron background in large-scale xenon detectors for dark matter searches". *Astroparticle Physics* **21** (6): 667–687. doi:10.1016/j.astropartphys.2004.05.001.

[10] Povh, B.; Rith, K.; Scholz, C.; Zetsche, F. (2002). *Particles and Nuclei: An Introduction to the Physical Concepts.* Berlin: Springer-Verlag. p. 73. ISBN 978-3-540-43823-6.

[11] J.-L. Basdevant, J. Rich, M. Spiro (2005). *Fundamentals in Nuclear Physics.* Springer. p. 155. ISBN 0-387-01672-4.

[12] Paul Allen Tipler, Ralph A. Llewellyn (2002). *Modern Physics* (4 ed.). Macmillan. p. 310. ISBN 0-7167-4345-0.

[13] Nakamura, K (2010). "Review of Particle Physics". *Journal of Physics G: Nuclear and Particle Physics* **37** (7A): 075021. Bibcode:2010JPhG...37g5021N. doi:10.1088/0954-3899/37/7A/075021. PDF with 2011 partial update for the 2012 edition The exact value of the mean lifetime is still uncertain, due to conflicting results from experiments. The Particle Data Group reports values up to six seconds apart (more than four standard deviations), commenting that "our 2006, 2008, and 2010 Reviews stayed with 885.7±0.8 s; but we noted that in light of SEREBROV 05 our value should be regarded as suspect until further experiments clarified matters. Since our 2010 Review, PICHLMAIER 10 has obtained a mean life of 880.7±1.8 s, closer to the value of SEREBROV 05 than to our average. And SEREBROV 10B[...] claims their values should be lowered by about 6 s, which would bring them into line with the two lower values. However, those reevaluations have not received an enthusiastic response from the experimenters in question; and in any case the Particle Data Group would have to await published changes (by those experimenters) of published values. At this point, we can think of nothing better to do than to average the seven best but discordant measurements, getting 881.5±1.5s. Note that the error includes a scale factor of 2.7. This is a jump of 4.2 old (and 2.8 new) standard deviations. This state of affairs is a particularly unhappy one, because the value is so important. We again call upon the experimenters to clear this up."

[14] R.K. Adair (1989). *The Great Design: Particles, Fields, and Creation.* Oxford University Press. p. 214.

[15] Cho, Adiran (2 April 2010). "Mass of the Common Quark Finally Nailed Down". *http://news.sciencemag.org".* American Association for the Advancement of Science. *Retrieved 27 September 2014.*

[16] W.N.Cottingham,D.A.Greenwood(1986).*An Introduction to Nuclear Physics.*Cambridge University Press.ISBN9780521657.

[17] E.Rutherford(1920). "Nuclear Constitution of Atoms".*Proceedings of the Royal Society A***97**(686):374–400.Bibcode:1920. doi:10.1098/rspa.1920.0040.

[18] "Wolfgang Pauli". *Sources in the History of Mathematics and Physical Sciences.* Sources in the History of Mathematics and Physical Sciences **6**: 105–144. 1985. doi:10.1007/978-3-540-78801-0_3. ISBN 978-3-540-13609-5. lchapter= ignored (help)

[19] Hendry, John, ed. (1984), *Cambridge Physics in the Thirties*, Adam Hilger Ltd, Bristol, ISBN 0852747616

[20] N.Feather(1960)."A history of neutrons and nuclei.Part1".*Contemporary Physics***1**(3):191–203.doi:10.1080/00107516008.

[21] Brown,Laurie M. (1978). "The idea of the neutrino".*Physics Today***31**(9):23.Bibcode:1978PhT....31i..23B.doi:10.1063/1.

[22] Friedlander G., Kennedy J.W. and Miller J.M. (1964) *Nuclear and Radiochemistry* (2nd edition), Wiley, pp. 22–23 and 38–39

[23] Stuewer, Roger H. (1985). "Niels Bohr and Nuclear Physics". In French, A. P.; Kennedy, P. J. *Niels Bohr: A Centenary Volume*. Harvard University Press. pp. 197–220. ISBN 0674624165.

[24] Pais, Abraham (1986). *Inward Bound*. Oxford: Oxford University Press. p. 299. ISBN 0198519974.

[25] Klein, O. (1929). "Die Reflexion von Elektronen an einem Potentialsprung nach der relativistischen Dynamik von Dirac". *Zeitschrift für Physik* **53** (3–4): 157–165. Bibcode:1929ZPhy...53..157K. doi:10.1007/BF01339716.

[26] Bothe, W.; Becker, H. (1930). "Künstliche Erregung von Kern-γ-Strahlen" [Artificial excitation of nuclear γ-radiation]. *Zeitschrift für Physik* **66** (5–6): 289–306. Bibcode:1930ZPhy...66..289B. doi:10.1007/BF01390908.

[27] Becker, H.; Bothe, W. (1932). "Die in Bor und Beryllium erregten γ-Strahlen" [Γ-rays excited in boron and beryllium]. *Zeitschrift für Physik* **76** (7–8): 421–438. Bibcode:1932ZPhy...76..421B. doi:10.1007/BF01336726.

[28] Joliot-Curie, Irène and Joliot, Frédéric (1932). "Émission de protons de grande vitesse par les substances hydrogénées sous l'influence des rayons γ très pénétrants" [Emission of high-speed protons by hydrogenated substances under the influence of very penetrating γ-rays]. *Comptes Rendus* **194**: 273.

[29] "Atop the Physics Wave: Rutherford Back in Cambridge, 1919–1937". *Rutherford's Nuclear World*. American Institute of Physics. 2011–2014. Retrieved 19 August 2014.

[30] Chadwick, J. (1933). "Bakerian Lecture. The Neutron". *Proceedings of the Royal Society A: Mathematical, Physical and Engineering Sciences* **142** (846): 1–25. Bibcode:1933RSPSA.142....1C. doi:10.1098/rspa.1933.0152.

[31] Heisenberg, W. (1932). "Über den Bau der Atomkerne. I". *Z. Phys.* **77**: 1–11. doi:10.1007/BF01342433.

[32] Heisenberg, W. (1932). "Über den Bau der Atomkerne. II". *Z. Phys.* **78** (3–4): 156–164. doi:10.1007/BF01337585.

[33] Heisenberg, W. (1933). "Über den Bau der Atomkerne. III". *Z. Phys.* **80** (9–10): 587–596. doi:10.1007/BF01335696.

[34] Iwanenko, D.D., The neutron hypothesis, Nature **129** (1932) 798.

[35] Miller A. I. *Early Quantum Electrodynamics: A Sourcebook*, Cambridge University Press, Cambridge, 1995, ISBN 0521568919, pp. 84–88.

[36] Wilson, Fred L. (1968). "Fermi's Theory of Beta Decay". *Am. J. Phys.* **36** (12): 1150–1160. Bibcode:1968AmJPh..36.1150W. doi:10.1119/1.1974382.

[37] Chadwick, J.; Goldhaber, M. (1934). "A nuclear photo-effect: disintegration of the diplon by gamma rays". *Nature* **134**: 237–238. doi:10.1038/134237a0.

[38] Chadwick,J.;Goldhaber,M. (1935)."A nuclear photoelectric effect"(PDF).*Proc.R.Soc.Lond***151**:479–493.doi:10.1098/rspa.

[39] Cooper, Dan (1999). *Enrico Fermi: And the Revolutions in Modern physics*. New York: Oxford University Press. ISBN 0-19-511762-X. OCLC 39508200.

[40] Hahn, O. (1958). "The Discovery of Fission". *Scientific American* **198** (2): 76. doi:10.1038/scientificamerican0258-76.

[41] Rife, Patricia (1999). *Lise Meitner and the dawn of the nuclear age*. Basel, Switzerland: Birkhäuser. ISBN 0-8176-3732-X.

[42] Hahn, O.; Strassmann, F. (10 February 1939). "Proof of the Formation of Active Isotopes of Barium from Uranium and Thorium Irradiated with Neutrons; Proof of the Existence of More Active Fragments Produced by Uranium Fission". *Die Naturwissenschaften* **27**: 89–95.

[43] "The Nobel Prize in Chemistry 1944". Nobel Foundation. Retrieved 2007-12-17.

[44] Bernstein, Jeremy (2001). *Hitler's uranium club: the secret recordings at Farm Hall*. New York: Copernicus. p. 281. ISBN 0-387-95089-3.

[45] "The Nobel Prize in Chemistry 1944: Presentation Speech". Nobel Foundation. Retrieved 2008-01-03.

[46] Sir James Chadwick's Discovery of Neutrons. ANS Nuclear Cafe. Retrieved on 2012-08-16.

[47] Particle Data Group Summary Data Table on Baryons. lbl.gov (2007). Retrieved on 2012-08-16.

[48] Basic Ideas and Concepts in Nuclear Physics: An Introductory Approach, Third Edition K. Heyde Taylor & Francis 2004. Print ISBN 978-0-7503-0980-6. eBook ISBN 978-1-4200-5494-1. DOI: 10.1201/9781420054941.ch5. full text

[49] Olive, K.A. et al. (2014). "Review of Particle Physics". *Chin. Phys. C* **38**: 090001. doi:10.1088/1674-1137/38/9/090001. |first2= missing |last2= in Authors list (help)

[50] "Pear-shaped particles probe big-bang mystery" (Press release). University of Sussex. 20 February 2006. Retrieved 2009-12-14.

[51] A cryogenic experiment to search for the EDM of the neutron. Hepwww.rl.ac.uk. Retrieved on 2012-08-16.

[52] Search for the neutron electric dipole moment: nEDM. Nedm.web.psi.ch (2001-09-12). Retrieved on 2012-08-16.

[53] SNS Neutron EDM Experiment. P25ext.lanl.gov. Retrieved on 2012-08-16.

[54] Measurement of the Neutron Electric Dipole Moment. Nrd.pnpi.spb.ru. Retrieved on 2012-08-16.

[55] Gell, Y.; Lichtenberg, D. B. (1969). "Quark model and the magnetic moments of proton and neutron". *Il Nuovo Cimento A*. Series 10 **61**: 27–40. Bibcode:1969NCimA..61...27G. doi:10.1007/BF02760010.

[56] Alvarez, L. W; Bloch, F. (1940). "A quantitative determination of the neutron magnetic moment in absolute nuclear magnetons". *Physical Review* **57**: 111–122. doi:10.1103/physrev.57.111.

[57] Miller,G.A. (2007). "Charge Densities of the Neutron and Proton".*Physical Review Letters***99**(11):112001.Bibcode:2007Ph. doi:10.1103/PhysRevLett.99.112001.

[58] Greene, GL et al. (1986). "New determination of the deuteron binding energy and the neutron mass". *Phys. Rev. Lett.* **56**: 819–822. Bibcode:1986PhRvL..56..819G. doi:10.1103/PhysRevLett.56.819.

[59] Byrne, J. *Neutrons, Nuclei, and Matter*, Dover Publications, Mineola, New York, 2011, ISBN 0486482383, pp. 18–19

[60] Spyrou, A. et al. (2012). "First Observation of Ground State Dineutron Decay: 16Be". *Physical Review Letters* **108** (10): 102501. Bibcode:2012PhRvL.108j2501S. doi:10.1103/PhysRevLett.108.102501. PMID 22463404.

[61] Llanes-Estrada, Felipe J.; Moreno Navarro, Gaspar (2011). "Cubic neutrons". arXiv:1108.1859v1 [nucl-th].

[62] Köhn, C., Ebert, U., Calculation of beams of positrons, neutrons and protons associated with terrestrial gamma-ray flashes, J. Geophys. Res. Atmos. (2015), vol. 23, doi:10.1002/2014JD022229

[63] Clowdsley, MS; Wilson, JW; Kim, MH; Singleterry, RC; Tripathi, RK; Heinbockel, JH; Badavi, FF; Shinn, JL (2001). "Neutron Environments on the Martian Surface" (PDF). *Physica Medica* **17** (Suppl 1): 94–6. PMID 11770546.

[64] Byrne, J. *Neutrons, Nuclei, and Matter*, Dover Publications, Mineola, New York, 2011, ISBN 0486482383, pp. 32–33.

[65] Science/Nature | Q&A: Nuclear fusion reactor. BBC News (2006-02-06). Retrieved on 2010-12-04.

[66] Byrne, J. *Neutrons, Nuclei, and Matter*, Dover Publications, Mineola, New York, 2011, ISBN 0486482383, p. 453.

[67] Kumakhov, M. A.; Sharov, V. A. (1992). "A neutron lens". *Nature* **357** (6377): 390–391. Bibcode:1992Natur.357..390K. doi:10.1038/357390a0.

[68] Physorg.com, "New Way of 'Seeing': A 'Neutron Microscope'". Physorg.com (2004-07-30). Retrieved on 2012-08-16.

[69] "NASA Develops a Nugget to Search for Life in Space". NASA.gov (2007-11-30). Retrieved on 2012-08-16.

[70] Hall EJ. Radiobiology for the Radiologist. Lippincott Williams & Wilkins; 5th edition (2000)

[71] Johns HE and Cunningham JR. The Physics of Radiology. Charles C Thomas 3rd edition 1978

[72] Tami Freeman (May 23, 2008). "Facing up to secondary neutrons". Medical Physics Web. Retrieved 2011-02-08.

[73] Heilbronn, L.; Nakamura, T; Iwata, Y; Kurosawa, T; Iwase, H; Townsend, LW (2005). "Expand+Overview of secondary neutron production relevant to shielding in space". *Radiation Protection Dosimetry* **116** (1–4): 140–143. doi:10.1093/rpd/nci033. PMID 16604615.

1.14   Further reading

- Annotated bibliography for neutrons from the Alsos Digital Library for Nuclear Issues

- Abraham Pais, *Inward Bound*, Oxford: Oxford University Press, 1986. ISBN 0198519974.

- Sin-Itiro Tomonaga, *The Story of Spin*, The University of Chicago Press, 1997

- Herwig Schopper, *Weak interactions and nuclear beta decay*, Publisher, North-Holland Pub. Co., 1966.

1.15   External links

- neutron properties at Particle Data Group, Lawrence Berkeley National Laboratory in Berkeley, CA. (pdgLive)

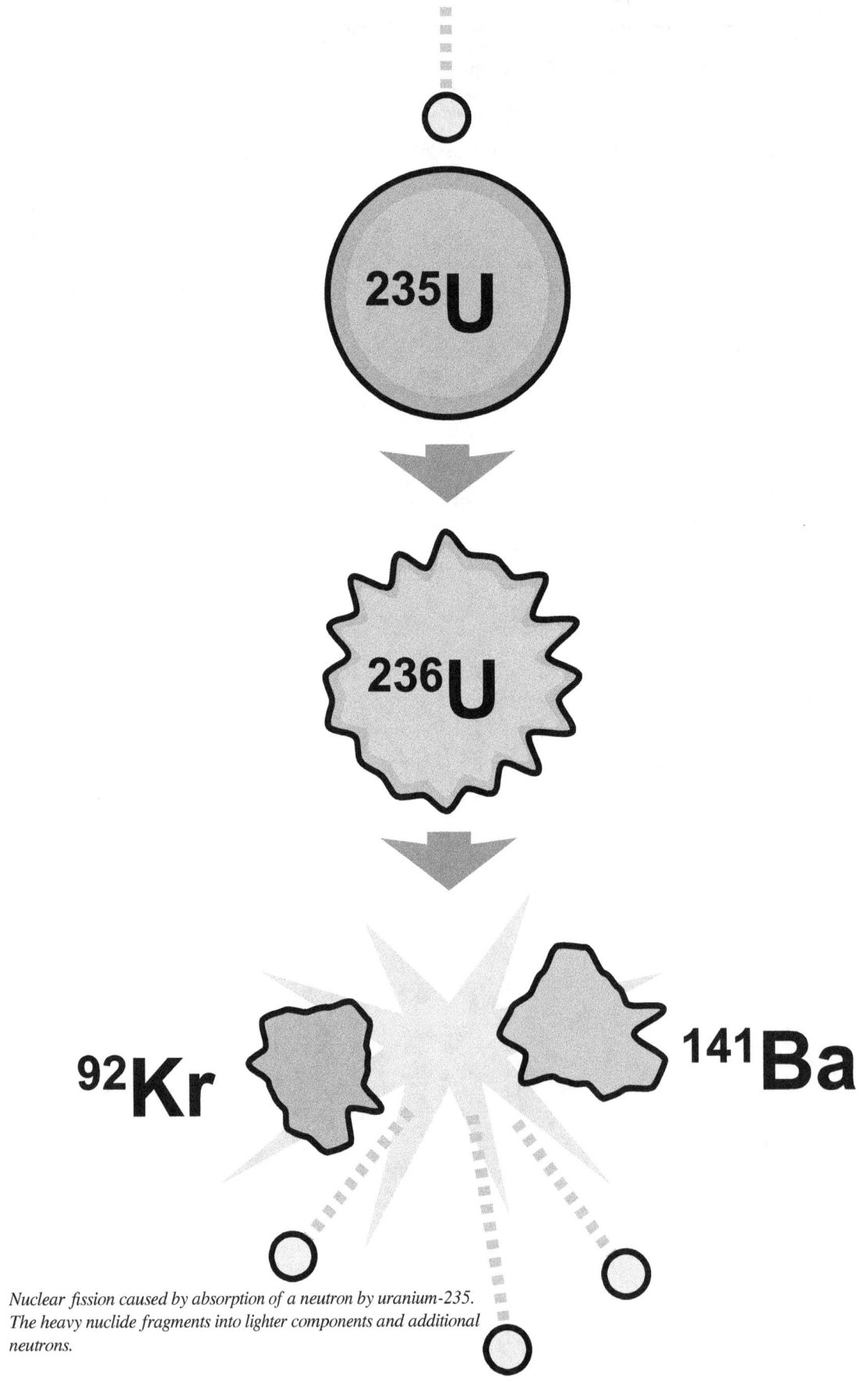

Nuclear fission caused by absorption of a neutron by uranium-235.
The heavy nuclide fragments into lighter components and additional
neutrons.

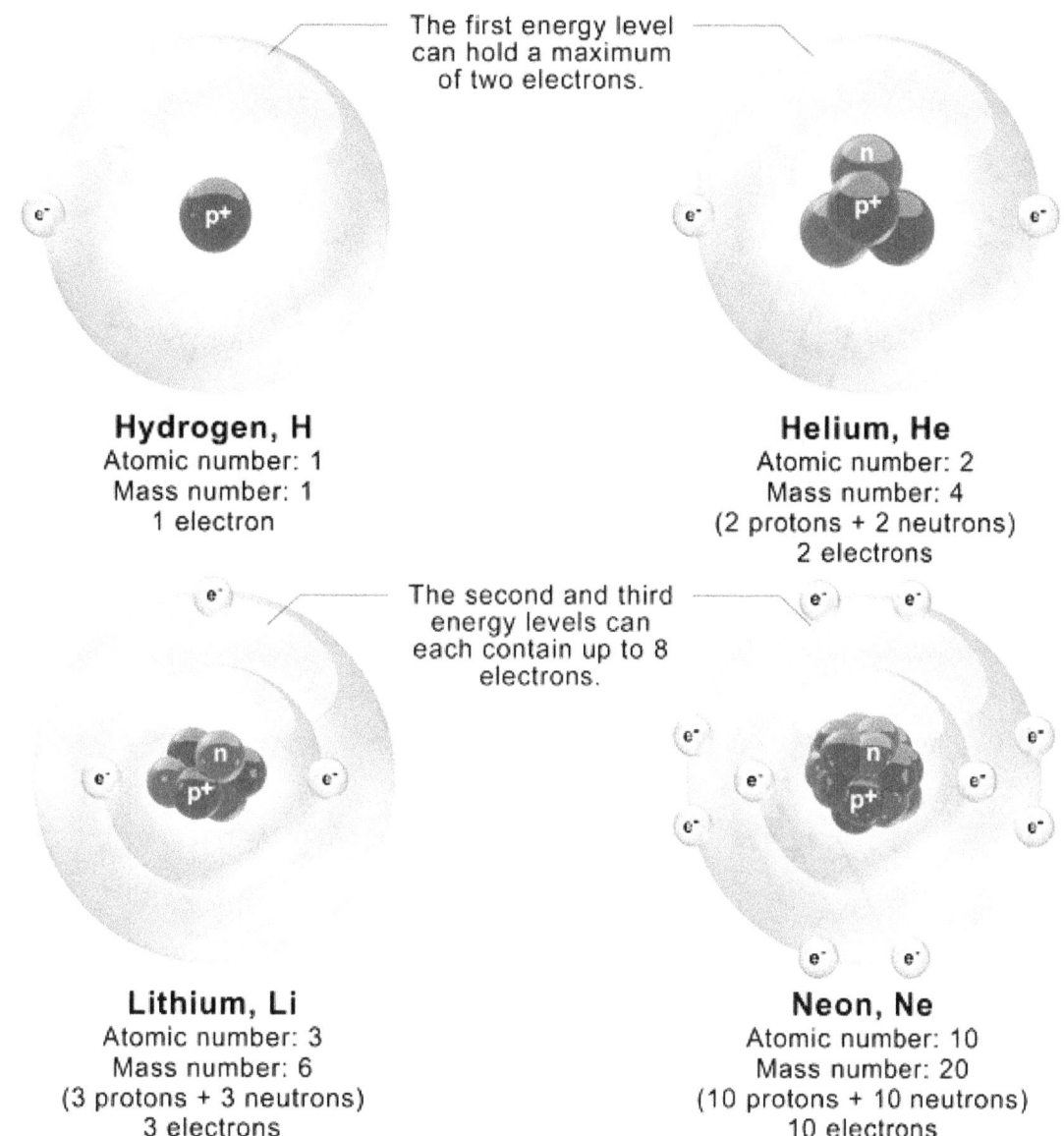

Models depicting the nucleus and electron energy levels in hydrogen, helium, lithium, and neon atoms. In reality, the diameter of the nucleus is about 100,000 times smaller than the diameter of the atom.

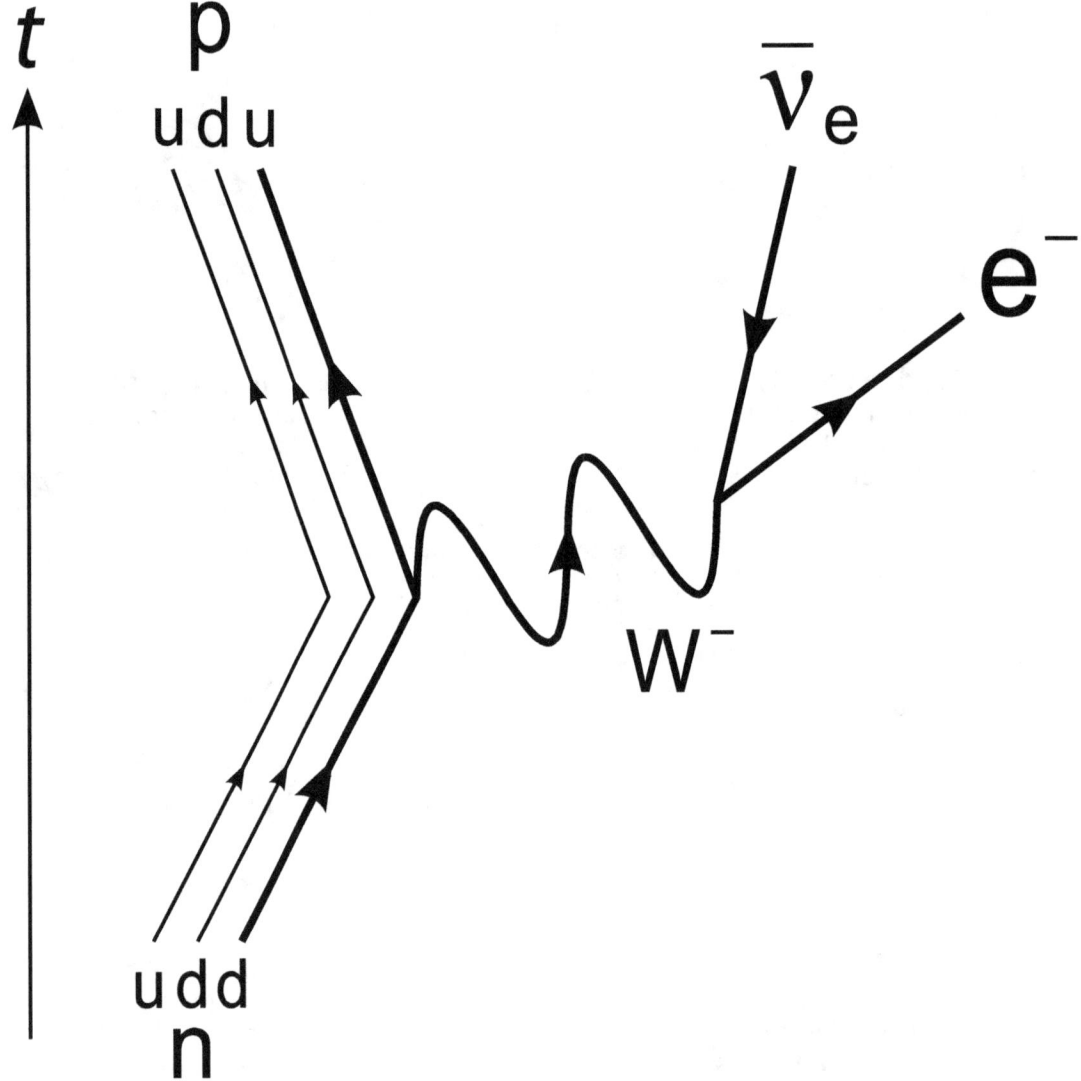

The Feynman diagram for beta decay of a neutron into a proton, electron, and electron antineutrino via an intermediate heavy W boson

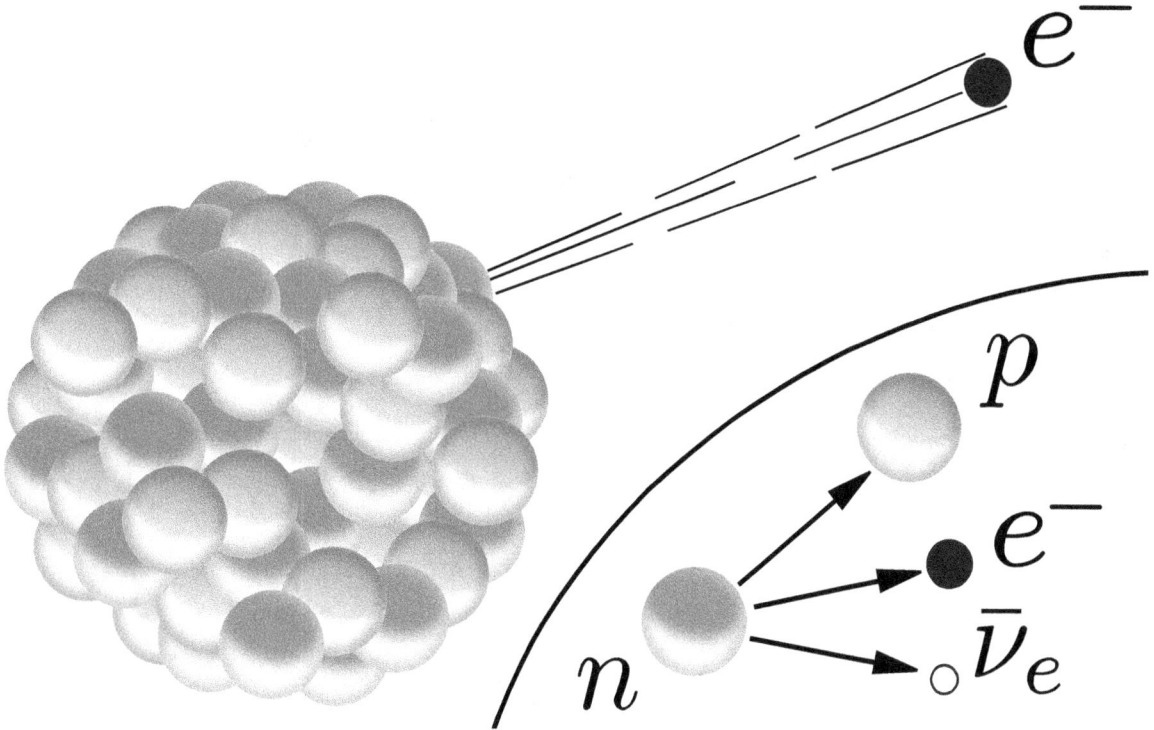

A schematic of the nucleus of an atom indicating β− radiation, the emission of a fast electron from the nucleus (the accompanying antineutrino is omitted). In the Rutherford model for the nucleus, red spheres were protons with positive charge and blue spheres were protons tightly bound to an electron with no net charge.
*The **inset** shows beta decay of a free neutron as it is understood today; an electron and antineutrino are created in this process.*

Institut Laue–Langevin (ILL) in Grenoble, France – a major neutron research facility.

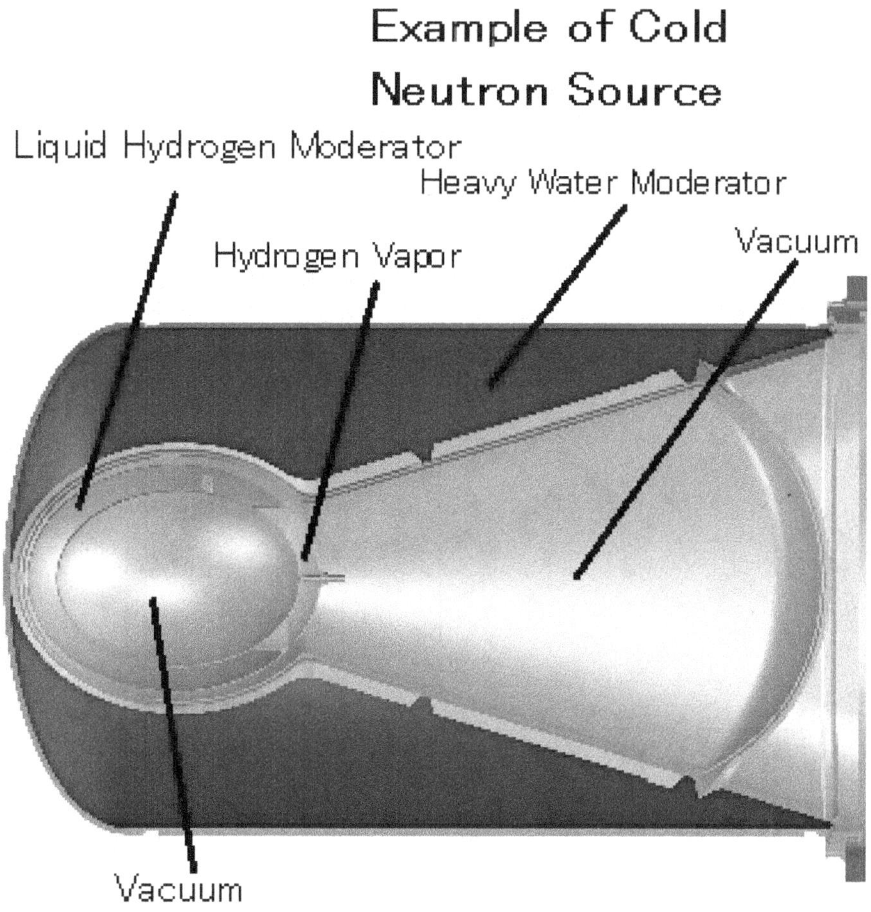

Example of Cold
Neutron Source

Liquid Hydrogen Moderator

Heavy Water Moderator

Hydrogen Vapor

Vacuum

Vacuum

Cold neutron source providing neutrons at about the temperature of liquid hydrogen

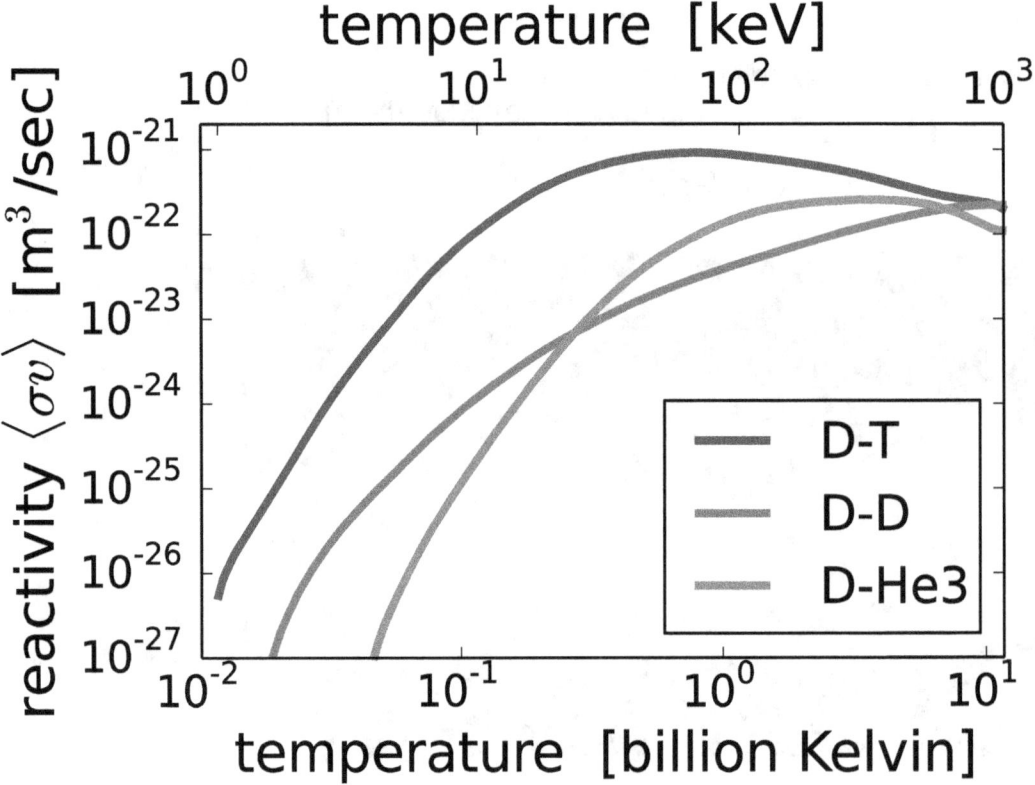

The fusion reaction rate increases rapidly with temperature until it maximizes and then gradually drops off. The DT rate peaks at a lower temperature (about 70 keV, or 800 million kelvins) and at a higher value than other reactions commonly considered for fusion energy.

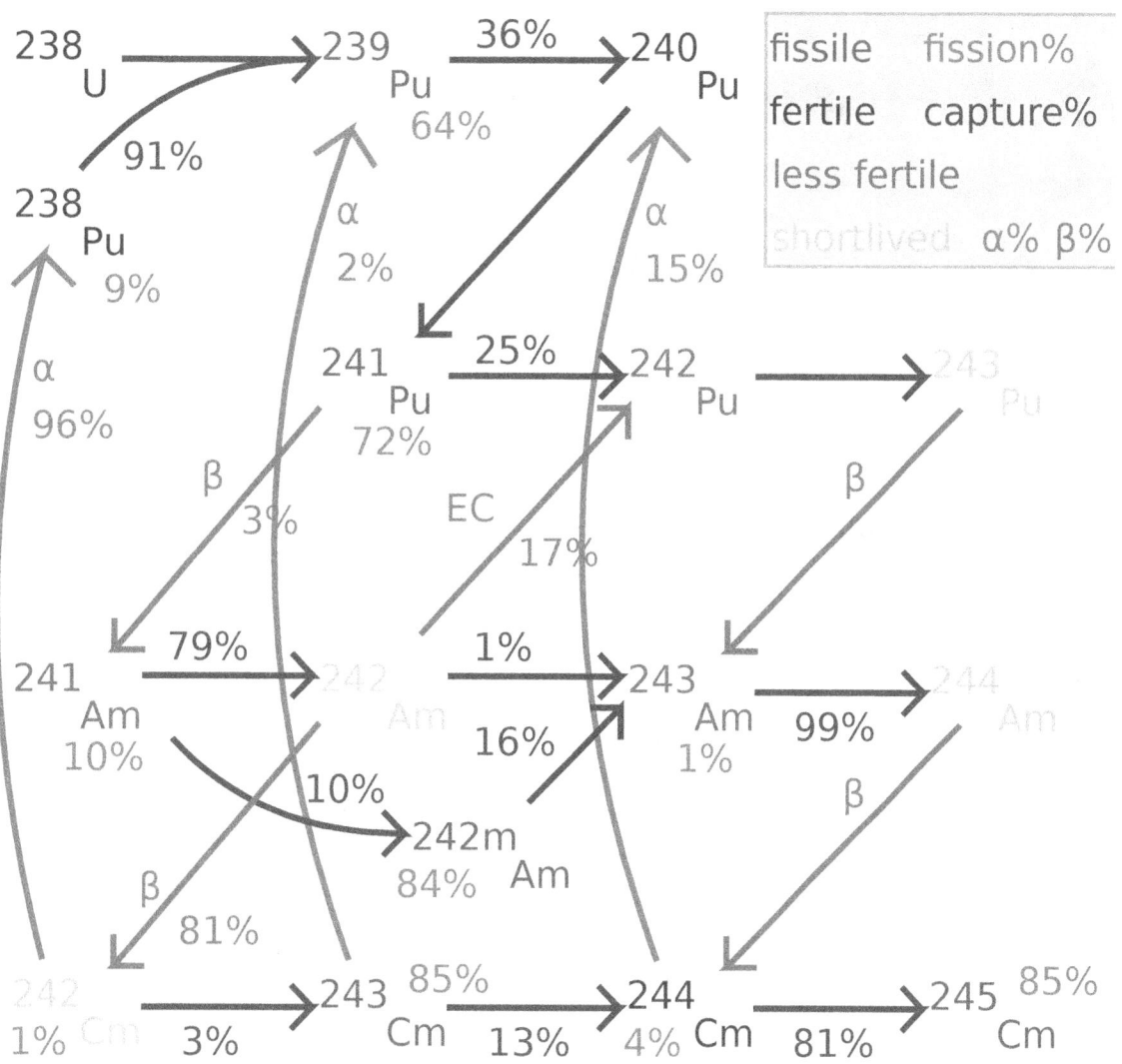

Transmutation flow in light water reactor, which is a thermal-spectrum reactor

Chapter 2

Discovery of the neutron

The story of the discovery of the neutron and its properties is central to the extraordinary developments in atomic physics that occurred in the first half of the 20th century, leading ultimately to the atomic bomb in 1945. The century began with Ernest Rutherford and Thomas Royds proving that alpha radiation is helium ions in 1908[1][2] and Rutherford's model for the atom in 1911,[3] in which atoms have their mass and positive charge concentrated in a very small nucleus.[4] The essential nature of the atomic nucleus was established with the discovery of the neutron by James Chadwick in 1932. By mid-century, these discoveries and subsequent developments had ushered in the atomic age.

2.1 Rutherford atom

See also: Plum pudding model

In the 1911 Rutherford model, the atom consisted of a small massive nucleus with positive charge surrounded by a much larger cloud of negatively charged electrons. This model had been developed from the extraordinary finding that alpha particles were on occasion scattered to high angle when passing through gold foil, indicating the alpha particles were occasionally reflecting from a small, but dense, component of atoms. Rutherford and others noted the disparity between the atomic number of an atom, or number of positive charges, and its mass computed in atomic mass units. The atomic number of an atom is usually about half its atomic mass. In 1920 Rutherford suggested that the disparity could be explained by the existence of a neutrally charged particle within the atomic nucleus.[5] Since at the time no such particle was known to exist, yet the mass of such a particle had to be about equal to that of the proton, Rutherford considered the required particle to be a neutral double consisting of an electron closely orbiting a proton.[5] The mass of protons is about 1800 times greater than that of electrons.

There were other motivations for the proton–electron model. As noted by Rutherford at the time, "We have strong reason for believing that the nuclei of atoms contain electrons as well as positively charged bodies...",[5] namely, it was known that beta radiation was electrons emitted from the nucleus.

Rutherford called these uncharged particles *neutrons*, apparently from the Latin root for *neutral* and the Greek ending *-on* (by imitation of *electron* and *proton*).[6][7] References to the word *neutron* in connection with the atom can be found in the literature as early as 1899, however.[8]

2.2 Problems of the nuclear electrons hypothesis

Throughout the 1920s, physicists assumed that the atomic nucleus was composed of protons and "nuclear electrons"[9][10] but there were obvious problems. Under this hypothesis, the nitrogen-14 (^{14}N) nucleus would be composed of 14 protons and 7 electrons, so that it would have a net charge of +7 elementary charge units and a mass of 14 atomic mass units. The nucleus was also orbited by another 7 electrons, termed "external electrons" by Rutherford,[5] to complete the ^{14}N atom. The Rutherford model was very influential, however, motivating the Bohr model for electrons orbiting the nucleus

25

James Chadwick discovered the neutron in 1932 while working at Cavendish Laboratory.

in 1913 and eventually leading to quantum mechanics by the mid-1920s.

By about 1930 it was generally recognized that it was difficult to reconcile the proton–electron model for nuclei with the Heisenberg uncertainty relation of quantum mechanics.[11][12] This relation, $\Delta x \cdot \Delta p \geq \tfrac{1}{2}\hbar$, implies that an electron confined

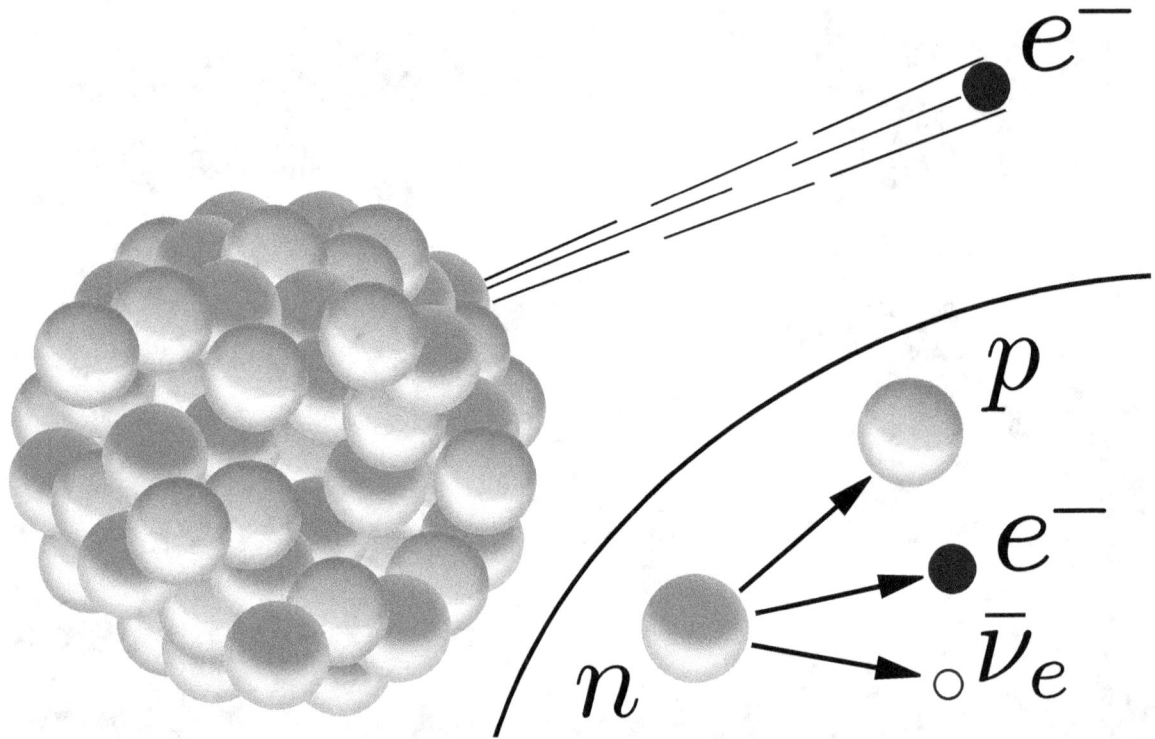

A schematic of the nucleus of an atom indicating β– radiation, the emission of a fast electron from the nucleus (the accompanying antineutrino is omitted). In the Rutherford model for the nucleus, red spheres were protons with positive charge and blue spheres were protons tightly bound to an electron with no net charge.
*The **inset** shows beta decay of a free neutron as it is understood today; an electron and antineutrino are created in this process.*

to a region the size of an atomic nucleus has an expected kinetic energy of 10–100 MeV.[12][13][14] This energy is larger than the binding energy of nucleons and larger than the observed energy of beta particles emitted from the nucleus.[12] While these considerations did not "prove" an electron could not exist in the nucleus, they were challenging for physicists to interpret. Indeed, Heisenberg at one time considered the possibility that the uncertainty relation was not applicable inside the nuclei.[15]

The Klein paradox,[16] discovered by Oskar Klein in 1928, presented further quantum mechanical objections to the notion of an electron confined within a nucleus.[11] Derived from the Dirac equation, this clear and precise paradox showed that a high-energy electron approaching a potential barrier has a high probability of passing through the barrier, or escaping, by transforming to a particle of negative mass. Apparently, an electron could not be confined within a nucleus by any potential well. The meaning of this paradox was intensely debated at the time.[11]

Observations of the energy levels of atoms and molecules were inconsistent with the nuclear spin expected from proton–electron hypothesis. Molecular spectroscopy of dinitrogen ($^{14}N_2$) showed that transitions originating from even rotational levels are more intense than those from odd levels, hence the even levels are more populated. According to quantum mechanics and the Pauli exclusion principle, the spin of the ^{14}N nucleus is therefore an integer multiple of $\hbar$ (the reduced Planck constant).[17][18] Both protons and electrons carry an intrinsic spin of ½ $\hbar$, and there is no way to arrange an odd number (14 protons + 7 electrons = 21) of spins ±½ $\hbar$ to give a spin integer multiple of $\hbar$.

The observed hyperfine structure of atomic spectra was inconsistent to the proton–electron hypothesis. This structure is caused by the influence of the nucleus on the dynamics of orbiting electrons. The magnetic moments of supposed "nuclear electrons" should produce hyperfine spectral line splittings similar to the Zeeman effect , but no such effects were observed.[11] This contradiction was somewhat mysterious,[9] until it was realized that there are no individual nuclear electrons in the nucleus.

2.3   Discovery of the neutron

In 1931, Walther Bothe and Herbert Becker in Giessen, Germany found that if the very energetic alpha particles emitted from polonium fell on certain light elements, specifically beryllium, boron, or lithium, an unusually penetrating radiation was produced. Since this radiation was not influenced by an electric field (neutrons have no charge), it was thought to be gamma radiation. The radiation was more penetrating than any gamma rays known, and the details of experimental results were difficult to interpret.[19][20] The following year Irène Joliot-Curie and Frédéric Joliot in Paris showed that if this unknown radiation fell on paraffin, or any other hydrogen-containing compound, it ejected protons of very high energy.[21] This observation was not in itself inconsistent with the assumed gamma ray nature of the new radiation, but detailed quantitative analysis of the data became increasingly difficult to reconcile with such a hypothesis. In Rome, the young physicist Ettore Majorana suggested that the manner in which the new radiation interacted with protons required a new neutral particle.[22]

On hearing of the Paris results in 1932, neither Rutherford nor James Chadwick at the Cavendish Laboratory in Cambridge were convinced by the gamma ray hypothesis.[9] Chadwick had searched for Rutherford's neutron by several experiments throughout the 1920s without success. Chadwick quickly performed a series of experiments showing that the gamma ray hypothesis was untenable. He repeated the creation of the radiation using beryllium, used better approaches to detection, and aimed the radiation at paraffin following the Paris experiment. Paraffin is high in hydrogen content, hence offers a target dense with protons; since neutrons and protons have almost equal mass, protons scatter energetically from neutrons. Chadwick measured the range of these protons, and also measured how the new radiation impacted the atoms of various gases.[23] He found that the new radiation consisted of not gamma rays, but uncharged particles with about the same mass as the proton; these particles were neutrons.[24][25] Chadwick won the Nobel Prize in Physics for this discovery in 1935.[26]

2.4   Proton–neutron model of the nucleus

Given the problems of the *proton–electron model*,[9][10] it was quickly accepted that the atomic nucleus is composed of protons and neutrons. Within months after the discovery of the neutron, Werner Heisenberg[27][28][29] and Dmitri Ivanenko[30] had proposed proton–neutron models for the nucleus.[31] Heisenberg's landmark papers approached the description of protons and neutrons in the nucleus through quantum mechanics. While Heisenberg's theory for protons and neutrons in the nucleus was a "major step toward understanding the nucleus as a quantum mechanical system,"[32] he still assumed the presence of nuclear electrons. In particular, Heisenberg assumed the neutron was a proton–electron composite, for which there is no quantum mechanical explanation. Heisenberg had no explanation for how lightweight electrons could be bound within the nucleus. Heisenberg introduced the first theory of nuclear exchange forces that bind the nucleons. He considered protons and neutrons to be different quantum states of the same particle, i.e., nucleons distinguished by the value of their nuclear isospin quantum numbers.

The proton–neutron model explained the puzzle of dinitrogen noticed independently by Ralph Kronig and Franco Rasetti. When ^{14}N was proposed to consist of 3 pairs each of protons and neutrons, with an additional unpaired neutron and proton each contributing a spin of $\frac{1}{2}$ ℏ in the same direction for a total spin of 1 ℏ, the model became viable. [33] [34][35] Soon, neutrons were used to naturally explain spin differences in many different nuclides in the same way.

If the proton–neutron model for the nucleus resolved many issues, it highlighted the problem of explaining the origins of beta radiation. No existing theory could account for how electrons could emanate from the nucleus. In 1934, Enrico Fermi published his classic paper describing the process of beta decay, in which the neutron decays to a proton by *creating* an electron and a (as yet undiscovered) neutrino.[36] The paper employed the analogy that photons, or electromagnetic radiation, were similarly created and destroyed in atomic processes. Ivanenko had suggested a similar analogy in 1932.[33][37] Fermi's theory requires the neutron to be a spin-½ particle. The theory preserved the principle of conservation of energy, which had been thrown into question by the continuous energy distribution of beta particles. The basic theory for beta decay proposed by Fermi was the first to show how particles could be created and destroyed. It established a general, basic theory for the interaction of particles by weak or strong forces.[36] While this influential paper has stood the test of time, the ideas within it were so new that when it was first submitted to the journal Nature in 1933 it was rejected as being too speculative.[32]

The question of whether the neutron was a composite particle of a proton and an electron persisted for a few years after

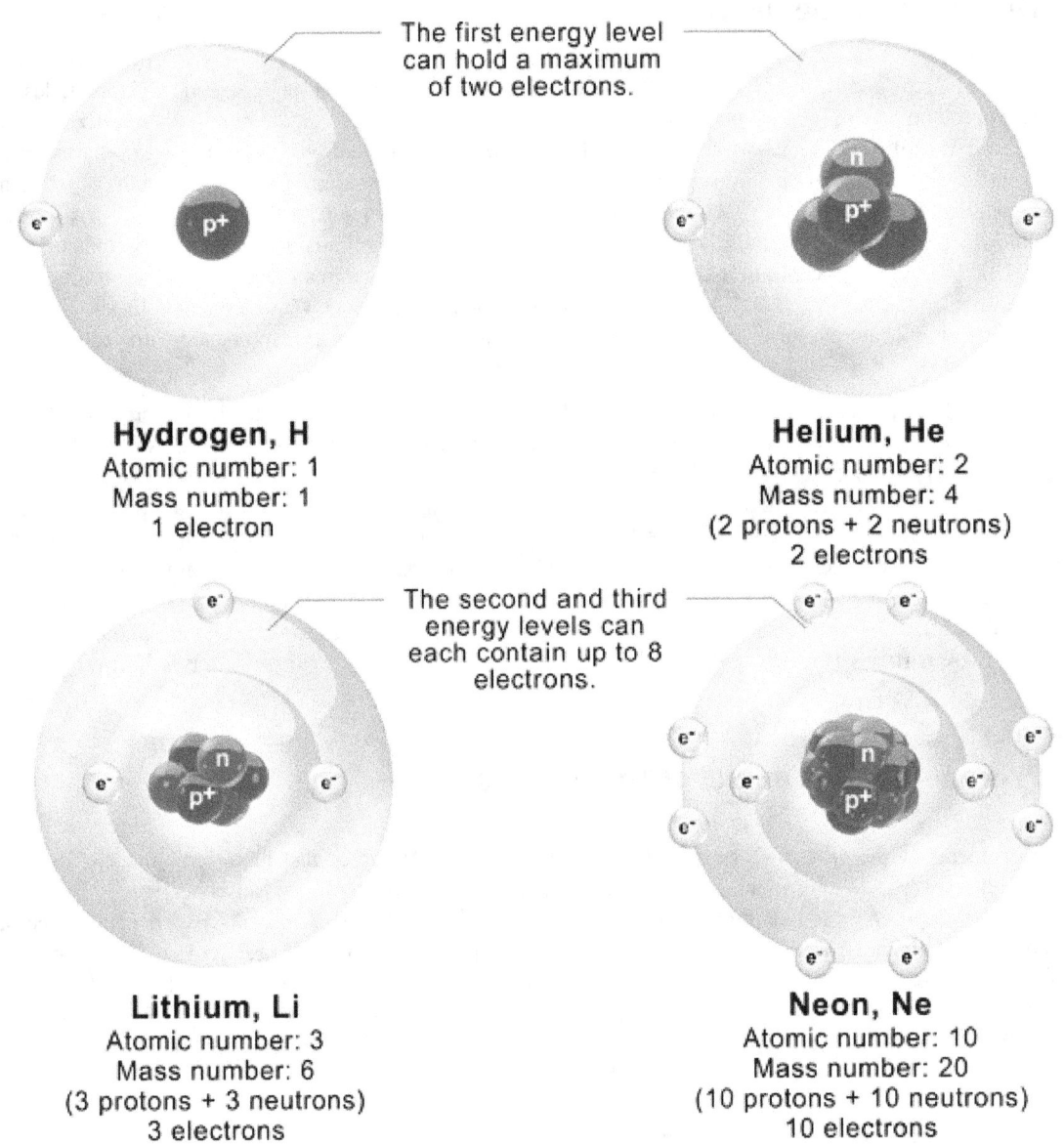

The first energy level can hold a maximum of two electrons.

The second and third energy levels can each contain up to 8 electrons.

Hydrogen, H
Atomic number: 1
Mass number: 1
1 electron

Helium, He
Atomic number: 2
Mass number: 4
(2 protons + 2 neutrons)
2 electrons

Lithium, Li
Atomic number: 3
Mass number: 6
(3 protons + 3 neutrons)
3 electrons

Neon, Ne
Atomic number: 10
Mass number: 20
(10 protons + 10 neutrons)
10 electrons

Models depicting the nucleus and electron energy levels in hydrogen, helium, lithium, and neon atoms. In reality, the diameter of the nucleus is about 100,000 times smaller than the diameter of the atom.

its discovery.[38][39] The issue was a legacy of the prevailing view from the 1920s that the only elementary particles were the proton and electron. The nature of the neutron was a primary topic of discussion at the 7th Solvay Conference held in October 1933, attended by Heisenberg, Niels Bohr, Lise Meitner, Ernest Lawrence, Fermi, Chadwick, and others.[32][40] As posed by Chadwick in his Bakerian Lecture in 1933, the primary question was the mass of the neutron relative to the proton. If the neutron's mass was less than the combined masses of a proton and an electron (1.0078 u), then the neutron could be a proton-electron composite because of the mass defect from the binding energy. If greater than the combined masses, then the neutron was elementary like the proton.[41] The question was challenging to answer because the electron's mass is only 0.05% of the proton's, hence precise measurements were required.

The difficulty of making the measurement is illustrated by the wide ranging values for the mass of the neutron obtained from 1932-1934. The accepted value today is 1.00866 u. In Chadwick's 1932 paper reporting on the discovery, he estimated the mass of the neutron to be between 1.005 u and 1.008 u.[42] By bombarding boron with alpha particles, Frédéric and Irène Joliot-Curie obtained a high value of 1.012 u, while Ernest Lawrence's team at the University of

Seventh Solvay Conference, 1933.

California measured the small value 1.0006 u using their new cyclotron.[43] In support of Fermi's theory and the neutron as an elementary particle, in 1935 Chadwick and his doctoral student Maurice Goldhaber, reported the first accurate measurement of the mass of the neutron. They used the 2.6 MeV gamma rays of ^{208}Tl (then known as thorium C") to photodisintegrate deuterium:

The energies of the resulting proton and neutron could be used to accurately determine the neutron's mass. Chadwick and Goldhaber found the neutron's mass to be slightly greater than the mass of the proton (1.0084 u or 1.0090 u, depending on precise values used for the proton and deuteron masses), and therefore predicted that an unbound neutron is unstable and would undergo beta decay.[44][45] The mass of the neutron was too large to be a proton-electron composite.[42]

2.5   Neutron physics in the 1930s

See also: Neutron magnetic moment

Soon after the discovery of the neutron, indirect evidence suggested the neutron had an unexpected non-zero value for its magnetic moment. Attempts to measure the neutron's magnetic moment originated with the discovery by Otto Stern in 1933 in Hamburg that the proton had an anomalously large magnetic moment.[46][47] By 1934 groups led by Stern, now in Pittsburgh, and I. I. Rabi in New York had independently deduced that the magnetic moment of the neutron was negative and unexpectedly large by measuring the magnetic moments of the proton and deuteron.[48][49][50] [39][51] Values for the magnetic moment of the neutron were also determined by Robert Bacher[52] at Ann Arbor (1933) and I.Y. Tamm and S.A. Altshuler[39][53] (1934) in the Soviet Union from studies of the hyperfine structure of atomic spectra. By the late 1930s accurate values for the magnetic moment of the neutron had been deduced by the Rabi group using measurements employing newly developed nuclear magnetic resonance techniques.[51] The large value for the proton's magnetic moment and the inferred negative value for the neutron's magnetic moment were unexpected and raised many questions.[39]

The discovery of the neutron immediately gave scientists a new tool for probing the properties of atomic nuclei. Alpha particles had been used over the previous decades in scattering experiments, but such particles, which are helium nuclei, have +2 charge. This charge makes it difficult for alpha particles to overcome the Coulomb repulsive force and interact directly with the nuclei of atoms. Since neutrons have no electric charge, they do not have to overcome this force to interact with nuclei. Almost coincident with its discovery, neutrons were used by Norman Feather, Chadwick's colleague and protege, in scattering experiments with nitrogen.[54] Feather was able to show that neutrons interacting with nitrogen nuclei scattered to protons or induced nitrogen to disintegrate to form boron with the emission of an alpha particle. Feather was therefore the first to show that neutrons produce nuclear disintegrations.

In Rome Enrico Fermi bombarded heavier elements with neutrons and found the products to be radioactive. By 1934 Fermi had used neutrons to induce radioactivity in 22 different elements, many of these elements of high atomic number. Noticing that other experiments with neutrons at his laboratory seemed to work better on a wooden table than a marble table, Fermi suspected that the protons of the wood were slowing the neutrons and so increasing the chance for the neutron to interact with nuclei. Fermi therefore passed neutrons through paraffin wax to slow them and found that the radioactivity of bombarded elements increased by a hundredfold. The cross section for interaction with nuclei is much larger for slow neutrons than for fast neutrons. In 1938 Fermi received the Nobel Prize in Physics *"for his demonstrations of the existence of new radioactive elements produced by neutron irradiation, and for his related discovery of nuclear reactions brought about by slow neutrons"*.[55]

Jointly with Lise Meitner and his pupil and assistant Fritz Strassmann, Otto Hahn furthered the research begun by Fermi and his team when he bombarded uranium with neutrons at his laboratory in Berlin. Between 1934 and 1938, Hahn, Meitner, and Strassmann found a great number of radioactive transmutation products from these experiments, all of which they regarded as transuranic.[56] The decisive experiment on 16–17 December 1938 (the celebrated "radium–barium–mesothorium–fractionation") produced puzzling results: the three isotopes consistently behaved not as radium, but as barium.[57] By January 1939 Hahn had concluded that he was seeing light platinoids, barium, lanthanum, and cerium. Hahn and his collaborators had observed nuclear fission, or the fractionation of uranium nuclei into light elements, induced by neutron bombardment. In their second publication on nuclear fission, Hahn and Strassmann predicted the existence and liberation of additional neutrons during the fission process.[58] Frédéric Joliot and his team proved this phenomena to be a chain reaction in March 1939. In 1945 Hahn received the 1944 Nobel Prize in Chemistry *"for his discovery of the fission of heavy atomic nuclei."* [59][60][61]

The discovery of nuclear fission at the end of 1938 marked a shift in the centers of nuclear research from Europe to the United States. Large numbers of scientists were migrating to the United States to escape the troubles in Europe and the looming war (See Jewish scientists and the Manhattan Project). The new centers of nuclear research were the universities in the United States, particularly Columbia University in New York and the University of Chicago where Enrico Fermi had relocated, and a new research facility at Los Alamos, New Mexico beginning in 1942, the new home of the Manhattan project.

2.6 See also

- Ionizing radiation

- List of particles

- Neutronium

- Neutron magnetic moment

- Neutron radiation and the Sievert radiation scale

- Nuclear reaction

- Thermal reactor

- Nucleosynthesis

2.7   References

[1] Campbell, John. "Rutherford – A Brief Biography". *Rutherford.org.nz*. Retrieved 4 March 2013.

[2] E. Rutherford and T. Royds (1908) "Spectrum of the radium emanation," *Philosophical Magazine*, Series 6, vol. 16, pages 313–317.

[3] Ernest Rutherford (1911). *The scattering of alpha and beta particles by matter and the structure of the atom*. Taylor & Francis. p. 688.

[4] M. S. Longair (2003). *Theoretical concepts in physics: an alternative view of theoretical reasoning in physics*. Cambridge University Press. pp. 377–378. ISBN 978-0-521-52878-8.

[5] E.Rutherford(1920). "Nuclear Constitution of Atoms".*Proceedings of the Royal Society A***97**(686):374.Bibcode:1920RSPSA..9. doi:10.1098/rspa.1920.0040.

[6] Wolfgang Pauli (1985). "Wolfgang Pauli Wissenschaftlicher Briefwechsel mit Bohr, Einstein, Heisenberg u.a.". Sources in the History of Mathematics and Physical Sciences **6**. p. 105. doi:10.1007/978-3-540-78801-0_3. ISBN 978-3-540-13609-5. |chapter= ignored (help)

[7] Hendry, John, ed. (1984), *Cambridge Physics in the Thirties*, Adam Hilger Ltd, Bristol, ISBN 0852747616

[8] N.Feather(1960)."A history of neutrons and nuclei.Part1".*Contemporary Physics***1**(3):191–203.Bibcode:1960ConPh...1..19. doi:10.1080/00107516008202611.

[9] Brown,Laurie M. (1978). "The idea of the neutrino".*Physics Today***31**(9):23.Bibcode:1978PhT....31i..23B.doi:10.1063/1.299.

[10] Friedlander G., Kennedy J.W. and Miller J.M. (1964) *Nuclear and Radiochemistry* (2nd edition), Wiley, pp. 22–23 and 38–39

[11] Stuewer, Roger H. (1985). "Niels Bohr and Nuclear Physics". In French, A. P.; Kennedy, P. J. *Niels Bohr: A Centenary Volume*. Harvard University Press. pp. 197–220. ISBN 0674624165.

[12] Pais, Abraham (1986). *Inward Bound*. Oxford: Oxford University Press. p. 299. ISBN 0198519974.

[13] Shultis, J. Kenneth; Faw, Richard E. (2007), *Fundamentals of Nuclear Science and Engineering*, CRC, 2nd edition, ISBN 1420051369

[14] In a nucleus of diameter R in the order of 10 fm, the uncertainty principle would require an electron to have a momentum p of the order of h/R. Such a momentum implies that the electron has a (relativistic) kinetic energy of 10–100 MeV.

[15] Tomonaga, chapter 9

[16] Klein, O. (1929). "Die Reflexion von Elektronen an einem Potentialsprung nach der relativistischen Dynamik von Dirac". *Zeitschrift für Physik* **53** (3–4): 157. Bibcode:1929ZPhy...53..157K. doi:10.1007/BF01339716.

[17] Atkins, P.W. and J. de Paula, P.W. (2006) "Atkins' Physical Chemistry" (8th edition), W.H. Freeman, p. 451

[18] Herzberg, G. (1950) *Spectra of Diatomic Molecules* (2nd edition), van Nostrand Reinhold, pp. 133–140

[19] Bothe, W.; Becker, H. (1930). "Künstliche Erregung von Kern-γ-Strahlen" [Artificial excitation of nuclear γ-radiation]. *Zeitschrift für Physik* **66** (5–6): 289. Bibcode:1930ZPhy...66..289B. doi:10.1007/BF01390908.

[20] Becker, H.; Bothe, W. (1932). "Die in Bor und Beryllium erregten γ-Strahlen" [Γ-rays excited in boron and beryllium]. *Zeitschrift für Physik* **76** (7–8): 421. Bibcode:1932ZPhy...76..421B. doi:10.1007/BF01336726.

[21] Joliot-Curie, Irène and Joliot, Frédéric (1932). "Émission de protons de grande vitesse par les substances hydrogénées sous l'influence des rayons γ très pénétrants" [Emission of high-speed protons by hydrogenated substances under the influence of very penetrating γ-rays]. *Comptes Rendus* **194**: 273.

[22] Ettore Majorana: genius and mystery, CERN courier.

[23] "Atop the Physics Wave: Rutherford Back in Cambridge, 1919–1937". *Rutherford's Nuclear World*. American Institute of Physics. 2011–2014. Retrieved 19 August 2014.

[24] Chadwick, James (1932). "Possible Existence of a Neutron". *Nature* **129** (3252): 312. Bibcode:1932Natur.129Q.312C. doi:10.1038/129312a0.

[25] Chadwick, J. (1933). "Bakerian Lecture. The Neutron". *Proceedings of the Royal Society A: Mathematical, Physical and Engineering Sciences* **142** (846): 1. Bibcode:1933RSPSA.142....1C. doi:10.1098/rspa.1933.0152.

[26] 1935 Nobel Prize in Physics. Nobelprize.org. Retrieved on 2012-08-16.

[27] Heisenberg, W. (1932). "Über den Bau der Atomkerne. I". *Z. Phys.* **77**: 1–11. Bibcode:1932ZPhy...77....1H. doi:10.10

[28] Heisenberg,W. (1932). "Über den Bau der Atomkerne.II".*Z.Phys.***78**(3–4):156–164.Bibcode:1932ZPhy...78..156H. doi:10.1007/BF01337585.

[29] Heisenberg, W. (1933). "Über den Bau der Atomkerne. III". *Z. Phys.* **80** (9–10): 587–596. Bibcode:1933ZPhy...80..587H. doi:10.1007/BF01335696.

[30] Iwanenko, D.D., The neutron hypothesis, Nature **129** (1932) 798.

[31] Miller A. I. *Early Quantum Electrodynamics: A Sourcebook*, Cambridge University Press, Cambridge, 1995, ISBN 0521568919, pp. 84–88.

[32] Brown, L.M.; Rechenberg, H. (1996). *The Origin of the Concept of Nuclear Forces.* Bristol and Philadelphia: Institute of Physics Publishing. ISBN 0750303735.

[33] Iwanenko, D. (1932). "Sur la constitution des noyaux atomiques". *Compt. Rend. Acad Sci. Paris* **195**: 439–441.

[34] Bacher, R.F.; Condon, E.U. (1932). "The Spin of the Neutron". *Physical Review* **41** (5): 683–685. Bibcode:1932PhRv...41..683G. doi:10.1103/PhysRev.41.683.

[35] Whaling, W. (2009). "Robert F. Bacher 1905–2004" (PDF). *Biographical Memoirs of the National Academy of Sciences.*

[36] Wilson, Fred L. (1968). "Fermi's Theory of Beta Decay". *Am. J. Phys.* **36** (12): 1150–1160. Bibcode:1968AmJPh..36.1150W. doi:10.1119/1.1974382.

[37] Iwanenko, D. (1932). "Neutronen und kernelektronen". *Physikalische Zeitschrift der Sowjetunion* **1**: 820–822.

[38] Kurie, F.N.D. (1933). "The Collisions of Neutrons with Protons". *Physical Review* **44** (6): 463. Bibcode:1933PhRv...44..463K. doi:10.1103/PhysRev.44.463.

[39] Breit, G.; Rabi, I.I. (1934). "On the interpretation of present values of nuclear moments". *Physical Review* **46** (3): 230. Bibcode:1934PhRv...46..230B. doi:10.1103/PhysRev.46.230.

[40] Sime, R.L. (1996). *Lise Meitner: A Life in Physics.* University of California Press. ISBN 0520089065.

[41] Chadwick, J. (1933). "Bakerian Lecture - The Neutron". *Proc. Roy. Soc.* **142** (846): 1–25. Bibcode:1933RSPSA.142....1C. doi:10.1098/rspa.1933.0152.

[42] Brown,A. (1997).*The Neutron and the Bomb:A Biography of Sir James Chadwick.*Oxford University Press.ISBN97801985399.

[43] Seidel, R.W. (1989). *Lawrence and his Laboratory: A History of the Lawrence Berkeley Laboratory.* University of California Press. ISBN 9780520064263.

[44] Chadwick, J.; Goldhaber, M. (1934). "A nuclear photo-effect: disintegration of the diplon by gamma rays". *Nature* **134** (3381): 237–238. Bibcode:1934Natur.134..237C. doi:10.1038/134237a0.

[45] Chadwick, J.; Goldhaber, M. (1935). "A nuclear photoelectric effect" (PDF). *Proc. R. Soc. Lond* **151** (873): 479–493. Bibcode:1935RSPSA.151..479C. doi:10.1098/rspa.1935.0162.

[46] Frisch, R.; Stern, O. (1933). "Über die magnetische Ablenkung von Wasserstoffmolekülen und das magnetische Moment des Protons. I / Magnetic Deviation of Hydrogen Molecules and the Magnetic Moment of the Proton. I.". *Z. Phys.* **84**: 4–16.

[47] Esterman, I.; Stern, O. (1933). "Über die magnetische Ablenkung von Wasserstoffmolekülen und das magnetische Moment des Protons. II / Magnetic Deviation of Hydrogen Molecules and the Magnetic Moment of the Proton. I.". *Z. Phys.* **85**: 17–24. Bibcode:1933ZPhy...85...17E. doi:10.1007/BF01330774.

[48] Esterman, I.; Stern, O. (1934). "Magnetic moment of the deuton". *Physical Review* **45**: 761(A109).

[49] Rabi, I.I.; Kellogg, J.M.; Zacharias, J.R. (1934). "The magnetic moment of the proton". *Physical Review* **46** (3): 157. Bibcode:1934PhRv...46..157R. doi:10.1103/PhysRev.46.157.

[50] Rabi, I.I.; Kellogg, J.M.; Zacharias, J.R. (1934). "The magnetic moment of the deuton". *Physical Review* **46** (3): 163. Bibcode:1934PhRv...46..163R. doi:10.1103/PhysRev.46.163.

[51] John S. Rigden (2000). *Rabi, Scientist and Citizen.* Harvard University Press. ISBN 9780674004351.

[52] Bacher,R.F. (1933)."Note on the Magnetic Moment of the Nitrogen Nucleus".*Physical Review***43**(12):1001.Bibcode:1933Ph. doi:10.1103/PhysRev.43.1001. Retrieved 2015-02-10.

[53] Tamm, I.Y.; Altshuler, S.A. (1934). "Magnetic Moment of the Neutron". *Doklady Akad. Nauk SSSR* **8**: 455. Retrieved 2015-01-30.

[54] N. Feather (1 June 1932). "The Collisions of Neutrons with Nitrogen Nuclei" (PDF). *Proceedings of the Royal Society A* **136** (830): 709–727. Bibcode:1932RSPSA.136..709F. doi:10.1098/rspa.1932.0113.

[55] Cooper, Dan (1999). *Enrico Fermi: And the Revolutions in Modern physics.* New York: Oxford University Press. ISBN 0-19-511762-X. OCLC 39508200.

[56] Hahn, O. (1958). "The Discovery of Fission". *Scientific American* **198** (2): 76. doi:10.1038/scientificamerican0258-76.

[57] Rife, Patricia (1999). *Lise Meitner and the dawn of the nuclear age.* Basel, Switzerland: Birkhäuser. ISBN 0-8176-3732-X.

[58] Hahn, O.; Strassmann, F. (10 February 1939). "Proof of the Formation of Active Isotopes of Barium from Uranium and Thorium Irradiated with Neutrons; Proof of the Existence of More Active Fragments Produced by Uranium Fission". *Die Naturwissenschaften* **27** (6): 89–95. Bibcode:1939NW.....27...89H. doi:10.1007/BF01488988.

[59] "The Nobel Prize in Chemistry 1944". Nobel Foundation. Retrieved 2007-12-17.

[60] Bernstein, Jeremy (2001). *Hitler's uranium club: the secret recordings at Farm Hall.* New York: Copernicus. p. 281. ISBN 0-387-95089-3.

[61] "The Nobel Prize in Chemistry 1944: Presentation Speech". Nobel Foundation. Retrieved 2008-01-03.

2.8   Further reading

- Annotated bibliography for neutrons from the Alsos Digital Library for Nuclear Issues

- Abraham Pais, *Inward Bound*, Oxford: Oxford University Press, 1986. ISBN 0198519974.

- Herwig Schopper, *Weak interactions and nuclear beta decay*, Publisher, North-Holland Pub. Co., 1966.

- Ruth Lewin Sime, *Lise Meitner: A Life in Physics*, Berkeley, University of California Press, 1996. ISBN 0520208609.

- Sin-Itiro Tomonaga, *The Story of Spin*, The University of Chicago Press, 1997

2.9   External links

Fermi and his students (the Via Panisperna boys) in the courtyard of Rome University's Physics Institute in Via Panisperna, about 1934. From Left to right: Oscar D'Agostino, Emilio Segrè, Edoardo Amaldi, Franco Rasetti and Fermi

Lise Meitner and Otto Hahn in their laboratory.

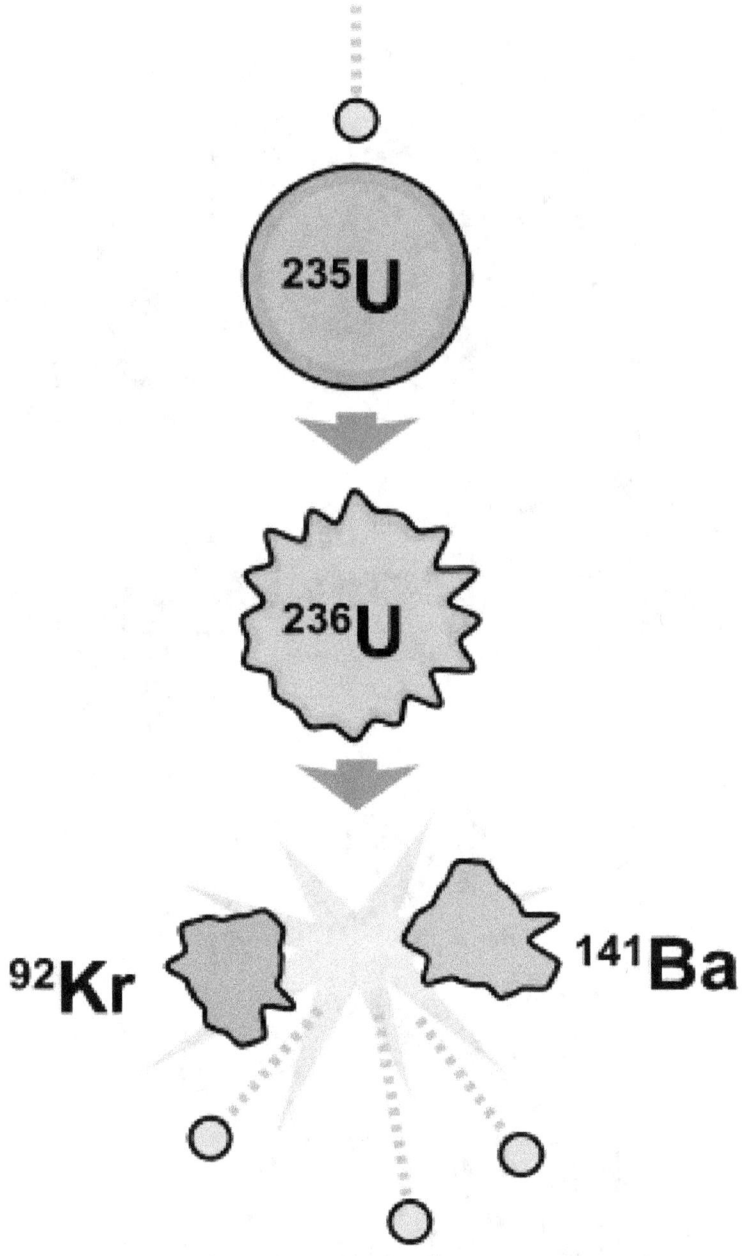

Nuclear fission caused by absorption of a neutron by uranium-235. The heavy nuclide fragments into lighter components and additional neutrons.

The Trinity test of the Manhattan Project in 1945 was the first atomic bomb.

Chapter 3

Standard Model

This article is about the Standard Model of particle physics. For other uses, see Standard model (disambiguation).

This article is a non-mathematical general overview of the Standard Model. For a mathematical description, see the article Standard Model (mathematical formulation).

For the Standard Model of Big Bang cosmology, Lambda-CDM model.

The **Standard Model** of particle physics is a theory concerning the electromagnetic, weak, and strong nuclear inter-

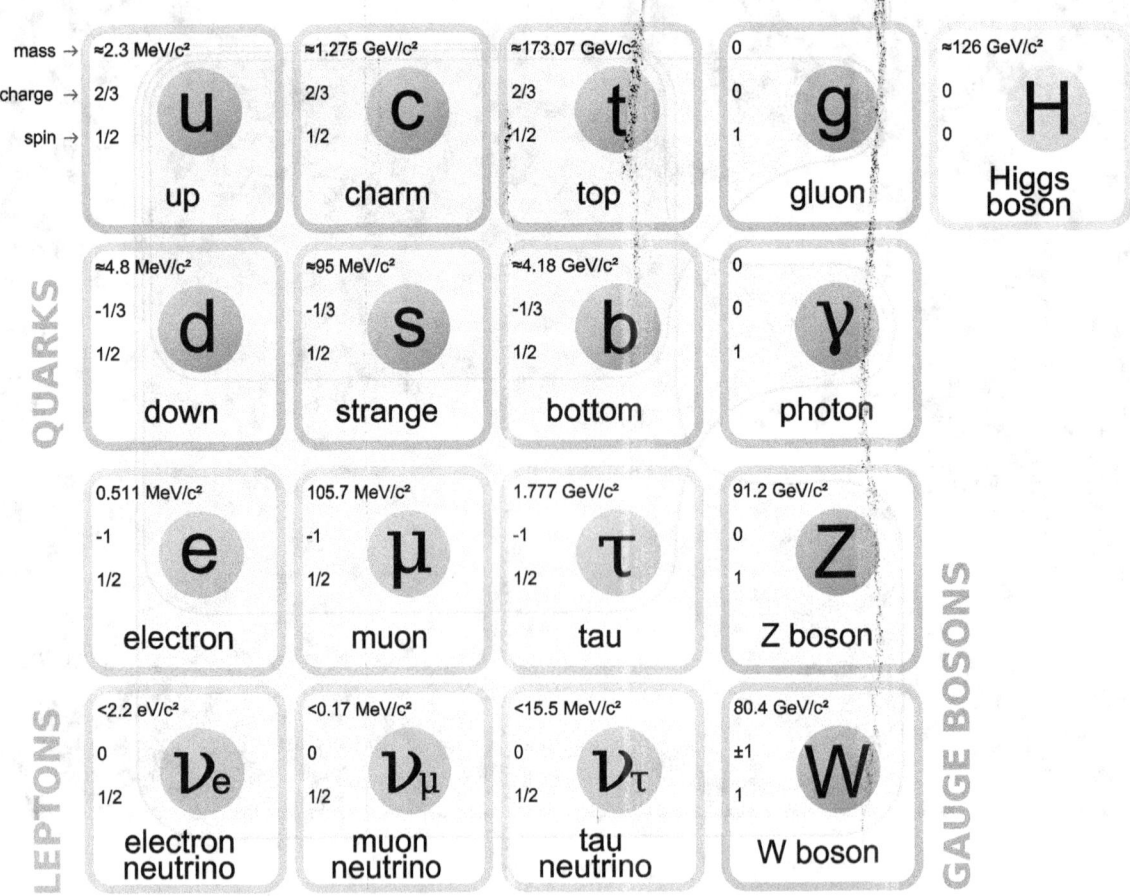

The Standard Model of elementary particles (more schematic depiction), with the three generations of matter, gauge bosons in the fourth column, and the Higgs boson in the fifth.

actions, as well as classifying all the subatomic particles known. It was developed throughout the latter half of the 20th century, as a collaborative effort of scientists around the world.[1] The current formulation was finalized in the mid-1970s upon experimental confirmation of the existence of quarks. Since then, discoveries of the top quark (1995), the tau neutrino (2000), and more recently the Higgs boson (2013), have given further credence to the Standard Model. Because of its success in explaining a wide variety of experimental results, the Standard Model is sometimes regarded as a "theory of almost everything".

Although the Standard Model is believed to be theoretically self-consistent[2] and has demonstrated huge and continued successes in providing experimental predictions, it does leave some phenomena unexplained and it falls short of being a complete theory of fundamental interactions. It does not incorporate the full theory of gravitation[3] as described by general relativity, or account for the accelerating expansion of the universe (as possibly described by dark energy). The model does not contain any viable dark matter particle that possesses all of the required properties deduced from observational cosmology. It also does not incorporate neutrino oscillations (and their non-zero masses).

The development of the Standard Model was driven by theoretical and experimental particle physicists alike. For theorists, the Standard Model is a paradigm of a quantum field theory, which exhibits a wide range of physics including spontaneous symmetry breaking, anomalies, non-perturbative behavior, etc. It is used as a basis for building more exotic models that incorporate hypothetical particles, extra dimensions, and elaborate symmetries (such as supersymmetry) in an attempt to explain experimental results at variance with the Standard Model, such as the existence of dark matter and neutrino oscillations.

3.1 Historical background

The first step towards the Standard Model was Sheldon Glashow's discovery in of a way to combine the electromagnetic andweak interactions. [4] In 1967 Steven Weinberg[5] and Abdus Salam[6] incorporated the Higgs mechanism[7][8][9] into Glashow's electroweak theory, giving it its modern form.

The Higgs mechanism is believed to give rise to the masses of all the elementary particles in the Standard Model. This includes the masses of the W and Z bosons, and the masses of the fermions, i.e. the quarks and leptons.

After the neutral weak currents caused by Z boson exchange were discovered at CERN in 1973,[10][11][12][13] the electroweak theory became widely accepted and Glashow, Salam, and Weinberg shared the 1979 Nobel Prize in Physics for discovering it. The W and Z bosons were discovered experimentally in 1981, and their masses were found to be as the Standard Model predicted.

The theory of the strong interaction, to which many contributed, acquired its modern form around 1973–74, when experiments confirmed that the hadrons were composed of fractionally charged quarks.

3.2 Overview

At present, matter and energy are best understood in terms of the kinematics and interactions of elementary particles. To date, physics has reduced the laws governing the behavior and interaction of all known forms of matter and energy to a small set of fundamental laws and theories. A major goal of physics is to find the "common ground" that would unite all of these theories into one integrated theory of everything, of which all the other known laws would be special cases, and from which the behavior of all matter and energy could be derived (at least in principle).[14]

3.3 Particle content

The Standard Model includes members of several classes of elementary particles (fermions, gauge bosons, and the Higgs boson), which in turn can be distinguished by other characteristics, such as color charge.

3.3.1 Fermions

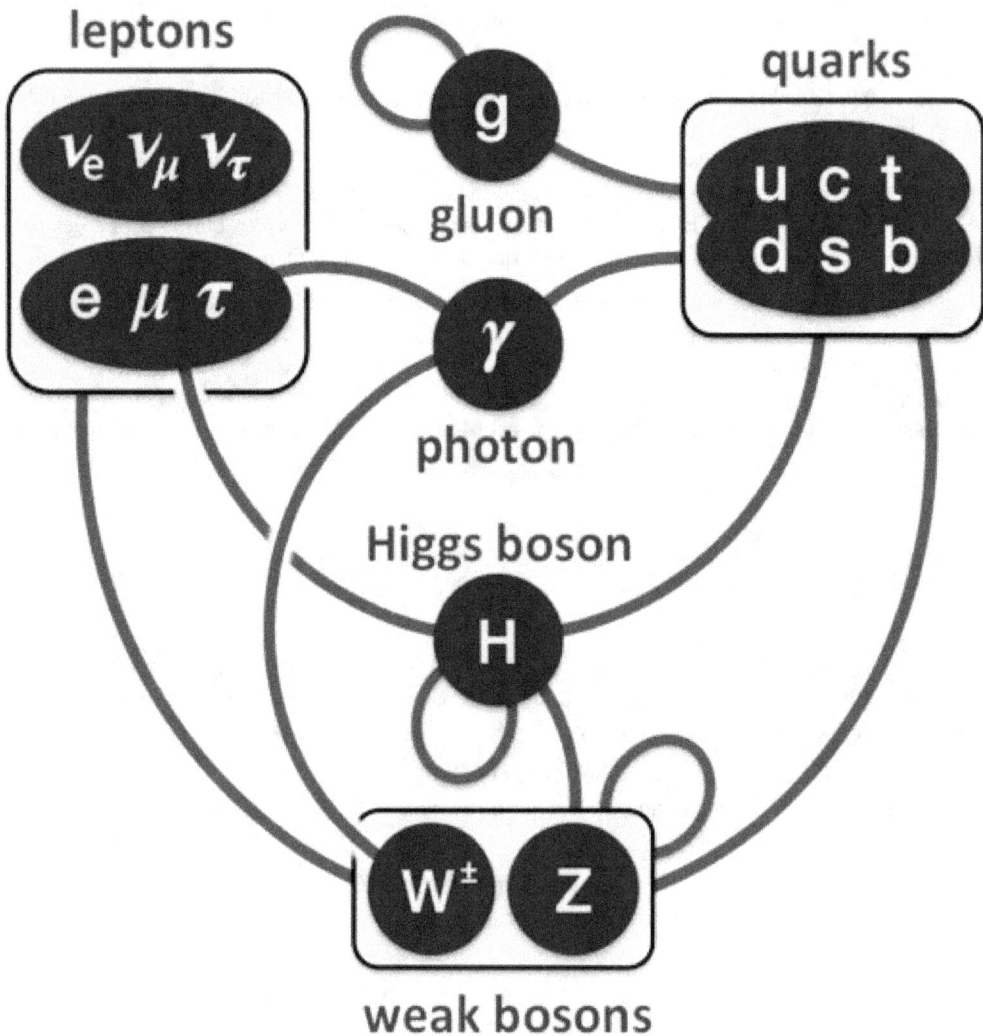

Summary of interactions between particles described by the Standard Model.

The Standard Model includes 12 elementary particles of spin-½ known as fermions. According to the spin-statistics theorem, fermions respect the Pauli exclusion principle. Each fermion has a corresponding antiparticle.

The fermions of the Standard Model are classified according to how they interact (or equivalently, by what charges they carry). There are six quarks (up, down, charm, strange, top, bottom), and six leptons (electron, electron neutrino, muon, muon neutrino, tau, tau neutrino). Pairs from each classification are grouped together to form a generation, with corresponding particles exhibiting similar physical behavior (see table).

The defining property of the quarks is that they carry color charge, and hence, interact via the strong interaction. A phenomenon called color confinement results in quarks being very strongly bound to one another, forming color-neutral composite particles (hadrons) containing either a quark and an antiquark (mesons) or three quarks (baryons). The familiar proton and the neutron are the two baryons having the smallest mass. Quarks also carry electric charge and weak isospin. Hence they interact with other fermions both electromagnetically and via the weak interaction.

The remaining six fermions do not carry colour charge and are called leptons. The three neutrinos do not carry electric charge either, so their motion is directly influenced only by the weak nuclear force, which makes them notoriously difficult to detect. However, by virtue of carrying an electric charge, the electron, muon, and tau all interact electromagnetically.

Each member of a generation has greater mass than the corresponding particles of lower generations. The first generation charged particles do not decay; hence all ordinary (baryonic) matter is made of such particles. Specifically, all atoms consist of electrons orbiting around atomic nuclei, ultimately constituted of up and down quarks. Second and third generations charged particles, on the other hand, decay with very short half lives, and are observed only in very high-energy environments. Neutrinos of all generations also do not decay, and pervade the universe, but rarely interact with baryonic matter.

3.3.2 Gauge bosons

In the Standard Model, gauge bosons are defined as force carriers that mediate the strong, weak, and electromagnetic fundamental interactions.

Interactions in physics are the ways that particles influence other particles. At a macroscopic level, electromagnetism allows particles to interact with one another via electric and magnetic fields, and gravitation allows particles with mass to attract one another in accordance with Einstein's theory of general relativity. The Standard Model explains such forces as resulting from matter particles exchanging other particles, generally referred to as *force mediating particles*. When a force-mediating particle is exchanged, at a macroscopic level the effect is equivalent to a force influencing both of them, and the particle is therefore said to have *mediated* (i.e., been the agent of) that force. The Feynman diagram calculations, which are a graphical representation of the perturbation theory approximation, invoke "force mediating particles", and when applied to analyze high-energy scattering experiments are in reasonable agreement with the data. However, perturbation theory (and with it the concept of a "force-mediating particle") fails in other situations. These include low-energy quantum chromodynamics, bound states, and solitons.

The gauge bosons of the Standard Model all have spin (as do matter particles). The value of the spin is 1, making them bosons. As a result, they do not follow the Pauli exclusion principle that constrains fermions: thus bosons (e.g. photons) do not have a theoretical limit on their spatial density (number per volume). The different types of gauge bosons are described below.

- Photons mediate the electromagnetic force between electrically charged particles. The photon is massless and is well-described by the theory of quantum electrodynamics.

- The W+, W−, and Z gauge bosons mediate the weak interactions between particles of different flavors (all quarks and leptons). They are massive, with the Z being more massive than the W±. The weak interactions involving the W± exclusively act on *left-handed* particles and *right-handed* antiparticles. Furthermore, the W± carries an electric charge of +1 and −1 and couples to the electromagnetic interaction. The electrically neutral Z boson interacts with both left-handed particles and antiparticles. These three gauge bosons along with the photons are grouped together, as collectively mediating the electroweak interaction.

- The eight gluons mediate the strong interactions between color charged particles (the quarks). Gluons are massless. The eightfold multiplicity of gluons is labeled by a combination of color and anticolor charge (e.g. red–antigreen).[nb 1] Because the gluons have an effective color charge, they can also interact among themselves. The gluons and their interactions are described by the theory of quantum chromodynamics.

The interactions between all the particles described by the Standard Model are summarized by the diagrams on the right of this section.

3.3.3 Higgs boson

Main article: Higgs boson

Standard Model Interactions
(Forces Mediated by Gauge Bosons)

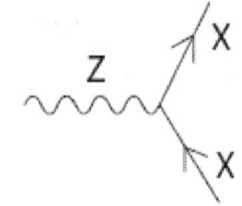

X is any fermion in
the Standard Model.

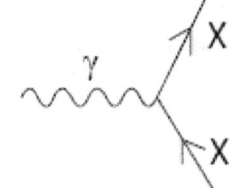

X is electrically charged.

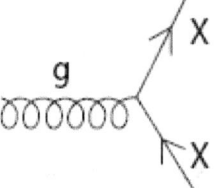

X is any quark.

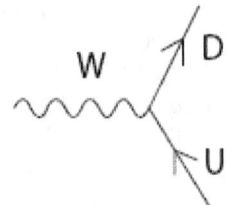

U is a up-type quark;
D is a down-type quark.

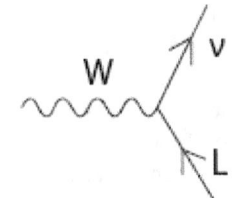

L is a lepton and ν is the
corresponding neutrino.

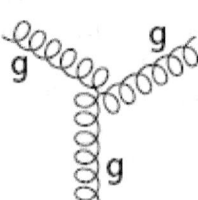

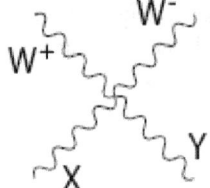

X is a photon or Z-boson.

X and Y are any two
electroweak bosons such
that charge is conserved.

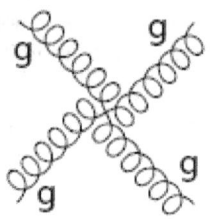

The above interactions form the basis of the standard model. Feynman diagrams in the standard model are built from these vertices. Modifications involving Higgs boson interactions and neutrino oscillations are omitted. The charge of the W bosons is dictated by the fermions they interact with; the conjugate of each listed vertex (i.e. reversing the direction of arrows) is also allowed.

The Higgs particle is a massive scalar elementary particle theorized by Robert Brout, François Englert, Peter Higgs, Gerald Guralnik, C. R. Hagen, and Tom Kibble in 1964 (see 1964 PRL symmetry breaking papers) and is a key building block in the Standard Model.[7][8][9][15] It has no intrinsic spin, and for that reason is classified as a boson (like the gauge bosons, which have integer spin).

The Higgs boson plays a unique role in the Standard Model, by explaining why the other elementary particles, except the photon and gluon, are massive. In particular, the Higgs boson explains why the photon has no mass, while the W and Z bosons are very heavy. Elementary particle masses, and the differences between electromagnetism (mediated by the photon) and the weak force (mediated by the W and Z bosons), are critical to many aspects of the structure of microscopic (and hence macroscopic) matter. In electroweak theory, the Higgs boson generates the masses of the leptons (electron, muon, and tau) and quarks. As the Higgs boson is massive, it must interact with itself.

Because the Higgs boson is a very massive particle and also decays almost immediately when created, only a very high-energy particle accelerator can observe and record it. Experiments to confirm and determine the nature of the Higgs boson using the Large Hadron Collider (LHC) at CERN began in early 2010, and were performed at Fermilab's Tevatron until its closure in late 2011. Mathematical consistency of the Standard Model requires that any mechanism capable of generating the masses of elementary particles become visible at energies above 1.4 TeV;[16] therefore, the LHC (designed to collide two 7 to 8 TeV proton beams) was built to answer the question of whether the Higgs boson actually exists.[17]

On 4 July 2012, the two main experiments at the LHC (ATLAS and CMS) both reported independently that they found a new particle with a mass of about 125 GeV/c^2 (about 133 proton masses, on the order of 10^{-25} kg), which is "consistent with the Higgs boson." Although it has several properties similar to the predicted "simplest" Higgs,[18] they acknowledged that further work would be needed to conclude that it is indeed the Higgs boson, and exactly which version of the Standard Model Higgs is best supported if confirmed.[19][20][21][22][23]

On 14 March 2013 the Higgs Boson was tentatively confirmed to exist.[24]

3.3.4   Total particle count

Counting particles by a rule that distinguishes between particles and their corresponding antiparticles, and among the many color states of quarks and gluons, gives a total of 61 elementary particles.[25]

3.4   Theoretical aspects

Main article: Standard Model (mathematical formulation)

3.4.1   Construction of the Standard Model Lagrangian

Technically, quantum field theory provides the mathematical framework for the Standard Model, in which a Lagrangian controls the dynamics and kinematics of the theory. Each kind of particle is described in terms of a dynamical field that pervades space-time. The construction of the Standard Model proceeds following the modern method of constructing most field theories: by first postulating a set of symmetries of the system, and then by writing down the most general renormalizable Lagrangian from its particle (field) content that observes these symmetries.

The global Poincaré symmetry is postulated for all relativistic quantum field theories. It consists of the familiar translational symmetry, rotational symmetry and the inertial reference frame invariance central to the theory of special relativity. The local SU(3)×SU(2)×U(1) gauge symmetry is an internal symmetry that essentially defines the Standard Model. Roughly, the three factors of the gauge symmetry give rise to the three fundamental interactions. The fields fall into different representations of the various symmetry groups of the Standard Model (see table). Upon writing the most general Lagrangian, one finds that the dynamics depend on 19 parameters, whose numerical values are established by experiment. The parameters are summarized in the table above (note: with the Higgs mass is at 125 GeV, the Higgs self-coupling strength $\lambda \sim 1/8$).

Quantum chromodynamics sector

Main article: Quantum chromodynamics

The quantum chromodynamics (QCD) sector defines the interactions between quarks and gluons, with SU(3) symmetry, generated by T^a. Since leptons do not interact with gluons, they are not affected by this sector. The Dirac Lagrangian of the quarks coupled to the gluon fields is given by

$$\mathcal{L}_{QCD} = i\overline{U}(\partial_\mu - ig_s G_\mu^a T^a)\gamma^\mu U + i\overline{D}(\partial_\mu - ig_s G_\mu^a T^a)\gamma^\mu D.$$

G_μ^a is the SU(3) gauge field containing the gluons, γ^μ are the Dirac matrices, D and U are the Dirac spinors associated with up- and down-type quarks, and g_s is the strong coupling constant.

Electroweak sector

Main article: Electroweak interaction

The electroweak sector is a Yang–Mills gauge theory with the simple symmetry group U(1)×SU(2)L,

$$\mathcal{L}_{\text{EW}} = \sum_\psi \bar{\psi}\gamma^\mu \left(i\partial_\mu - g'\frac{1}{2}Y_{\text{W}}B_\mu - g\frac{1}{2}\vec{\tau}_{\text{L}}\vec{W}_\mu \right)\psi$$

where $B\mu$ is the U(1) gauge field; Y_{W} is the weak hypercharge—the generator of the U(1) group; $\vec{W}_\mu$ is the three-component SU(2) gauge field; $\vec{\tau}_{\text{L}}$ are the Pauli matrices—infinitesimal generators of the SU(2) group. The subscript L indicates that they only act on left fermions; g' and g are coupling constants.

Higgs sector

Main article: Higgs mechanism

In the Standard Model, the Higgs field is a complex scalar of the group SU(2)L:

$$\varphi = \frac{1}{\sqrt{2}}\left(\begin{array}{c} \varphi^+ \\ \varphi^0 \end{array} \right),$$

where the indices + and 0 indicate the electric charge (Q) of the components. The weak isospin (Y_{W}) of both components is 1.

Before symmetry breaking, the Higgs Lagrangian is:

$$\mathcal{L}_{\text{H}} = \varphi^\dagger \left(\partial^\mu - \frac{i}{2}\left(g'Y_{\text{W}}B^\mu + g\vec{\tau}\vec{W}^\mu \right) \right)\left(\partial_\mu + \frac{i}{2}\left(g'Y_{\text{W}}B_\mu + g\vec{\tau}\vec{W}_\mu \right) \right)\varphi - \frac{\lambda^2}{4}\left(\varphi^\dagger\varphi - v^2 \right)^2,$$

which can also be written as:

$$\mathcal{L}_{\text{H}} = \left| \left(\partial_\mu + \frac{i}{2}\left(g'Y_{\text{W}}B_\mu + g\vec{\tau}\vec{W}_\mu \right) \right)\varphi \right|^2 - \frac{\lambda^2}{4}\left(\varphi^\dagger\varphi - v^2 \right)^2.$$

3.5 Fundamental forces

Main article: Fundamental interaction

The Standard Model classified all four fundamental forces in nature. In the Standard Model, a force is described as an exchange of bosons between the objects affected, such as a photon for the electromagnetic force and a gluon for the strong interaction. Those particles are called force carriers.[26]

3.6 Tests and predictions

The Standard Model (SM) predicted the existence of the W and Z bosons, gluon, and the top and charm quarks before these particles were observed. Their predicted properties were experimentally confirmed with good precision. To give an idea of the success of the SM, the following table compares the measured masses of the W and Z bosons with the masses predicted by the SM:

The SM also makes several predictions about the decay of Z bosons, which have been experimentally confirmed by the Large Electron-Positron Collider at CERN.

In May 2012 BaBar Collaboration reported that their recently analyzed data may suggest possible flaws in the Standard Model of particle physics.[28][29] These data show that a particular type of particle decay called "B to D-star-tau-nu" happens more often than the Standard Model says it should. In this type of decay, a particle called the B-bar meson decays into a D meson, an antineutrino and a tau-lepton. While the level of certainty of the excess (3.4 sigma) is not enough to claim a break from the Standard Model, the results are a potential sign of something amiss and are likely to impact existing theories, including those attempting to deduce the properties of Higgs bosons.[30]

On December 13, 2012, physicists reported the constancy, over space and time, of a basic physical constant of nature that supports the *standard model of physics*. The scientists, studying methanol molecules in a distant galaxy, found the change $(\Delta\mu/\mu)$ in the proton-to-electron mass ratio μ to be equal to "$(0.0 \pm 1.0) \times 10^{-7}$ at redshift z = 0.89" and consistent with "a null result".[31][32]

3.7 Challenges

See also: Physics beyond the Standard Model

Self-consistency of the Standard Model (currently formulated as a non-abelian gauge theory quantized through path-integrals) has not been mathematically proven. While regularized versions useful for approximate computations (for example lattice gauge theory) exist, it is not known whether they converge (in the sense of S-matrix elements) in the limit that the regulator is removed. A key question related to the consistency is the Yang–Mills existence and mass gap problem.

Experiments indicate that neutrinos have mass, which the classic Standard Model did not allow.[33] To accommodate this finding, the classic Standard Model can be modified to include neutrino mass.

If one insists on using only Standard Model particles, this can be achieved by adding a non-renormalizable interaction of leptons with the Higgs boson.[34] On a fundamental level, such an interaction emerges in the seesaw mechanism where heavy right-handed neutrinos are added to the theory. This is natural in the left-right symmetric extension of the Standard Model[35][36] and in certain grand unified theories.[37] As long as new physics appears below or around 10^{14} GeV, the neutrino masses can be of the right order of magnitude.

Theoretical and experimental research has attempted to extend the Standard Model into a Unified field theory or a Theory of everything, a complete theory explaining all physical phenomena including constants. Inadequacies of the Standard Model that motivate such research include:

- It does not attempt to explain gravitation, although a theoretical particle known as a graviton would help explain it, and unlike for the strong and electroweak interactions of the Standard Model, there is no known way of describing general relativity, the canonical theory of gravitation, consistently in terms of quantum field theory. The reason for this is, among other things, that quantum field theories of gravity generally break down before reaching the Planck scale. As a consequence, we have no reliable theory for the very early universe;

- Some consider it to be *ad hoc* and inelegant, requiring 19 numerical constants whose values are unrelated and arbitrary. Although the Standard Model, as it now stands, can explain why neutrinos have masses, the specifics of neutrino mass are still unclear. It is believed that explaining neutrino mass will require an additional 7 or 8 constants, which are also arbitrary parameters;

- The Higgs mechanism gives rise to the hierarchy problem if some new physics (coupled to the Higgs) is present at high energy scales. In these cases in order for the weak scale to be much smaller than the Planck scale, severe fine tuning of the parameters is required; there are, however, other scenarios that include quantum gravity in which such fine tuning can be avoided.[38]There are also issues of Quantum triviality, which suggests that it may not be possible to create a consistent quantum field theory involving elementary scalar particles.

- It should be modified so as to be consistent with the emerging "Standard Model of cosmology." In particular, the Standard Model cannot explain the observed amount of cold dark matter (CDM) and gives contributions to dark energy which are many orders of magnitude too large. It is also difficult to accommodate the observed predominance of matter over antimatter (matter/antimatter asymmetry). The isotropy and homogeneity of the visible universe over large distances seems to require a mechanism like cosmic inflation, which would also constitute an extension of the Standard Model.

- The existence of ultra-high-energy cosmic rays are difficult to explain under the Standard Model.

Currently, no proposed Theory of Everything has been widely accepted or verified.

3.8 See also

- Fundamental interaction:

  - Quantum electrodynamics

  - Strong interaction: Color charge, Quantum chromodynamics, Quark model

  - Weak interaction: Electroweak theory, Fermi theory of beta decay, Weak hypercharge, Weak isospin

- Gauge theory: Nontechnical introduction to gauge theory

- Generation

- Higgs mechanism: Higgs boson, Higgsless model

- J. C. Ward

- J. J. Sakurai Prize for Theoretical Particle Physics

- Lagrangian

- Open questions: BTeV experiment, CP violation, Neutrino masses, Quark matter, Quantum triviality

- Penguin diagram

- Quantum field theory

- Standard Model: Mathematical formulation of, Physics beyond the Standard Model

3.9 Notes and references

[1] Technically, there are nine such color–anticolor combinations. However, there is one color-symmetric combination that can be constructed out of a linear superposition of the nine combinations, reducing the count to eight.

3.10   References

[1] R. Oerter (2006). *The Theory of Almost Everything: The Standard Model, the Unsung Triumph of Modern Physics* (Kindle ed.). Penguin Group. p. 2. ISBN 0-13-236678-9.

[2] In fact, there are mathematical issues regarding quantum field theories still under debate (see e.g. Landau pole), but the predictions extracted from the Standard Model by current methods applicable to current experiments are all self-consistent. For a further discussion see e.g. Chapter 25 of R. Mann (2010). *An Introduction to Particle Physics and the Standard Model*. CRC Press. ISBN 978-1-4200-8298-2.

[3] Sean Carroll, Ph.D., Cal Tech, 2007, The Teaching Company, *Dark Matter, Dark Energy: The Dark Side of the Universe*, Guidebook Part 2 page 59, Accessed Oct. 7, 2013, "...Standard Model of Particle Physics: The modern theory of elementary particles and their interactions ... It does not, strictly speaking, include gravity, although it's often convenient to include gravitons among the known particles of nature..."

[4] S.L.Glashow(1961). "Partial-symmetries of weak interactions".*Nuclear Physics***22**(4):579–588.Bibcode:1961NucPh..22. doi:10.1016/0029-5582(61)90469-2.

[5] S. Weinberg (1967). "A Model of Leptons". *Physical Review Letters* **19** (21): 1264–1266. Bibcode:1967PhRvL..19.1264W. doi:10.1103/PhysRevLett.19.1264.

[6] A. Salam (1968). N. Svartholm, ed. *Elementary Particle Physics: Relativistic Groups and Analyticity*. Eighth Nobel Symposium. Stockholm: Almquvist and Wiksell. p. 367.

[7] F. Englert, R. Brout (1964). "Broken Symmetry and the Mass of Gauge Vector Mesons". *Physical Review Letters* **13** (9): 321–323. Bibcode:1964PhRvL..13..321E. doi:10.1103/PhysRevLett.13.321.

[8] P.W. Higgs (1964). "Broken Symmetries and the Masses of Gauge Bosons". *Physical Review Letters* **13** (16): 508–509. Bibcode:1964PhRvL..13..508H. doi:10.1103/PhysRevLett.13.508.

[9] G.S. Guralnik, C.R. Hagen, T.W.B. Kibble (1964). "Global Conservation Laws and Massless Particles". *Physical Review Letters* **13** (20): 585–587. Bibcode:1964PhRvL..13..585G. doi:10.1103/PhysRevLett.13.585.

[10] F.J.Hasert et al. (1973). "Search for elastic muon-neutrino electron scattering".*Physics Letters B***46**(1):121.Bibcode:1973PhL. doi:10.1016/0370-2693(73)90494-2.

[11] F.J. Hasert et al. (1973). "Observation of neutrino-like interactions without muon or electron in the Gargamelle neutrino experiment". *Physics Letters B* **46** (1): 138. Bibcode:1973PhLB...46..138H. doi:10.1016/0370-2693(73)90499-1.

[12] F.J. Hasert et al. (1974). "Observation of neutrino-like interactions without muon or electron in the Gargamelle neutrino experiment". *Nuclear Physics B* **73** (1): 1. Bibcode:1974NuPhB..73....1H. doi:10.1016/0550-3213(74)90038-8.

[13] D. Haidt (4 October 2004). "The discovery of the weak neutral currents". *CERN Courier*. Retrieved 8 May 2008.

[14] "Details can be worked out if the situation is simple enough for us to make an approximation, which is almost never, but often we can understand more or less what is happening." from *The Feynman Lectures on Physics*, Vol 1. pp. 2–7

[15] G.S. Guralnik (2009). "The History of the Guralnik, Hagen and Kibble development of the Theory of Spontaneous Symmetry Breaking and Gauge Particles". *International Journal of Modern Physics A* **24** (14): 2601–2627. arXiv:0907.3466. Bibcode:2009IJMPA..24.2601G. doi:10.1142/S0217751X09045431.

[16] B.W. Lee, C. Quigg, H.B. Thacker (1977). "Weak interactions at very high energies: The role of the Higgs-boson mass". *Physical Review D* **16** (5): 1519–1531. Bibcode:1977PhRvD..16.1519L. doi:10.1103/PhysRevD.16.1519.

[17] "Huge $10 billion collider resumes hunt for 'God particle'". CNN. 11 November 2009. Retrieved 2010-05-04.

[18] M. Strassler (10 July 2012). "Higgs Discovery: Is it a Higgs?". Retrieved 2013-08-06.

[19] "CERN experiments observe particle consistent with long-sought Higgs boson". CERN. 4 July 2012. Retrieved 2012-07-04.

[20] "Observation of a New Particle with a Mass of 125 GeV". CERN. 4 July 2012. Retrieved 2012-07-05.

[21] "ATLAS Experiment". ATLAS. 1 January 2006. Retrieved 2012-07-05.

[22] "Confirmed: CERN discovers new particle likely to be the Higgs boson". *YouTube*. Russia Today. 4 July 2012. Retrieved 2013-08-06.

[23] D. Overbye (4 July 2012). "A New Particle Could Be Physics' Holy Grail". *New York Times*. Retrieved 2012-07-04.

[24] "New results indicate that new particle is a Higgs boson". CERN. 14 March 2013. Retrieved 2013-08-06.

[25] S. Braibant, G. Giacomelli, M. Spurio (2009). *Particles and Fundamental Interactions: An Introduction to Particle Physics*. Springer. pp. 313–314. ISBN 978-94-007-2463-1.

[26] http://home.web.cern.ch/about/physics/standard-model Official CERN website

[27] http://www.pha.jhu.edu/~{ }dfehling/particle.gif

[28] "BABAR Data in Tension with the Standard Model". SLAC. 31 May 2012. Retrieved 2013-08-06.

[29] BaBar Collaboration (2012). "Evidence for an excess of $B \to D^{(*)} \tau^- \nu\tau$ decays". *Physical Review Letters* **109** (10): 101802. arXiv:1205.5442. Bibcode:2012PhRvL.109j1802L. doi:10.1103/PhysRevLett.109.101802.

[30] "BaBar data hint at cracks in the Standard Model". *e! Science News*. 18 June 2012. Retrieved 2013-08-06.

[31] J. Bagdonaite et al. (2012). "A Stringent Limit on a Drifting Proton-to-Electron Mass Ratio from Alcohol in the Early Universe". *Science* **339** (6115): 46. Bibcode:2013Sci...339...46B. doi:10.1126/science.1224898.

[32] C. Moskowitz (13 December 2012). "Phew! Universe's Constant Has Stayed Constant". Space.com. Retrieved 2012-12-14.

[33] "Particle chameleon caught in the act of changing". CERN. 31 May 2010. Retrieved 2012-07-05.

[34] S.Weinberg(1979). "Baryon and Lepton Nonconserving Processes".*Physical Review Letters***43**(21):1566.Bibcode:1979Ph. doi:10.1103/PhysRevLett.43.1566.

[35] P.Minkowski(1977). "μ→eγat a Rate of One Out of10₉Muon Decays?".*Physics Letters B***67**(4):421.Bibcode:1977PhLB.1M. doi:10.1016/0370-2693(77)90435-X.

[36] R. N. Mohapatra, G. Senjanovic (1980). "Neutrino Mass and Spontaneous Parity Nonconservation". *Physical Review Letters* **44** (14): 912–915. Bibcode:1980PhRvL..44..912M. doi:10.1103/PhysRevLett.44.912.

[37] M. Gell-Mann, P. Ramond and R. Slansky (1979). F. van Nieuwenhuizen and D. Z. Freedman, ed. *Supergravity*. North Holland. pp. 315–321. ISBN 0-444-85438-X.

[38] Salvio,Strumia(2014-03-17)."Agravity".*JHEP1406(2014)080*.arXiv:1403.4226.Bibcode:2014JHEP...06..080S.doi:)080.

3.11 Further reading

- R. Oerter (2006). *The Theory of Almost Everything: The Standard Model, the Unsung Triumph of Modern Physics*. Plume.

- B.A. Schumm (2004). *Deep Down Things: The Breathtaking Beauty of Particle Physics*. Johns Hopkins University Press. ISBN 0-8018-7971-X.

- "The Standard Model of Particle Physics Interactive Graphic".

Introductory textbooks

- I. Aitchison, A. Hey (2003). *Gauge Theories in Particle Physics: A Practical Introduction*. Institute of Physics. ISBN 978-0-585-44550-2.

- W. Greiner, B. Müller (2000). *Gauge Theory of Weak Interactions*. Springer. ISBN 3-540-67672-4.

- G.D. Coughlan, J.E. Dodd, B.M. Gripaios (2006). *The Ideas of Particle Physics: An Introduction for Scientists*. Cambridge University Press.

- D.J. Griffiths (1987). *Introduction to Elementary Particles*. John Wiley & Sons. ISBN 0-471-60386-4.

- G.L. Kane (1987). *Modern Elementary Particle Physics*. Perseus Books. ISBN 0-201-11749-5.

Advanced textbooks

- T.P. Cheng, L.F. Li (2006). *Gauge theory of elementary particle physics*. Oxford University Press. ISBN 0-19-851961-3. Highlights the gauge theory aspects of the Standard Model.

- J.F. Donoghue, E. Golowich, B.R. Holstein (1994). *Dynamics of the Standard Model*. Cambridge University Press. ISBN 978-0-521-47652-2. Highlights dynamical and phenomenological aspects of the Standard Model.

- L. O'Raifeartaigh (1988). *Group structure of gauge theories*. Cambridge University Press. ISBN 0-521-34785-8.

- Nagashima Y. Elementary Particle Physics: Foundations of the Standard Model, Volume 2. (Wiley 2013) 920 рапуы

- Schwartz, M.D. Quantum Field Theory and the Standard Model (Cambridge University Press 2013) 952 pages

- Langacker P. The standard model and beyond. (CRC Press, 2010) 670 pages Highlights group-theoretical aspects of the Standard Model.

Journal articles

- E.S.Abers,B.W.Lee(1973). "Gauge theories".*Physics Reports***9**:1–141.Bibcode:1973PhR.....9....1A.doi:10.101-1573(73)90027-6.

- M. Baak et al. (2012). "The Electroweak Fit of the Standard Model after the Discovery of a New Boson at the LHC". *The European Physical Journal C* **72** (11). arXiv:1209.2716. Bibcode:2012EPJC...72.2205B. doi:10.1140/epjc/s10052-012-2205-9.

- Y. Hayato et al. (1999). "Search for Proton Decay through $p \rightarrow \nu K^+$ in a Large Water Cherenkov Detector". *Physical Review Letters***83**(8):1529.arXiv:hep-ex/9904020.Bibcode:1999PhRvL..83.1529H.doi:10.1103/Phy.

- S.F. Novaes (2000). "Standard Model: An Introduction". arXiv:hep-ph/0001283 [hep-ph].

- D.P. Roy (1999). "Basic Constituents of Matter and their Interactions — A Progress Report". arXiv:hep-ph/9912523 [hep-ph].

- F. Wilczek (2004). "The Universe Is A Strange Place". *Nuclear Physics B - Proceedings Supplements* **134**: 3. arXiv:astro-ph/0401347. Bibcode:2004NuPhS.134....3W. doi:10.1016/j.nuclphysbps.2004.08.001.

3.12   External links

- "The Standard Model explained in Detail by CERN's John Ellis" omega tau podcast.

- "LHC sees hint of lightweight Higgs boson" "New Scientist".

- "Standard Model may be found incomplete," *New Scientist*.

- "Observation of the Top Quark" at Fermilab.

- "The Standard Model Lagrangian." After electroweak symmetry breaking, with no explicit Higgs boson.

- "Standard Model Lagrangian" with explicit Higgs terms. PDF, PostScript, and LaTeX versions.

- "The particle adventure." Web tutorial.

- Nobes, Matthew (2002) "Introduction to the Standard Model of Particle Physics" on Kuro5hin: Part 1, Part 2, Part 3a, Part 3b.

- "The Standard Model" The Standard Model on the CERN web site explains how the basic building blocks of matter interact, governed by four fundamental forces.

Chapter 4

Nuclear force

This article is about the force that holds nucleons together in a nucleus. For the force that holds quarks together in a nucleon, see Strong interaction.

The **nuclear force** (or **nucleon–nucleon interaction** or **residual strong force**) is the force between protons and

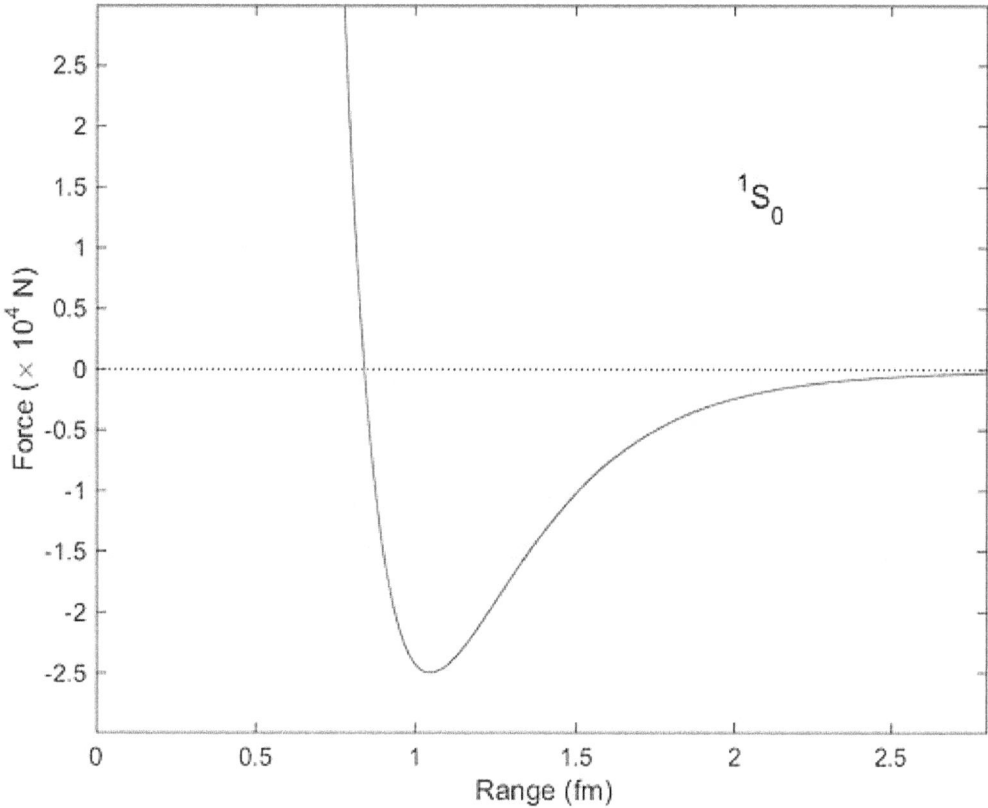

Force (in units of 10,000 N) between two nucleons as a function of distance as computed from the Reid potential (1968).[1] The spins of the neutron and proton are aligned, and they are in the S angular momentum state. The attractive (negative) force has a maximum at a distance of about 1 fm with a force of about 25,000 N. Particles much closer than a distance of 0.8 fm experience a large repulsive (positive) force. Particles separated by a distance greater than 1 fm are still attracted (Yukawa potential), but the force falls as an exponential function of distance.

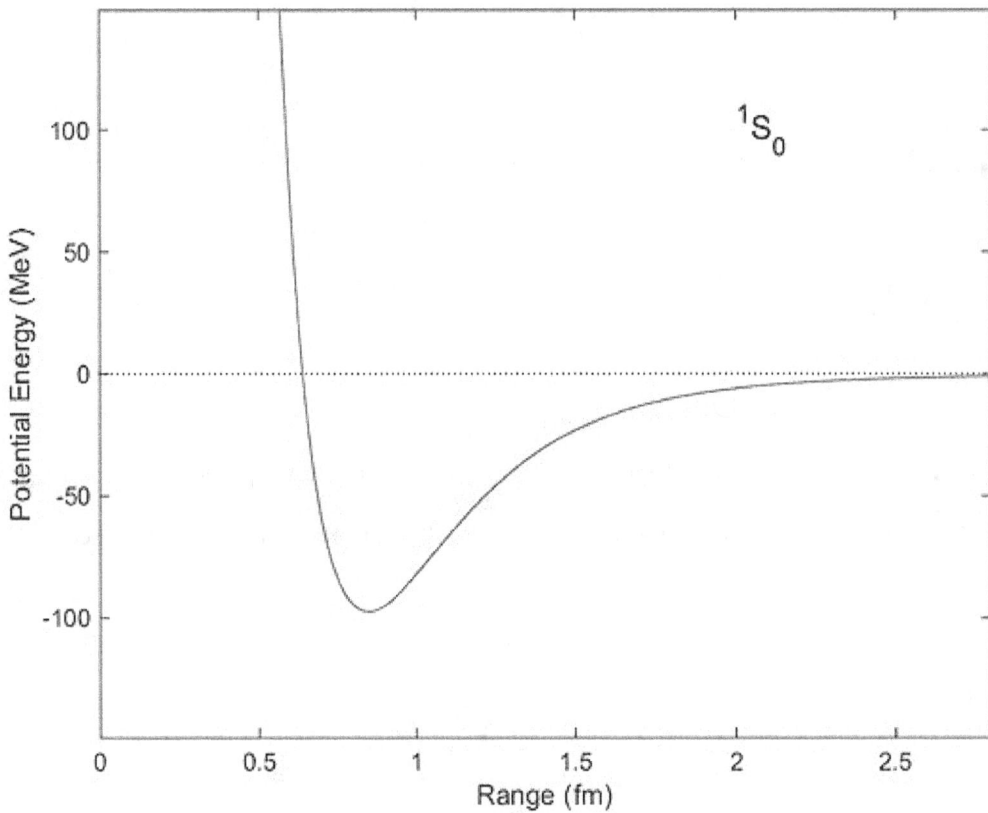

Corresponding potential energy (in units of MeV) of two nucleons as a function of distance as computed from the Reid potential. The potential well is a minimum at a distance of about 0.8 fm. With this potential nucleons can become bound with a negative "binding energy."

neutrons, subatomic particles that are collectively called nucleons. The nuclear force is responsible for binding protons and neutrons into atomic nuclei. Neutrons and protons are affected by the nuclear force almost identically. Since protons have charge +1 e, they experience a Coulomb repulsion that tends to push them apart, but at short range the nuclear force is sufficiently attractive as to overcome the electromagnetic repulsive force. The mass of a nucleus is less than the sum total of the individual masses of the protons and neutrons which form it. The difference in mass between bound and unbound nucleons is known as the mass defect. Energy is released when nuclei break apart, and it is this energy that used in nuclear power and nuclear weapons.[2][3]

The nuclear force is powerfully attractive between nucleons at distances of about 1 femtometer (fm, or 1.0×10^{-15} metres) between their centers, but rapidly decreases to insignificance at distances beyond about 2.5 fm. At distances less than 0.7 fm, the nuclear force becomes repulsive. This repulsive component is responsible for the physical size of nuclei, since the nucleons can come no closer than the force allows. By comparison, the size of an atom, measured in angstroms (Å, or 1.0×10^{-10} m), is five orders of magnitude larger. The nuclear force is not simple, however, since it depends on the nucleon spins, has a tensor component, and may depend on the relative momentum of the nucleons.[4]

A quantitative description of the nuclear force relies on partially empirical equations that model the internucleon potential energies, or potentials. (Generally, forces within a system of particles can be more simply modeled by describing the system's potential energy; the negative gradient of a potential is equal to the vector force.) The constants for the equations are phenomenological, that is, determined by fitting the equations to experimental data. The internucleon potentials attempt to describe the properties of nucleon–nucleon interaction. Once determined, any given potential can be used in, e.g., the Schrödinger equation to determine the quantum mechanical properties of the nucleon system.

The discovery of the neutron in 1932 revealed that atomic nuclei were made of protons and neutrons, held together by an attractive force. By 1935 the nuclear force was conceived to be transmitted by particles called mesons. This theoretical development included a description of the Yukawa potential, an early example of a nuclear potential. Mesons, predicted by theory, were discovered experimentally in 1947. By the 1970s, the quark model had been developed, which showed that the mesons and nucleons were composed of quarks and gluons. By this new model, the nuclear force, resulting from the exchange of mesons between neighboring nucleons, is a residual effect of the strong force.

4.1   Description

The nuclear force is only felt between particles composed of quarks, or hadrons. At small separations between nucleons (less than ~ 0.7 fm between their centers, depending upon spin alignment) the force becomes repulsive, which keeps the nucleons at a certain average separation, even if they are of different types. This repulsion arises from the Pauli exclusion force for identical nucleons (such as two neutrons or two protons). A Pauli exclusion force also occurs between quarks of the same type within nucleons, when the nucleons are different (a proton and a neutron, for example). The nuclear force also has a "tensor" component which depends on whether or not the spins (angular momentum vectors) of the nucleons are aligned (point in the same direction) or anti-aligned (i.e., point in opposite directions in space).

At distances larger than 0.7 fm the force becomes attractive between spin-aligned nucleons, becoming maximal at a center–center distance of about 0.9 fm. Beyond this distance the force drops exponentially, until beyond about 2.0 fm separation, the force is negligible. Nucleons have a radius of about 0.8 fm.[5]

At short distances (less than 1.7 fm or so), the nuclear force is stronger than the Coulomb force between protons; it thus overcomes the repulsion of protons inside the nucleus. However, the Coulomb force between protons has a much larger range due to its decay as the inverse square of charge separation, and Coulomb repulsion thus becomes the only significant force between protons when their separation exceeds about 2 to 2.5 fm.

For two particles that are the same (such as two neutrons or two protons) the force is not enough to bind the particles, since the spin vectors of two particles of the same type must point in opposite directions when the particles are near each other and are (save for spin) in the same quantum state. This requirement for fermions stems from the Pauli exclusion principle. For fermion particles of different types (such as a proton and neutron), particles may be close to each other and have aligned spins without violating the Pauli exclusion principle, and the nuclear force may bind them (in this case, into a deuteron), since the nuclear force is much stronger for spin-aligned particles. But if the particles' spins are anti-aligned the nuclear force is too weak to bind them, even if they are of different types.

To disassemble a nucleus into unbound protons and neutrons requires work against the nuclear force. Conversely, energy is released when a nucleus is created from free nucleons or other nuclei: the nuclear binding energy. Because of mass–energy equivalence (i.e. Einstein's famous formula $E = mc^2$), releasing this energy causes the mass of the nucleus to be lower than the total mass of the individual nucleons, leading to the so-called "mass defect".[6]

The nuclear force is nearly independent of whether the nucleons are neutrons or protons. This property is called *charge independence*. The force depends on whether the spins of the nucleons are parallel or antiparallel, and it has a noncentral or *tensor* component. This part of the force does not conserve orbital angular momentum, which is a constant of motion under central forces.

The symmetry resulting in the strong force, proposed by Werner Heisenberg, is that protons and neutrons are identical in every respect, other than their charge. This is not completely true, because neutrons are a tiny bit heavier, but it is an approximate symmetry. Protons and neutrons are therefore viewed as the same particle, but with different isospin quantum number. The strong force is invariant under SU(2) transformations, just as particles with "regular spin" are. Isospin and "regular" spin are related under this SU(2) symmetry group. There are only strong attractions when the total isospin is 0, as is confirmed by experiment.[7]

The information on nuclear force are obtained by scattering experiments and the study of light nuclei binding energy.

The nuclear force occurs by the exchange of virtual light mesons, such as the virtual pions, as well as two types of virtual mesons with spin (vector mesons), the rho mesons and the omega mesons. The vector mesons account for the spin-dependence of the nuclear force in this "virtual meson" picture.

The nuclear force is separate from what historically was known as the weak nuclear force. The weak interaction is one

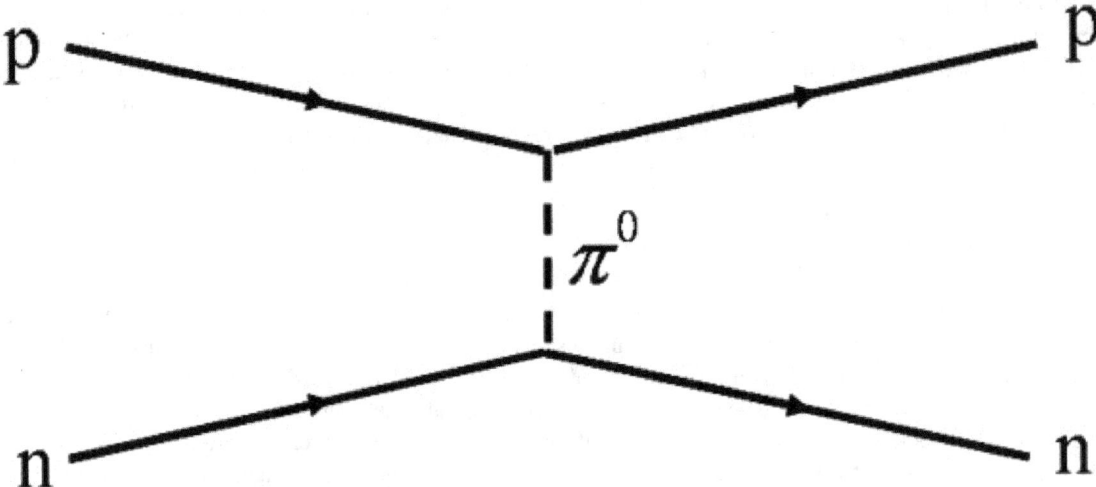

A Feynman diagram of a strong proton–neutron interaction mediated by a neutral pion. Time proceeds from left to right.

of the four fundamental interactions, and it refers to such processes as beta decay. The weak force plays no role in the interaction of nucleons, though it is responsible for the decay of neutrons to protons and vice versa.

4.2 History

The nuclear force has been at the heart of nuclear physics ever since the field was born in 1932 with the discovery of the neutron by James Chadwick. The traditional goal of nuclear physics is to understand the properties of atomic nuclei in terms of the 'bare' interaction between pairs of nucleons, or nucleon–nucleon forces (NN forces).

Within months after the discovery of the neutron, Werner Heisenberg[8][9][10] and Dmitri Ivanenko[11] had proposed proton–neutron models for the nucleus.[12] Heisenberg approached the description of protons and neutrons in the nucleus through quantum mechanics, an approach that was not at all obvious at the time. Heisenberg's theory for protons and neutrons in the nucleus was a "major step toward understanding the nucleus as a quantum mechanical system."[13] Heisenberg introduced the first theory of nuclear exchange forces that bind the nucleons. He considered protons and neutrons to be different quantum states of the same particle, i.e., nucleons distinguished by the value of their nuclear isospin quantum numbers.

One of the earliest models for the nucleus was the liquid drop model developed in the 1930s. One property of nuclei is that the average binding energy per nucleon is approximately the same for all stable nuclei, which is similar to a liquid drop. The liquid drop model treated the nucleus as a drop of incompressible nuclear fluid, with nucleons behaving like molecules in a liquid. The model was first proposed by George Gamow and then developed by Niels Bohr, Werner Heisenberg and Carl Friedrich von Weizsäcker. This crude model did not explain all the properties of the nucleus, but it did explain the spherical shape of most nuclei. The model also gave good predictions for the nuclear binding energy of nuclei.

In 1934, Hideki Yukawa made the earliest attempt to explain the nature of the nuclear force. According to his theory, massive bosons (mesons) mediate the interaction between two nucleons. Although, in light of quantum chromodynamics (QCD), meson theory is no longer perceived as fundamental, the meson-exchange concept (where hadrons are treated as elementary particles) continues to represent the best working model for a quantitative *NN* potential. The Yukawa potential (also called a screened Coulomb potential) is a potential of the form

$$V_{\text{Yukawa}}(r) = -g^2 \frac{e^{-\mu r}}{r},$$

where g is a magnitude scaling constant, i.e., the amplitude of potential, μ is the Yukawa particle mass, r is the radial distance to the particle. The potential is monotone increasing, implying that the force is always attractive. The constants are determined empirically. The Yukawa potential depends only on the distance between particles, r, hence it models a central force.

Throughout the 1930s a group at Columbia University led by I. I. Rabi developed magnetic resonance techniques to determine the magnetic moments of nuclei. These measurements led to the discovery in 1939 that the deuteron also possessed an electric quadrupole moment.[14][15] This electrical property of the deuteron had been interfering with the measurements by the Rabi group. The deuteron, composed of a proton and a neutron, is one of the simplest nuclear systems. The discovery meant that the physical shape of the deuteron was not symmetric, which provided valuable insight into the nature of the nuclear force binding nucleons. In particular, the result showed that the nuclear force was not a central force, but had a tensor character.[1] Hans Bethe identified the discovery of the deuteron's quadrupole moment as one of the important events during the formative years of nuclear physics.[14]

Historically, the task of describing the nuclear force phenomenologically was formidable. The first semi-empirical quantitative models came in the mid-1950s,[1] such as the Woods–Saxon potential (1954). There was substantial progress in experiment and theory related to the nuclear force in the 1960s and 1970s. One influential model was the Reid potential (1968).[1] In recent years, experimenters have concentrated on the subtleties of the nuclear force, such as its charge dependence, the precise value of the πNN coupling constant, improved phase shift analysis, high-precision NN data, high-precision NN potentials, NN scattering at intermediate and high energies, and attempts to derive the nuclear force from QCD.

4.3   The nuclear force as a residual of the strong force

The nuclear force is a residual effect of the more fundamental strong force, or strong interaction. The strong interaction is the attractive force that binds the elementary particles called quarks together to form the nucleons themselves. This more powerful force is mediated by particles called gluons. Gluons hold quarks together with a force like that of electric charge, but of far greater strength. Quarks, gluons and their dynamics are mostly confined within nucleons, but residual influences extend slightly beyond nucleon boundaries to give rise to the nuclear force.

The nuclear forces arising between nucleons are analogous to the forces in chemistry between neutral atoms or molecules called London forces. Such forces between atoms are much weaker than the attractive electrical forces that hold the atoms themselves together (i.e., that bind electrons to the nucleus), and their range between atoms is shorter, because they arise from small separation of charges inside the neutral atom. Similarly, even though nucleons are made of quarks in combinations which cancel most gluon forces (they are "color neutral"), some combinations of quarks and gluons nevertheless leak away from nucleons, in the form of short-range nuclear force fields that extend from one nucleon to another nearby nucleon. These nuclear forces are very weak compared to direct gluon forces ("color forces" or strong forces) inside nucleons, and the nuclear forces extend only over a few nuclear diameters, falling exponentially with distance. Nevertheless, they are strong enough to bind neutrons and protons over short distances, and overcome the electrical repulsion between protons in the nucleus.

Sometimes, the nuclear force is called the **residual strong force**, in contrast to the strong interactions which arise from QCD. This phrasing arose during the 1970s when QCD was being established. Before that time, the *strong nuclear force* referred to the inter-nucleon potential. After the verification of the quark model, *strong interaction* has come to mean QCD.

4.4   Nucleon–nucleon potentials

Two-nucleon systems such as the deuteron, the nucleus of a deuterium atom, as well as proton–proton or neutron–proton scattering are ideal for studying the NN force. Such systems can be described by attributing a *potential* (such as the Yukawa potential) to the nucleons and using the potentials in a Schrödinger equation. The form of the potential is derived phenomenologically, although for the long-range interaction, meson-exchange theories help to construct the potential. The parameters of the potential are determined by fitting to experimental data such as the deuteron binding energy or NN

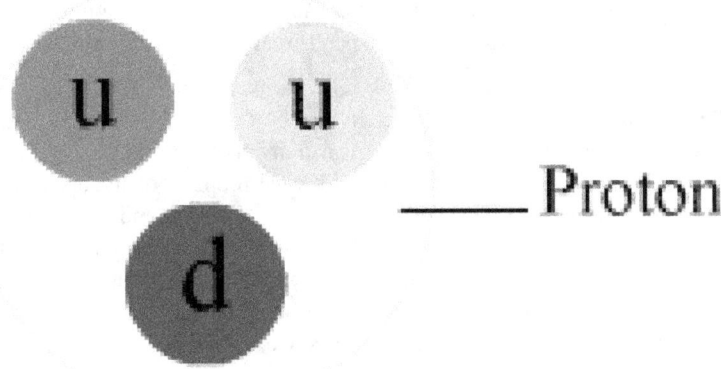

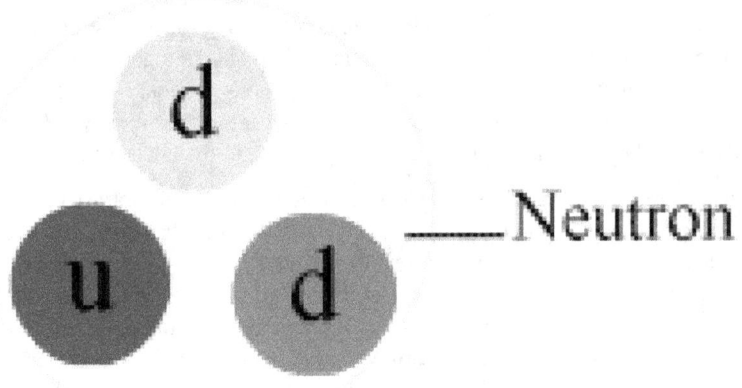

An animation of the interaction. The colored double circles are gluons. Anticolors are shown as per this diagram (larger version).

elastic scattering cross sections (or, equivalently in this context, so-called *NN* phase shifts).

The most widely used *NN* potentials are the Paris potential, the Argonne AV18 potential ,[16] the CD-Bonn potential and the Nijmegen potentials.

A more recent approach is to develop effective field theories for a consistent description of nucleon–nucleon and three-nucleon forces. In particular, chiral symmetry breaking can be analyzed in terms of an effective field theory (called chiral perturbation theory) which allows perturbative calculations of the interactions between nucleons with pions as exchange particles.

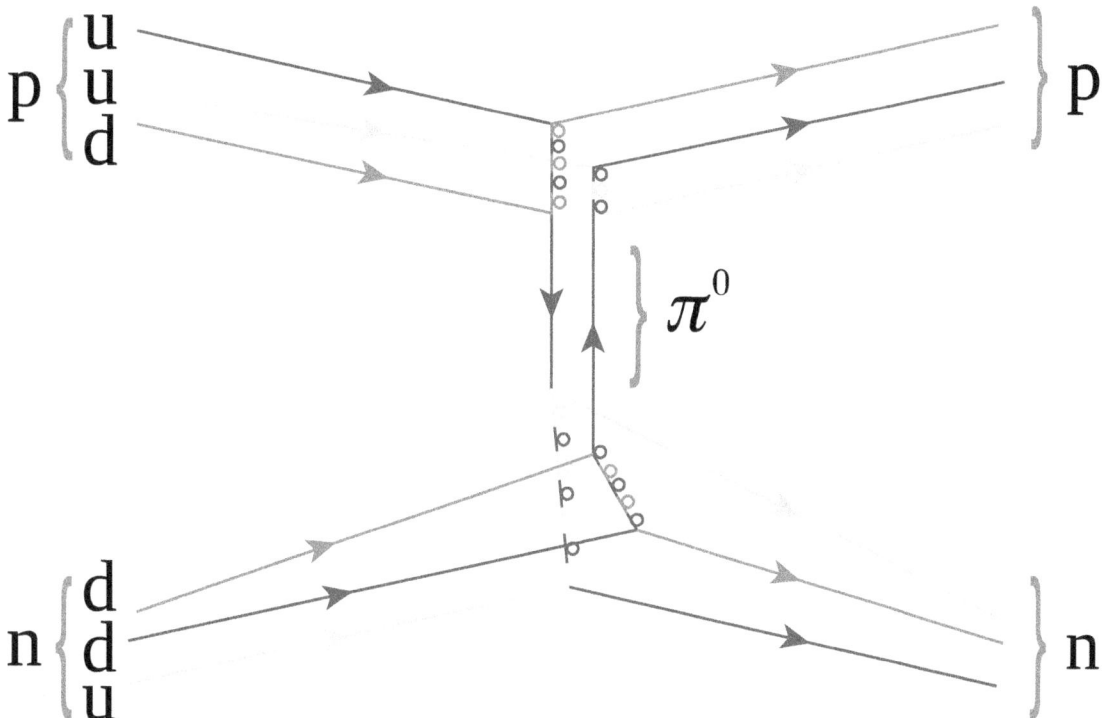

The same diagram as that above with the individual quark constituents shown, to illustrate how the fundamental *strong interaction gives rise to the* **nuclear force**. *Straight lines are quarks, while multi-colored loops are gluons (the carriers of the fundamental force). Other gluons, which bind together the proton, neutron, and pion "in-flight," are not shown.*

4.4.1   From nucleons to nuclei

The ultimate goal of nuclear physics would be to describe all nuclear interactions from the basic interactions between nucleons. This is called the *microscopic* or *ab initio* approach of nuclear physics. There are two major obstacles to overcome before this dream can become reality:

- Calculations in many-body systems are difficult and require advanced computation techniques.

- There is evidence that three-nucleon forces (and possibly higher multi-particle interactions) play a significant role. This means that three-nucleon potentials must be included into the model.

This is an active area of research with ongoing advances in computational techniques leading to better first-principles calculations of the nuclear shell structure. Two- and three-nucleon potentials have been implemented for nuclides up to $A = 12$.

4.4.2   Nuclear potentials

A successful way of describing nuclear interactions is to construct one potential for the whole nucleus instead of considering all its nucleon components. This is called the *macroscopic* approach. For example, scattering of neutrons from nuclei can be described by considering a plane wave in the potential of the nucleus, which comprises a real part and an imaginary part. This model is often called the **optical model** since it resembles the case of light scattered by an opaque glass sphere.

Nuclear potentials can be *local* or *global*: local potentials are limited to a narrow energy range and/or a narrow nuclear mass range, while global potentials, which have more parameters and are usually less accurate, are functions of the energy and the nuclear mass and can therefore be used in a wider range of applications.

4.5 See also

- Strong interaction

- Standard Model

4.6 References

[1] Reid,R.V. (1968). "Local phenomenological nucleon–nucleon potentials".*Annals of Physics***50**:411–448.Bibcode:19R. doi:10.1016/0003-4916(68)90126-7.

[2] Binding Energy, Mass Defect, Furry Elephant physics educational site, retr 2012 7 1

[3] Chapter 4 NUCLEAR PROCESSES, THE STRONG FORCE, M. Ragheb 1/30/2013, University of Illinois

[4] Kenneth S. Krane (1988). *Introductory Nuclear Physics.* Wiley & Sons. ISBN 0-471-80553-X.

[5] Povh, B.; Rith, K.; Scholz, C.; Zetsche, F. (2002). *Particles and Nuclei: An Introduction to the Physical Concepts.* Berlin: Springer-Verlag. p. 73. ISBN 978-3-540-43823-6.

[6] Stern, Dr. Swapnil Nikam (February 11, 2009). "Nuclear Binding Energy". *"From Stargazers to Starships".* NASA website. Retrieved 2010-12-30.

[7] Griffiths, David, Introduction to Elementary Particles

[8] Heisenberg, W. (1932). "Über den Bau der Atomkerne. I". *Z. Phys.* **77**: 1–11. doi:10.1007/BF01342433.

[9] Heisenberg, W. (1932). "Über den Bau der Atomkerne. II". *Z. Phys.* **78** (3–4): 156–164. doi:10.1007/BF01337585.

[10] Heisenberg, W. (1933). "Über den Bau der Atomkerne. III". *Z. Phys.* **80** (9–10): 587–596. doi:10.1007/BF01335696.

[11] Iwanenko, D.D., The neutron hypothesis, Nature **129** (1932) 798.

[12] Miller A. I. *Early Quantum Electrodynamics: A Sourcebook*, Cambridge University Press, Cambridge, 1995, ISBN 0521568919, pp. 84–88.

[13] Brown, L.M.; Rechenberg, H. (1996). *The Origin of the Concept of Nuclear Forces.* Bristol and Philadelphia: Institute of Physics Publishing. ISBN 0750303735.

[14] John S. Rigden (1987). *Rabi, Scientist and Citizen.* New York: Basic Books, Inc. pp. 99–114. ISBN 9780674004351. Retrieved May 9, 2015.

[15] Kellogg, J.M.; Rabi, I.I.; Ramsey, N.F.; Zacharias, J.R. (1939). "An electrical quadrupole moment of the deuteron". *Physical Review* **55**: 318–319. Bibcode:1939PhRv...55..318K. doi:10.1103/physrev.55.318. Retrieved May 9, 2015.

[16] Wiringa, R. B.; Stoks, V. G. J.; Schiavilla, R. (1995). "Accurate nucleon–nucleon potential with charge-independence breaking". *Physical Review C* **51**: 38. arXiv:nucl-th/9408016. Bibcode:1995PhRvC..51...38W. doi:10.1103/PhysRevC.51.38.

4.7 Bibliography

- Gerald Edward Brown and A. D. Jackson, *The Nucleon–Nucleon Interaction*, (1976) North-Holland Publishing, Amsterdam ISBN 0-7204-0335-9

- R. Machleidt and I. Slaus, "The nucleon–nucleon interaction", *J. Phys.* G **27** (2001) R69 *(topical review).*

- E.A. Nersesov, *Fundamentals of atomic and nuclear physics*, (1990), Mir Publishers, Moscow, ISBN 5-06-001249-2

- P. Navrátil and W.E. Ormand, "Ab initio shell model with a genuine three-nucleon force for the p-shell nuclei", Phys. Rev. C **68**, 034305 (2003).

Chapter 5

Binding energy

Binding energy is the energy required to disassemble a whole system into separate parts. A bound system typically has a lower potential energy than the sum of its constituent parts; this is what keeps the system together. Often this means that energy is released upon the creation of a bound state. This definition corresponds to a *positive* binding energy.

5.1 General idea

In general, binding energy represents the mechanical work that must be done against the forces which hold an object together, disassembling the object into component parts separated by sufficient distance that further separation requires negligible additional work.

At the atomic level the **atomic binding energy** of the atom derives from electromagnetic interaction and is the energy required to disassemble an atom into free electrons and a nucleus.[1] Electron binding energy is a measure of the energy required to free electrons from their atomic orbits. This is more commonly known as ionization energy.[2]

At the molecular level, bond energy and bond-dissociation energy are measures of the binding energy between the atoms in a chemical bond.

At the nuclear level, binding energy is also equal to the energy liberated when a nucleus is created from other nucleons or nuclei.[3][4] This nuclear binding energy (binding energy of nucleons into a nuclide) is derived from the nuclear force (residual strong interaction) and is the energy required to disassemble a nucleus into the same number of free, unbound neutrons and protons it is composed of, so that the nucleons are far/distant enough from each other so that the nuclear force can no longer cause the particles to interact.[5] Mass excess is a related concept which compares the mass number of a nucleus with its true measured mass.[6]

In astrophysics, the gravitational binding energy of a celestial body is the energy required to expand the material to infinity.

In bound systems, if the binding energy is removed from the system, it must be subtracted from the mass of the unbound system, simply because this energy *has* mass. Thus, if energy is removed (or emitted) from the system at the time it is bound, the loss of energy from the system will also result in the loss of the mass of the energy, from the system.[7] System mass is not conserved in this process because the system is "open" (i.e., is not an isolated system to mass or energy input or loss) during the binding process.

5.2 Mass-energy relation

Main articles: Mass–energy equivalence and Mass in special relativity

Classically a bound system is at a lower energy level than its unbound constituents, and its mass must be less than the total mass of its unbound constituents. For systems with low binding energies, this "lost" mass after binding may be fractionally small. For systems with high binding energies, however, the missing mass may be an easily measurable fraction. This missing mass may be lost during the process of binding as energy in the form of heat or light, with the removed energy corresponding to removed mass through Einstein's equation $E = mc^2$. Note that in the process of binding, the constituents of the system might enter higher energy states of the nucleus/atom/molecule, but these types of energy also have mass, and it is necessary that they be removed from the system before its mass may decrease. Once the system cools to normal temperatures and returns to ground states in terms of energy levels, there is less mass remaining in the system than there was when it first combined and was at high energy. In that case, the removed heat represents exactly the mass "deficit", and the heat itself retains the mass which was lost (from the point of view of the initial system). This mass appears in any other system which absorbs the heat and gains thermal energy.[8]

As an illustration, consider two objects attracting each other in space through their gravitational field. The attraction force accelerates the objects and they gain some speed toward each other converting the potential (gravity) energy into kinetic (movement) energy. When either the particles 1) pass through each other without interaction or 2) elastically repel during the collision, the gained kinetic energy (related to speed), starts to revert into potential form driving the collided particles apart. The decelerating particles will return to the initial distance and beyond into infinity or stop and repeat the collision (oscillation takes place). This shows that the system, which loses no energy, does not combine (bind) into a solid object, parts of which oscillate at short distances. Therefore, in order to bind the particles, the kinetic energy gained due to the attraction must be dissipated (by resistive force). Complex objects in collision ordinarily undergo inelastic collision, transforming some kinetic energy into internal energy (heat content, which is atomic movement), which is further radiated in the form of photons—the light and heat. Once the energy to escape the gravity is dissipated in the collision, the parts will oscillate at closer, possibly atomic, distance, thus looking like one solid object. This lost energy, necessary to overcome the potential barrier in order to separate the objects, is the binding energy. If this binding energy were retained in the system as heat, its mass would not decrease. However, binding energy lost from the system (as heat radiation) would itself have mass, and directly represents the "mass deficit" of the cold, bound system.

Closely analogous considerations apply in chemical and nuclear considerations. Exothermic chemical reactions in closed systems do not change mass, but become less massive once the heat of reaction is removed, though this mass change is much too small to measure with standard equipment. In nuclear reactions, however, the fraction of mass that may be removed as light or heat, i.e., binding energy, is often a much larger fraction of the system mass. It may thus be measured directly as a mass difference between rest masses of reactants and (cooled) products. This is because nuclear forces are comparatively stronger than the Coulombic forces associated with the interactions between electrons and protons, that generate heat in chemistry.

5.2.1 Mass change

Mass change (decrease) in bound systems, particularly atomic nuclei, has also been termed *mass defect*, *mass deficit*, or mass *packing fraction*.

The difference between the unbound system calculated mass and experimentally measured mass of nucleus (mass change) is denoted as Δm. It can be calculated as follows:

Mass change = (unbound system calculated mass) − (measured mass of system)

i.e., (sum of masses of protons and neutrons) − (measured mass of nucleus)

After nuclear reactions that result in an excited nucleus, the energy that must be radiated or otherwise removed as binding energy for a single nucleus to produce the unexcited state may be in any of several forms. This may be electromagnetic waves, such as gamma radiation, the kinetic energy of an ejected particle, such as an electron, in internal conversion decay, or partly as the rest mass of one or more emitted particles, such as the particles of beta decay. No mass deficit can in theory appear until this radiation or this energy has been emitted, and is no longer part of the system.

When nucleons bind together to form a nucleus, they must lose a small amount of mass, i.e., there is a change in mass, in order to stay bound. This mass change must be released as various types of photon or other particle energy as above, according to the relation $E = mc^2$. Thus, after binding energy has been removed, **binding energy = mass change $\times c^2$**.

This energy is a measure of the forces that hold the nucleons together, and it represents energy that must be supplied again from the environment, if the nucleus were to be broken up into individual nucleons.

The energy given off during either nuclear fusion or nuclear fission is the difference between the binding energies of the "fuel", i.e., the initial nuclide(s), and the fission or fusion products. In practice, this energy may also be calculated from the substantial mass differences between the fuel and products, which uses previous measurement of the atomic masses of known nuclides, which always have the same mass for each species. This mass difference appears once evolved heat and radiation have been removed, which is a given requirement for measuring the (rest) masses of the (non-excited) nuclides involved in such calculations.

5.3 See also

- Chemical bond

- Electron binding energy

- Semi-empirical mass formula

- William Prout

- Virial mass

5.4 References

[1] "Nuclear Power Binding Energy". Retrieved 16 May 2015.

[2] IUPAC, *Compendium of Chemical Terminology*, 2nd ed. (the "Gold Book") (1997). Online corrected version: (2006–) "Ionization energy".

[3] *Britannica Online Encyclopaedia* - "nuclear binding energy". Accessed 8 September 2010. http://www.britannica.com/EBchec/topic/65615/binding-energy

[4] Nuclear Engineering - "Binding Energy". Bill Garland, McMaster University. Accessed 8 September 2010. http://www.nuceng.ca/igna/binding_energy.htm

[5] *Atomic Alchemy: Nuclear Processes* - "Binding Energy". About. Accessed 7 September 2010. http://library.thinkquest.org/17940/texts/binding_energy/binding_energy.html

[6] Krane, K. S (1987). *Introductory Nuclear Physics*. John Wiley & Sons. ISBN 0-471-80553-X.

[7] *HyperPhysics* - "Nuclear Binding Energy". *C.R. Nave*, Georgia State University. Accessed 7 September 2010. http://hyperphysics.phy-astr.gsu.edu/hbase/nucene/nucbin.html

[8] E. F. Taylor and J. A. Wheeler, *Spacetime Physics*, W.H. Freeman and Co., NY. 1992. ISBN 0-7167-2327-1, see pp. 248-9 for discussion of mass remaining constant after detonation of nuclear bombs, until heat is allowed to escape.

5.5 External links

- Nuclear Binding energy

- Mass and Nuclide Stability

- Experimental atomic mass data compiled Nov. 2003

Chapter 6

Neutron emission

Neutron emission is a type of radioactive decay of atoms containing excess neutrons, in which a neutron is simply ejected from the nucleus. Neutron emission is one of the ways an atom reaches its stability. An atom is unstable, therefore radioactive, when the forces in the nucleus are unbalanced. The instability of the nucleus results from the nuclei having extra neutrons or extra protons. Two examples of isotopes that emit neutrons are beryllium-13 (mean life 2.7×10^{-21} s) and helium-5 (7×10^{-22} s). Commonly, it is abbreviated with a lower case n.

As only a neutron is lost in this process, the atom does not gain or lose any protons, and so it does not become an atom of a different element. Instead, the atom will become a new isotope of the original element, such as beryllium-13 becoming beryllium-12 after emitting one of its neutrons.[1]

6.1 Neutron emission in fission

Neutron emission usually happens from nuclei that are in an excited state, such as the excited ^{17}O* produced from the beta decay of ^{17}N. The neutron emission process itself is controlled by the nuclear force and therefore is extremely fast, sometimes referred to as "nearly instantaneous". This process allows unstable atoms to become more stable. The ejection of the neutron may be as a product of the movement of many nucleons, but it is ultimately mediated by the repulsive action of the nuclear force that exists at extremely short-range distances between nucleons. The life time of an ejected neutron inside the nucleus before it is emitted is usually comparable to the flight time of a typical neutron before it leaves the small nuclear "potential well", or about 10^{-23} seconds.[2]

6.1.1 Induced fission

A synonym for such neutron emission is "prompt neutron" production, of the type that is best known to occur simultaneously with induced nuclear fission. Induced fission happens only when a nucleus is bombarded with neutrons, gamma rays, or other carriers of energy. Many heavy isotopes, most notably californium-252, also emit prompt neutrons among the products of a similar spontaneous radioactive decay process, spontaneous fission.

6.1.2 Spontaneous fission

Spontaneous fission happens when an atom's nucleus splits into two smaller nuclei and generally one or more neutron.

6.1.3 Delayed neutrons in reactor control

Main article: Nuclear Reactor Control

64

Most neutron emission outside prompt neutron production associated with fission (either induced or spontaneous), is from neutron-heavy isotopes produced as fission products. These neutrons are sometimes emitted with a delay, giving them the term delayed neutrons, but the actual delay in their production is a delay waiting for the beta decay of fission products to produce the excited-state nuclear precursors that immediately undergo prompt neutron emission. Thus, the delay in neutron emission is not from the neutron-production process, but rather its precursor beta decay, which is controlled by the weak force, and thus requires a far longer time. The beta decay half lives for the precursors to delayed neutron-emitter radioisotopes, are typically fractions of a second to tens of seconds.

Nevertheless, the delayed neutrons emitted by neutron-rich fission products aid control of nuclear reactors by making reactivity change far more slowly than it would if it were controlled by prompt neutrons alone. About 0.65% of neutrons are released in a nuclear chain reaction in a delayed way due to the mechanism of neutron emission, and it is this fraction of neutrons that allows a nuclear reactor to be controlled on human reaction time-scales, without proceeding to a prompt critical state, and runaway melt down.

6.2   See also

- Neutron radiation

- Neutron drip line

- Proton emission

6.3   References

[1] "Neutron Emission" (webpage). Retrieved 2014-10-30.

[2] "Neutron emission lifetime and why" (PDF). Retrieved 2012-09-17.

6.4   External links

- "Why Are Some Atoms Radioactive?" EPA. Environmental Protection Agency, n.d. Web. 31 Oct. 2014.

-  **The LIVEChart of Nuclides - IAEA** with filter on delayed neutron emission decay

- **Nuclear Structure and Decay Data - IAEA** with query on Neutron Separation Energy

Chapter 7

Pauli exclusion principle

The **Pauli exclusion principle** is the quantum mechanical principle that states that two identical fermions (particles with half-integer spin) cannot occupy the same quantum state simultaneously. In the case of electrons, it can be stated as follows: it is impossible for two electrons of a poly-electron atom to have the same values of the four quantum numbers (n, ℓ, $m\ell$ and ms). For two electrons residing in the same orbital, n, ℓ, and $m\ell$ are the same, so ms must be different and the electrons have opposite spins. This principle was formulated by Austrian physicist Wolfgang Pauli in 1925.

A more rigorous statement is that the total wave function for two identical fermions is antisymmetric with respect to exchange of the particles. This means that the wave function changes its sign if the space *and* spin co-ordinates of any two particles are interchanged.

Integer spin particles, bosons, are not subject to the Pauli exclusion principle: any number of identical bosons can occupy the same quantum state, as with, for instance, photons produced by a laser and Bose–Einstein condensate.

7.1   Overview

The Pauli exclusion principle governs the behavior of all fermions (particles with "half-integer spin"), while bosons (particles with "integer spin") are not subject to it. Fermions include elementary particles such as quarks (the constituent particles of protons and neutrons), electrons and neutrinos. In addition, protons and neutrons (subatomic particles composed from three quarks) and some atoms are fermions, and are therefore subject to the Pauli exclusion principle as well. Atoms can have different overall "spin", which determines whether they are fermions or bosons — for example helium-3 has spin 1/2 and is therefore a fermion, in contrast to helium-4 which has spin 0 and is a boson.[1]:123–125 As such, the Pauli exclusion principle underpins many properties of everyday matter, from its large-scale stability, to the chemical behavior of atoms.

"Half-integer spin" means that the intrinsic angular momentum value of fermions is $\hbar = h/2\pi$ (reduced Planck's constant) times a half-integer (1/2, 3/2, 5/2, etc.). In the theory of quantum mechanics fermions are described by antisymmetric states. In contrast, particles with integer spin (called bosons) have symmetric wave functions; unlike fermions they may share the same quantum states. Bosons include the photon, the Cooper pairs which are responsible for superconductivity, and the W and Z bosons. (Fermions take their name from the Fermi–Dirac statistical distribution that they obey, and bosons from their Bose–Einstein distribution).

7.2   History

In the early 20th century it became evident that atoms and molecules with even numbers of electrons are more chemically stable than those with odd numbers of electrons. In the 1916 article "The Atom and the Molecule" by Gilbert N. Lewis, for example, the third of his six postulates of chemical behavior states that the atom tends to hold an even number of electrons in the shell and especially to hold eight electrons which are normally arranged symmetrically at the eight corners

Wolfgang Pauli

of a cube (see: cubical atom).[2] In 1919 chemist Irving Langmuir suggested that the periodic table could be explained if the electrons in an atom were connected or clustered in some manner. Groups of electrons were thought to occupy a set of electron shells around the nucleus.[3] In 1922, Niels Bohr updated his model of the atom by assuming that certain numbers of electrons (for example 2, 8 and 18) corresponded to stable "closed shells".[4]:203

Pauli looked for an explanation for these numbers, which were at first only empirical. At the same time he was trying to explain experimental results of the Zeeman effect in atomic spectroscopy and in ferromagnetism. He found an essential clue in a 1924 paper by Edmund C. Stoner, which pointed out that for a given value of the principal quantum number (*n*), the number of energy levels of a single electron in the alkali metal spectra in an external magnetic field, where all degenerate energy levels are separated, is equal to the number of electrons in the closed shell of the noble gases for the same value of *n*. This led Pauli to realize that the complicated numbers of electrons in closed shells can be reduced to the simple rule of *one* electron per state, if the electron states are defined using four quantum numbers. For this purpose he introduced a new two-valued quantum number, identified by Samuel Goudsmit and George Uhlenbeck as electron spin.[5]

7.3 Connection to quantum state symmetry

The Pauli exclusion principle with a single-valued many-particle wavefunction is equivalent to requiring the wavefunction to be antisymmetric. An antisymmetric two-particle state is represented as a sum of states in which one particle is in state $|x\rangle$ and the other in state $|y\rangle$:

$$|\psi\rangle = \sum_{x,y} A(x,y)|x,y\rangle,$$

and antisymmetry under exchange means that $A(x,y) = -A(y,x)$. This implies $A(x,y) = 0$ when *x=y*, which is Pauli exclusion. It is true in any basis, since unitary changes of basis keep antisymmetric matrices antisymmetric, although strictly speaking, the quantity $A(x,y)$ is not a matrix but an antisymmetric rank-two tensor.

Conversely, if the diagonal quantities $A(x,x)$ are zero *in every basis*, then the wavefunction component

$$A(x,y) = \langle\psi|x,y\rangle = \langle\psi|(|x\rangle \otimes |y\rangle)$$

is necessarily antisymmetric. To prove it, consider the matrix element

$$\langle\psi|\Big((|x\rangle + |y\rangle) \otimes (|x\rangle + |y\rangle)\Big).$$

This is zero, because the two particles have zero probability to both be in the superposition state $|x\rangle + |y\rangle$. But this is equal to

$$\langle\psi|x,x\rangle + \langle\psi|x,y\rangle + \langle\psi|y,x\rangle + \langle\psi|y,y\rangle.$$

The first and last terms on the right side are diagonal elements and are zero, and the whole sum is equal to zero. So the wavefunction matrix elements obey:

$$\langle\psi|x,y\rangle + \langle\psi|y,x\rangle = 0,$$

or

$$A(x,y) = -A(y,x).$$

7.3.1 Pauli principle in advanced quantum theory

According to the spin-statistics theorem, particles with integer spin occupy symmetric quantum states, and particles with half-integer spin occupy antisymmetric states; furthermore, only integer or half-integer values of spin are allowed by the principles of quantum mechanics. In relativistic quantum field theory, the Pauli principle follows from applying a rotation operator in imaginary time to particles of half-integer spin.

In one dimension, bosons, as well as fermions, can obey the exclusion principle. A one-dimensional Bose gas with delta-function repulsive interactions of infinite strength is equivalent to a gas of free fermions. The reason for this is that, in one dimension, exchange of particles requires that they pass through each other; for infinitely strong repulsion this cannot happen. This model is described by a quantum nonlinear Schrödinger equation. In momentum space the exclusion principle is valid also for finite repulsion in a Bose gas with delta-function interactions,[6] as well as for interacting spins and Hubbard model in one dimension, and for other models solvable by Bethe ansatz. The ground state in models solvable by Bethe ansatz is a Fermi sphere.

7.4 Consequences

7.4.1 Atoms and the Pauli principle

The Pauli exclusion principle helps explain a wide variety of physical phenomena. One particularly important consequence of the principle is the elaborate electron shell structure of atoms and the way atoms share electrons, explaining the variety of chemical elements and their chemical combinations. An electrically neutral atom contains bound electrons equal in number to the protons in the nucleus. Electrons, being fermions, cannot occupy the same quantum state as other electrons, so electrons have to "stack" within an atom, i.e. have different spins while at the same electron orbital as described below.

An example is the neutral [helium] atom, which has two bound electrons, both of which can occupy the lowest-energy (*1s*) states by acquiring opposite spin; as spin is part of the quantum state of the electron, the two electrons are in different quantum states and do not violate the Pauli principle. However, the spin can take only two different values (eigenvalues). In a lithium atom, with three bound electrons, the third electron cannot reside in a *1s* state, and must occupy one of the higher-energy *2s* states instead. Similarly, successively larger elements must have shells of successively higher energy. The chemical properties of an element largely depend on the number of electrons in the outermost shell; atoms with different numbers of shells but the same number of electrons in the outermost shell have similar properties, which gives rise to the periodic table of the elements.[7]:214–218

7.4.2 Solid state properties and the Pauli principle

In conductors and semiconductors, there are very large numbers of molecular orbitals which effectively form a continuous band structure of energy levels. In strong conductors (metals) electrons are so degenerate that they cannot even contribute much to the thermal capacity of a metal.[8]:133–147 Many mechanical, electrical, magnetic, optical and chemical properties of solids are the direct consequence of Pauli exclusion.

7.4.3 Stability of matter

The stability of the electrons in an atom itself is unrelated to the exclusion principle, but is described by the quantum theory of the atom. The underlying idea is that close approach of an electron to the nucleus of the atom necessarily increases its kinetic energy, an application of the uncertainty principle of Heisenberg.[9] However, stability of large systems with many electrons and many nucleons is a different matter, and requires the Pauli exclusion principle.[10]

It has been shown that the Pauli exclusion principle is responsible for the fact that ordinary bulk matter is stable and occupies volume. This suggestion was first made in 1931 by Paul Ehrenfest, who pointed out that the electrons of each atom cannot all fall into the lowest-energy orbital and must occupy successively larger shells. Atoms therefore occupy a volume and cannot be squeezed too closely together.[11]

A more rigorous proof was provided in 1967 by Freeman Dyson and Andrew Lenard, who considered the balance of attractive (electron–nuclear) and repulsive (electron–electron and nuclear–nuclear) forces and showed that ordinary matter would collapse and occupy a much smaller volume without the Pauli principle.[12][13]

The consequence of the Pauli principle here is that electrons of the same spin are kept apart by a repulsive exchange interaction, which is a short-range effect, acting simultaneously with the long-range electrostatic or Coulombic force. This effect is partly responsible for the everyday observation in the macroscopic world that two solid objects cannot be in the same place at the same time.

7.4.4 Astrophysics and the Pauli principle

Dyson and Lenard did not consider the extreme magnetic or gravitational forces which occur in some astronomical objects. In 1995 Elliott Lieb and coworkers showed that the Pauli principle still leads to stability in intense magnetic fields such as in neutron stars, although at a much higher density than in ordinary matter.[14] It is a consequence of general relativity that, in sufficiently intense gravitational fields, matter collapses to form a black hole.

Astronomy provides a spectacular demonstration of the effect of the Pauli principle, in the form of white dwarf and neutron stars. In both types of body, atomic structure is disrupted by large gravitational forces, leaving the constituents supported by "degeneracy pressure" alone. This exotic form of matter is known as degenerate matter. In white dwarfs atoms are held apart by electron degeneracy pressure. In neutron stars, subject to even stronger gravitational forces, electrons have merged with protons to form neutrons. Neutrons are capable of producing an even higher degeneracy pressure, albeit over a shorter range. This can stabilize neutron stars from further collapse, but at a smaller size and higher density than a white dwarf. Neutrons are the most "rigid" objects known; their Young modulus (or more accurately, bulk modulus) is 20 orders of magnitude larger than that of diamond. However, even this enormous rigidity can be overcome by the gravitational field of a massive star or by the pressure of a supernova, leading to the formation of a black hole.[15]:286–287

7.5 See also

- Exchange force

- Exchange interaction

- Exchange symmetry

- Hund's rule

- Fermi hole

- Pauli effect

7.6 References

[1] Kenneth S. Krane (5 November 1987). *Introductory Nuclear Physics*. Wiley. ISBN 978-0-471-80553-3.

[2]

[3] Langmuir, Irving (1919). "The Arrangement of Electrons in Atoms and Molecules" (PDF). *Journal of the American Chemical Society* **41** (6): 868–934. doi:10.1021/ja02227a002. Retrieved 2008-09-01.

[4] Shaviv, Glora. *The Life of Stars: The Controversial Inception and Emergence of the Theory of Stellar Structure* (2010 ed.). Springer. ISBN 978-3642020872.

[5] Straumann, Norbert (2004). "The Role of the Exclusion Principle for Atoms to Stars: A Historical Account". *Invited talk at the 12th Workshop on Nuclear Astrophysics*.

[6] A. Izergin and V. Korepin, Letter in Mathematical Physics vol 6, page 283, 1982

[7] Griffiths, David J. (2004), *Introduction to Quantum Mechanics (2nd ed.)*, Prentice Hall, ISBN 0-13-111892-7

[8] Kittel, Charles (2005), *Introduction to Solid State Physics* (8th ed.), USA: John Wiley & Sons, Inc., ISBN 978-0-471-41526-8

[9] Elliot J. Lieb *The Stability of Matter and Quantum Electrodynamics*

[10] This realization is attributed by Lieb and by GL Sewell (2002). *Quantum Mechanics and Its Emergent Macrophysics*. Princeton University Press. ISBN 0-691-05832-6. to FJ Dyson and A Lenard: *Stability of Matter, Parts I and II* (*J. Math. Phys.*, **8**, 423–434 (1967); *J. Math. Phys.*, **9**, 698–711 (1968)).

[11] As described by FJ Dyson (J.Math.Phys. **8**, 1538–1545 (1967)), Ehrenfest made this suggestion in his address on the occasion of the award of the Lorentz Medal to Pauli.

[12] FJ Dyson and A Lenard: *Stability of Matter, Parts I and II* (*J. Math. Phys.*, **8**, 423–434 (1967); *J. Math. Phys.*, **9**, 698–711 (1968))

[13] Dyson, Freeman (1967). "Ground-State Energy of a Finite System of Charged Particles". *J. Math. Phys.* **8** (8): 1538–1545. Bibcode:1967JMP.....8.1538D. doi:10.1063/1.1705389.

[14] Lieb, E. H.; Loss, M.; Solovej, J. P. (1995). "Stability of Matter in Magnetic Fields". *Phys. Rev. Letters* **75** (6): 985–9. arXiv:cond-mat/9506047. Bibcode:1995PhRvL..75..985L. doi:10.1103/PhysRevLett.75.985.

[15] Martin Bojowald (5 November 2012). *The Universe: A View from Classical and Quantum Gravity*. John Wiley & Sons. ISBN 978-3-527-66769-7.

- Dill, Dan (2006). "Chapter 3.5, Many-electron atoms: Fermi holes and Fermi heaps". *Notes on General Chemistry (2nd ed.)*. W. H. Freeman. ISBN 1-4292-0068-5.

- Liboff, Richard L. (2002). *Introductory Quantum Mechanics*. Addison-Wesley. ISBN 0-8053-8714-5.

- Massimi, Michela (2005). *Pauli's Exclusion Principle*. Cambridge University Press. ISBN 0-521-83911-4.

- Tipler, Paul; Llewellyn, Ralph (2002). *Modern Physics (4th ed.)*. W. H. Freeman. ISBN 0-7167-4345-0.

7.7 External links

- Nobel Lecture: Exclusion Principle and Quantum Mechanics Pauli's own account of the development of the Exclusion Principle.

Chapter 8

Inverse beta decay

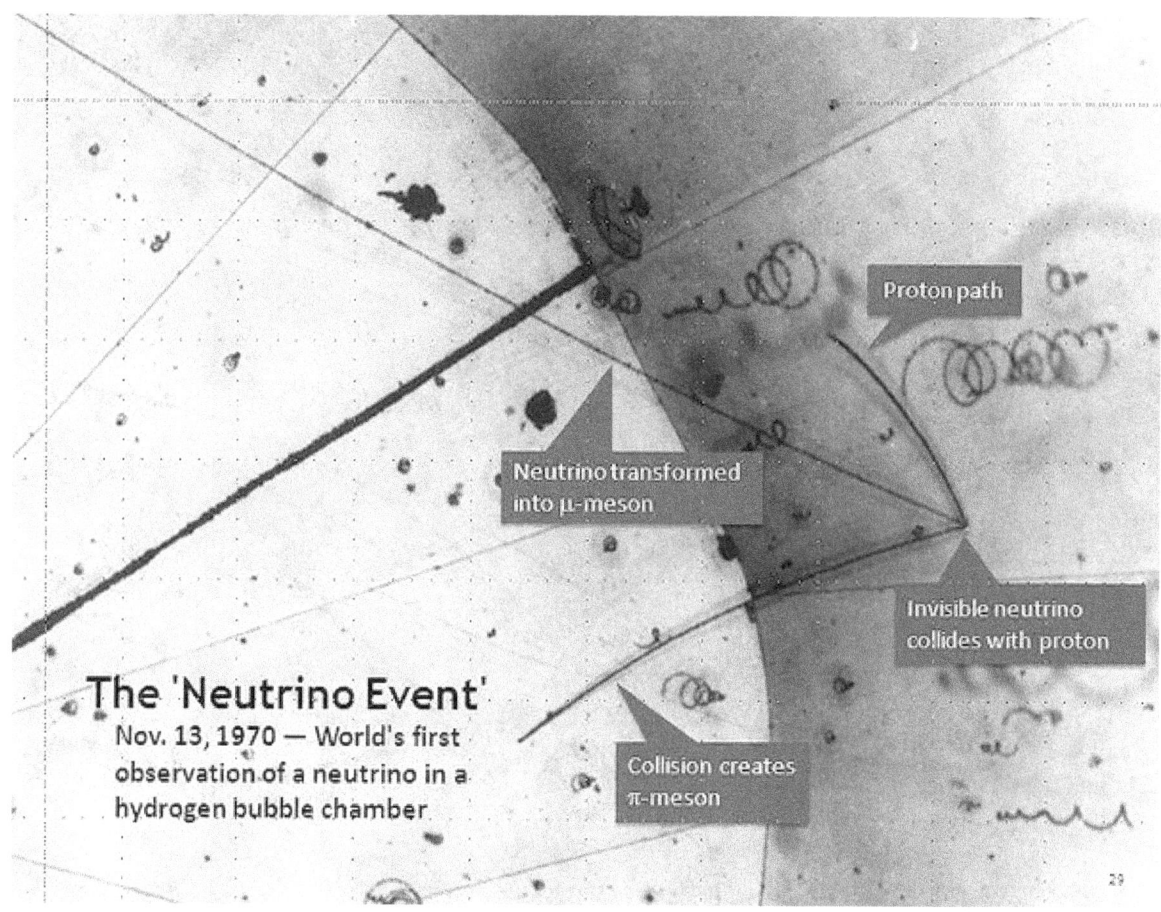

Inverse beta decay process involving a muon antineutrino.

Inverse beta decay is a somewhat vague term referring to one of several processes related to beta decay.

Inverse beta decay originally referred to the process $\bar{\nu}_e + p \to e^+ + n$,

(electron antineutrino scattering off a proton into a positron and a neutron) in which the existence of the antineutrino was decisively verified in the Cowan–Reines neutrino experiment. Understanding this process is important to our understanding of the mechanism of a supernova explosion.

Inverse beta decay may also sometimes refer to the process $e^- + p \to \nu_e + n$ normally called electron capture.

Chapter 9

Beta decay

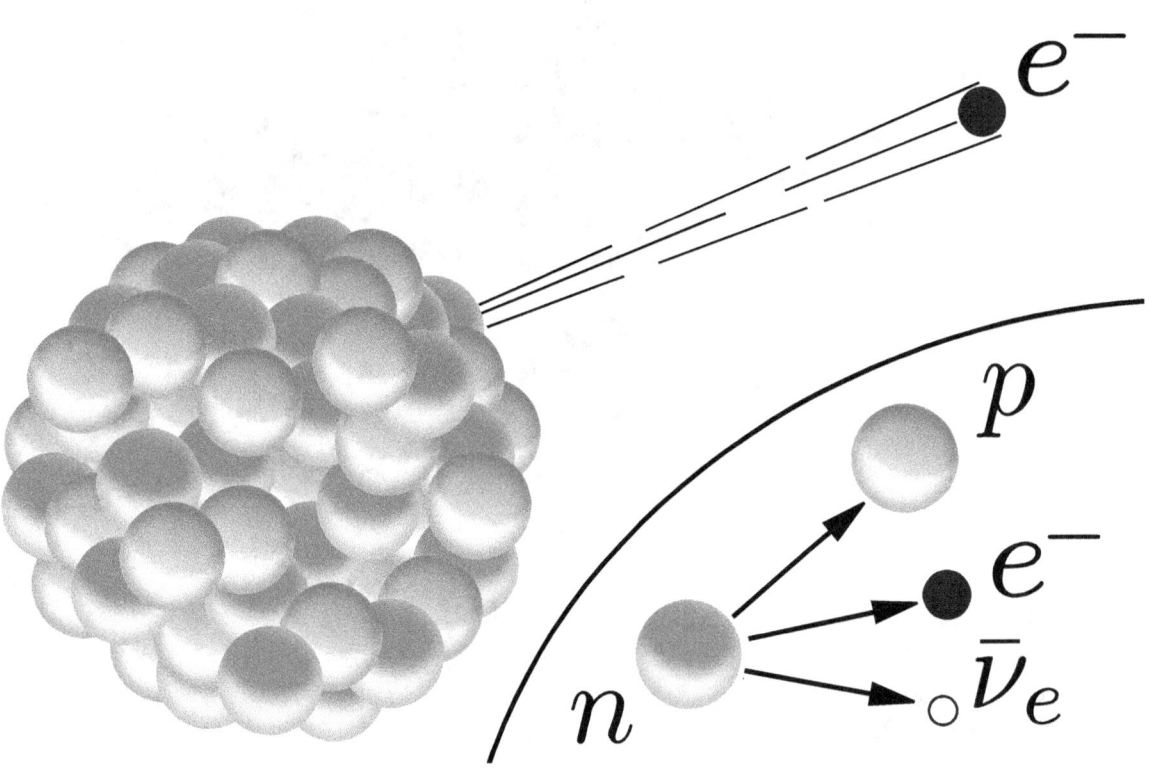

β− decay in an atomic nucleus (the accompanying antineutrino is omitted). The inset shows beta decay of a free neutron. In both processes, the intermediate emission of a virtual W− boson (which then decays to electron and antineutrino) is not shown.

In nuclear physics, **beta decay** (β-decay) is a type of radioactive decay in which a proton is transformed into a neutron, or vice versa, inside an atomic nucleus. This process allows the atom to move closer to the optimal ratio of protons and neutrons. As a result of this transformation, the nucleus emits a detectable beta particle, which is an electron or positron.[1]

Beta decay is mediated by the weak force. There are two types of beta decay, known as *beta minus* and *beta plus*. Beta minus (β⁻) decay produces an electron and electron antineutrino, while beta plus (β⁺) decay produces a positron and electron neutrino; β⁺ decay is thus also known as positron emission.[2]

An example of electron emission (β⁻ decay) is the decay of carbon-14 into nitrogen-14:

$$^{14}_{6}C \rightarrow {}^{14}_{7}N + e^- + \nu_e$$

In this form of decay, the original element becomes a new chemical element in a process known as nuclear transmutation. This new element has an unchanged mass number A, but an atomic number Z that is increased by one. As in all nuclear decays, the decaying element (in this case $^{14}_{6}C$) is known as the *parent nuclide* while the resulting element (in this case $^{14}_{7}N$) is known as the *daughter nuclide*. The emitted electron or positron is known as a beta particle.

An example of positron emission (β^+ decay) is the decay of magnesium-23 into sodium-23:

$$^{23}_{12}Mg \rightarrow {}^{23}_{11}Na + e^+ + \nu_e$$

In contrast to β^- decay, β^+ decay is accompanied by the emission of an electron neutrino and a positron. β^+ decay also results in nuclear transmutation, with the resulting element having an atomic number that is decreased by one.

Electron capture is sometimes included as a type of beta decay, because the basic nuclear process, mediated by the weak force, is the same. In electron capture, an inner atomic electron is captured by a proton in the nucleus, transforming it into a neutron, and an electron neutrino is released. An example of electron capture is the decay of krypton-81 into bromine-81:

$$^{81}_{36}Kr + e^- \rightarrow {}^{81}_{35}Br + \nu_e$$

Electron capture is a competing (simultaneous) decay process for all nuclei that can undergo β^+ decay. The converse, however, is not true: electron capture is the *only* type of decay that is allowed in proton-rich nuclides that do not have sufficient energy to emit a positron and neutrino.[3]

9.1 β^- decay

In $\beta-$ decay, the weak interaction converts an atomic nucleus into a nucleus with atomic number increased by one, while emitting an electron (e−) and an electron antineutrino (ν_e).

The generic equation is:

$$^{A}_{Z}X \rightarrow {}^{A}_{Z+1}X' + e^- + \nu_e [1]$$

where A and Z are the mass number and atomic number of the decaying nucleus, and X and X' are the initial and final elements, respectively.

Another example is when the free neutron (^1_0n) decays by $\beta-$ decay into a proton (p):

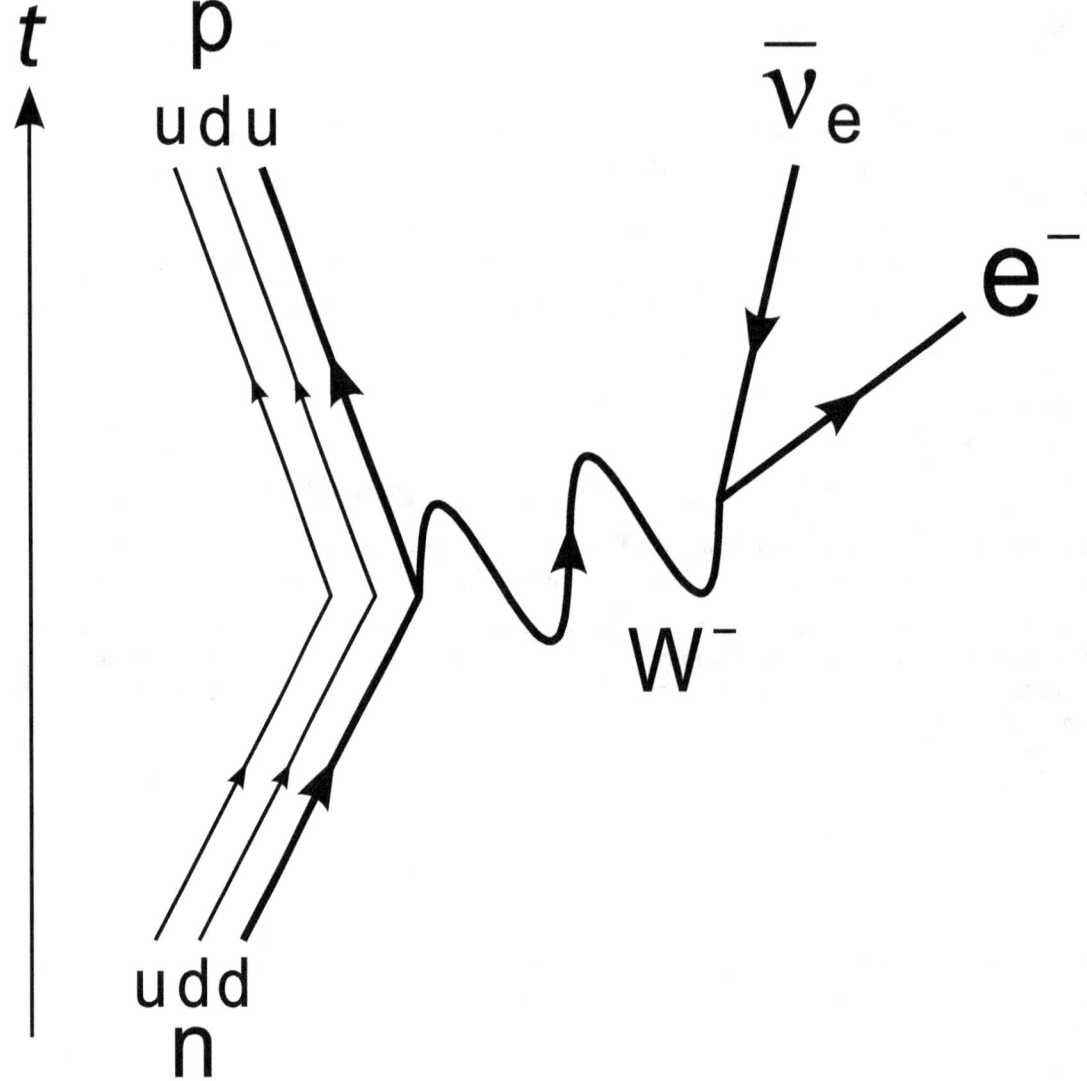

The Feynman diagram for β− decay of a neutron into a proton, electron, and electron antineutrino via an intermediate W− boson.

n → p + e− + ν
e.

At the fundamental level (as depicted in the Feynman diagram on the left), this is caused by the conversion of the negatively charged ($-\frac{1}{3}$ e) down quark to the positively charged ($+\frac{2}{3}$ e) up quark by emission of a W− boson; the W− boson subsequently decays into an electron and an electron antineutrino:

d → u + e− + ν
e.

The beta spectrum is a continuous spectrum: the total decay energy is divided between the electron and the antineutrino. In the figure to the right, this is shown, by way of example, for an electron of 0.4 MeV energy. In this example, the antineutrino then gets the remainder: 0.76 MeV, since the total decay energy is assumed to be 1.16 MeV.

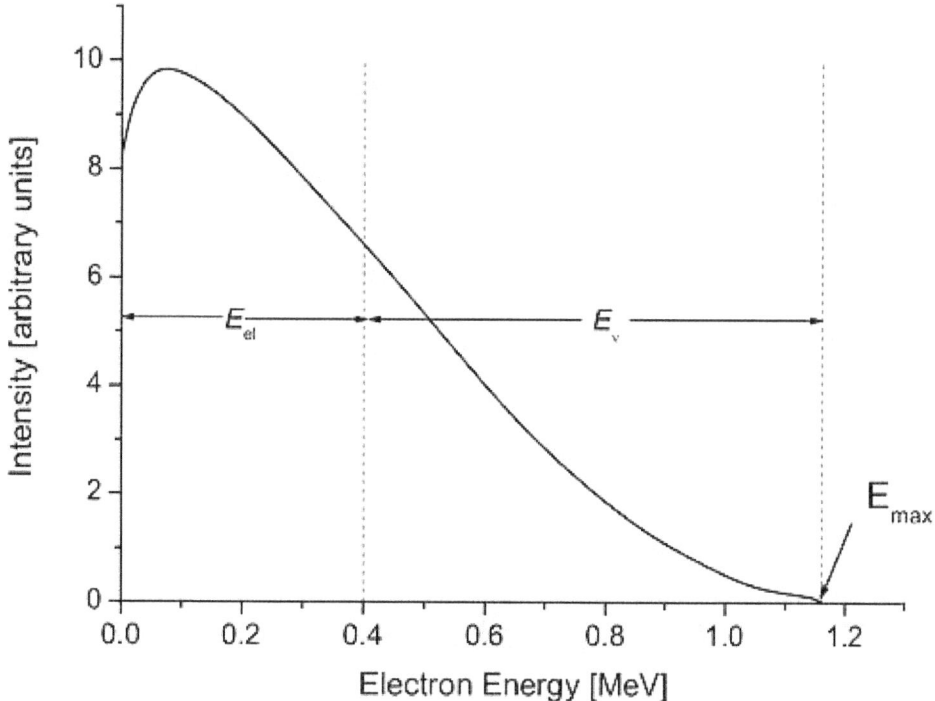

A beta spectrum, showing a typical division of energy between electron and antineutrino

β− decay generally occurs in neutron-rich nuclei.[4]

9.2 β⁺ decay

Main article: Positron emission

In β+ decay, or "positron emission", the weak interaction converts an atomic nucleus into a nucleus with atomic number decreased by one, while emitting a positron (e+) and an electron neutrino (ν
e). The generic equation is:

A
ZX → A
Z−1X' + e+ + ν
e[1]

This may be considered as the decay of a proton inside the nucleus to a neutron

p → n + e+ + ν
e[1]

However, β+ decay cannot occur in an isolated proton because it requires energy due to the mass of the neutron being greater than the mass of the proton. β+ decay can only happen inside nuclei when the daughter nucleus has a greater binding energy (and therefore a lower total energy) than the mother nucleus. The difference between these energies goes into the reaction of converting a proton into a neutron, a positron and a neutrino and into the kinetic energy of these particles. In an opposite process to negative beta decay, the weak interaction converts a proton into a neutron by converting an up quark into a down quark by having it emit a W+ or absorb a W−.

9.3 Electron capture (K-capture)

Main article: Electron capture

In all cases where β+ decay of a nucleus is allowed energetically, so is electron capture, the process in which the same nucleus captures an atomic electron with the emission of a neutrino:

$$^A_Z X + e^- \rightarrow {}^A_{Z-1} X' + \nu_e$$

The emitted neutrino is mono-energetic. In proton-rich nuclei where the energy difference between initial and final states is less than $2m_e c^2$, β+ decay is not energetically possible, and electron capture is the sole decay mode.[3]

If the captured electron comes from the innermost shell of the atom, the K-shell, which has the highest probability to interact with the nucleus, the process is called K-capture.[5] If it comes from the L-shell, the process is called L-capture, etc.

9.4 Competition of beta decay types

Three types of beta decay in competition are illustrated by the single isotope copper-64 (29 protons, 35 neutrons), which has a half-life of about 12.7 hours. This isotope has one unpaired proton and one unpaired neutron, so either the proton or the neutron can decay. This particular nuclide (though not all nuclides in this situation) is almost equally likely to decay through proton decay by positron emission (18%) or electron capture (43%), as through neutron decay by electron emission (39%).

9.5 Helicity (polarization) of neutrinos, electrons and positrons emitted in beta decay

After the discovery of parity non-conservation (see history below), it was found that, in beta decay, electrons are emitted mostly with negative helicity, i.e., they move, naively speaking, like left-handed screws driven into a material (they have negative longitudinal polarization).[6] Conversely, positrons have mostly positive helicity, i.e., they move like right-handed screws. Neutrinos (emitted in positron decay) have positive helicity, while antineutrinos (emitted in electron decay) have negative helicity.[7]

The higher the energy of the particles, the higher their polarization.

9.6 Energy release

The Q value is defined as the total energy released in a given nuclear decay. In beta decay, Q is therefore also the sum of the kinetic energies of the emitted beta particle, neutrino, and recoiling nucleus. (Because of the large mass of the nucleus compared to that of the beta particle and neutrino, the kinetic energy of the recoiling nucleus can generally be neglected.) Beta particles can therefore be emitted with any kinetic energy ranging from 0 to Q.[1] A typical Q is around 1 MeV, but can range from a few keV to a few tens of MeV.

Since the rest mass of the electron is 511 keV, the most energetic beta particles are ultrarelativistic, with speeds very close to the speed of light.

9.6.1 β⁻ decay

Consider the generic equation for beta decay

A
ZX → A
Z+1X' + e− + ν
e.

The Q value for this decay is

$$Q = \left[m_N \left({}^A_Z\text{X} \right) - m_N \left({}^A_{Z+1}\text{X}' \right) - m_e - m_{\overline{\nu}_e} \right] c^2$$

where $m_N \left({}^A_Z\text{X} \right)$ is the mass of the nucleus of the A
ZX atom, m_e is the mass of the electron, and $m_{\overline{\nu}_e}$ is the mass of the electron antineutrino. In other words, the total energy released is the mass energy of the initial nucleus, minus the mass energy of the final nucleus, electron, and antineutrino. The mass of the nucleus mN is related to the standard atomic mass m by

$$m \left({}^A_Z\text{X} \right) c^2 = m_N \left({}^A_Z\text{X} \right) c^2 + Z m_e c^2 - \sum_{i=1}^Z B_i$$

That is, the total atomic mass is the mass of the nucleus, plus the mass of the electrons, minus the binding energy B_i of each electron. Substituting this into our original equation, while neglecting the nearly-zero antineutrino mass and difference in electron binding energy, which is very small for high-Z atoms, we have

$$Q = \left[m \left({}^A_Z\text{X} \right) - m \left({}^A_{Z+1}\text{X}' \right) \right] c^2$$

This energy is carried away as kinetic energy by the electron and neutrino.

Because the reaction will proceed only when the Q-value is positive, β⁻ decay can occur when the mass of atom A
ZX is greater than the mass of atom A
Z+1X'.[8]

9.6.2 β⁺ decay

The equations for β⁺ decay are similar, with the generic equation

A
ZX → A
Z−1X' + e+ + ν
e

giving

$$Q = \left[m_N \left({}^A_Z X \right) - m_N \left({}^A_{Z-1} X' \right) - m_e - m_{\nu_e} \right] c^2$$

However, in this equation, the electron masses do not cancel, and we are left with

$$Q = \left[m \left({}^A_Z X \right) - m \left({}^A_{Z-1} X' \right) - 2m_e \right] c^2$$

Because the reaction will proceed only when the Q-value is positive, β^+ decay can occur when the mass of atom A ZX exceeds that of A
Z−1X' by at least twice the mass of the electron.[8]

9.6.3   Electron capture

The analogous calculation for electron capture must take into account the binding energy of the electrons. This is because the atom will be left in an excited state after capturing the electron, and the binding energy of the captured innermost electron is significant. Using the generic equation for electron capture

A
ZX + e− → A
Z−1X' + ν
e

we have

$$Q = \left[m_N \left({}^A_Z X \right) + m_e - m_N \left({}^A_{Z-1} X' \right) - m_{\nu_e} \right] c^2$$

which simplifies to

$$Q = \left[m \left({}^A_Z X \right) - m \left({}^A_{Z-1} X' \right) \right] c^2 - B_n$$

where B_n is the binding energy of the captured electron.

Because the binding energy of the electron is much less than the mass of the electron, nuclei that can undergo β^+ decay can always also undergo electron capture, but the reverse is not true.[8]

9.7   Nuclear transmutation

If the proton and neutron are part of an atomic nucleus, these decay processes transmute one chemical element into another. For example:

Beta decay does not change the number A of nucleons in the nucleus, but changes only its charge Z. Thus the set of all nuclides with the same A can be introduced; these *isobaric* nuclides may turn into each other via beta decay. Among them, several nuclides (at least one for any given mass number A) are beta stable, because they present local minima of the mass excess: if such a nucleus has (A, Z) numbers, the neighbour nuclei (A, Z−1) and (A, Z+1) have higher mass excess and can beta decay into (A, Z), but not vice versa. For all odd mass numbers A, there is only one known beta-stable isobar. For even A, there are up to three different beta-stable isobars experimentally known; for example, 96

40Zr, 96
42Mo, and 96
44Ru are all beta-stable. There are about 355 known beta-decay stable nuclides total.[9]

Usually, unstable nuclides are clearly either "neutron rich" or "proton rich", with the former undergoing beta decay and the latter undergoing electron capture (or more rarely, due to the higher energy requirements, positron decay). However, in a few cases of odd-proton, odd-neutron radionuclides, it may be energetically favorable for the radionuclide to decay to an even-proton, even-neutron isobar either by undergoing beta-positive or beta-negative decay. An often-cited example is 64
29Cu, which decays by positron emission/electron capture 61% of the time to 64
28Ni, and 39% of the time by (negative) beta decay to 64
30Zn.[10]

Most naturally occurring isotopes on Earth are beta stable. Those that are not have half-lives ranging from under a second to periods of time significantly greater than the age of the universe. One common example of a long-lived isotope is the odd-proton odd-neutron nuclide 40
19K, which undergoes all three types of beta decay ($\beta-$, $\beta+$ and electron capture) with a half-life of 1.277×10^9 years.[11]

9.8 Double beta decay

Main article: Double beta decay

Some nuclei can undergo double beta decay ($\beta\beta$ decay) where the charge of the nucleus changes by two units. Double beta decay is difficult to study, as the process has an extremely long half-life. In nuclei for which both β decay and $\beta\beta$ decay are possible, the rarer $\beta\beta$ decay process is effectively impossible to observe. However, in nuclei where β decay is forbidden but $\beta\beta$ decay is allowed, the process can be seen and a half-life measured.[12] Thus, $\beta\beta$ decay is usually studied only for beta stable nuclei. Like single beta decay, double beta decay does not change A; thus, at least one of the nuclides with some given A has to be stable with regard to both single and double beta decay.

"Ordinary" double beta decay results in the emission of two electrons and two antineutrinos. If neutrinos are Majorana particles (i.e., they are their own antiparticles), then a decay known as neutrinoless double beta decay will occur. Most neutrino physicists believe that neutrinoless double beta decay has never been observed.[12]

9.9 Bound-state β^- decay

A very small minority of free neutron decays (about four per million) are so-called "two-body decays", in which the proton, electron and antineutrino are produced, but the electron fails to gain the 13.6 eV energy necessary to escape the proton, and therefore simply remains bound to it, as a neutral hydrogen atom.[13] In this type of beta decay, in essence all of the neutron decay energy is carried off by the antineutrino.

For fully ionized atoms (bare nuclei), it is possible in likewise manner for electrons to fail to escape the atom, and to be emitted from the nucleus into low-lying atomic bound states (orbitals). This can not occur for neutral atoms whose low-lying bound states are already filled by electrons.

The phenomenon in fully ionized atoms was first observed for ^{163}Dy^{66+} in 1992 by Jung et al. of the Darmstadt Heavy-Ion Research group. Although neutral ^{163}Dy is a stable isotope, the fully ionized ^{163}Dy^{66+} undergoes β decay into the K and L shells with a half-life of 47 days.[14]

Another possibility is that a fully ionized atom undergoes greatly accelerated β decay, as observed for ^{187}Re by Bosch et al., also at Darmstadt. Neutral ^{187}Re does undergo β decay with a half-life of 42×10^9 years, but for fully ionized ^{187}Re^{75+} this is shortened by a factor of 10^9 to only 32.9 years.[15] For comparison the variation of decay rates of other nuclear processes due to chemical environment is less than 1%.

9.10 Forbidden transitions

Beta decays can be classified according to the L-value of the emitted radiation. When $L > 0$, the decay is referred to as "forbidden". Nuclear selection rules require high L-values to be accompanied by changes in nuclear spin (J) and parity (π). The selection rules for the Lth forbidden transitions are:

$$\Delta J = L - 1, L, L + 1; \Delta \pi = (-1)^L,$$

where $\Delta \pi = 1$ or -1 corresponds to no parity change or parity change, respectively. The special case of a transition between isobaric analogue states, where the structure of the final state is very similar to the structure of the initial state, is referred to as "superallowed" for beta decay, and proceeds very quickly. The following table lists the ΔJ and $\Delta \pi$ values for the first few values of L:

9.11 Fermi transitions

A **Fermi transition** is a beta decay in which the spins of the emitted electron (positron) and anti-neutrino (neutrino) couple to total spin $S = 0$, leading to an angular momentum change $\Delta J = 0$ between the initial and final states of the nucleus (assuming an allowed transition $\Delta L = 0$). In the non-relativistic limit, the nuclear part of the operator for a Fermi transition is given by

$$\mathcal{O}_F = G_V \sum_a \hat{\tau}_{a\pm}$$

with G_V the weak vector coupling constant, $\tau_\pm$ the isospin raising and lowering operators, and a running over all protons and neutrons in the nucleus.

9.12 Gamow-Teller transitions

A **Gamow-Teller transition** is a beta decay in which the spins of the emitted electron (positron) and anti-neutrino (neutrino) couple to total spin $S = 1$, leading to an angular momentum change $\Delta J = 0, \pm 1$ between the initial and final states of the nucleus (assuming an allowed transition). In this case, the nuclear part of the operator is given by

$$\mathcal{O}_{GT} = G_A \sum_a \hat{\sigma}_a \hat{\tau}_{a\pm}$$

with G_A the weak axial-vector coupling constant, and σ the spin Pauli matrices, which can produce a spin-flip in the decaying nucleon.

9.13 Beta emission spectrum

Beta decay can be considered as a perturbation as described in quantum mechanics, and thus Fermi's Golden Rule can be applied. This leads to an expression for the kinetic energy spectrum $N(T)$ of emitted betas as follows:[16]

$$N(T) = C_L(T)F(Z, T)pE(Q - T)^2$$

where T is the kinetic energy, CL is a shape function that depends on the forbiddenness of the decay (it is constant for allowed decays), $F(Z, T)$ is the Fermi Function (see below) with Z the charge of the final-state nucleus, $E = T + mc^2$

is the total energy, $p = \sqrt{(E/c)^2 - (mc)^2}$ is the momentum, and Q is the Q value of the decay. The kinetic energy of the emitted neutrino is given approximately by Q minus the kinetic energy of the beta.

As an example, the beta decay spectrum of ^{210}Bi (originally called RaE) is shown to the right.

9.13.1   Fermi function

The Fermi function that appears in the beta spectrum formula accounts for the Coulomb attraction / repulsion between the emitted beta and the final state nucleus. Approximating the associated wavefunctions to be spherically symmetric, the Fermi function can be analytically calculated to be:[17]

$$F(Z,T) = \frac{2(1+S)}{\Gamma(1+2S)^2}(2p\rho)^{2S-2}e^{\pi\eta}|\Gamma(S+i\eta)|^2,$$

where $S = \sqrt{1 - \alpha^2 Z^2}$ (α is the fine-structure constant), $\eta = \pm\, \alpha ZE/pc$ (+ for electrons, − for positrons), $\varrho = rN/\hbar$ (rN is the radius of the final state nucleus), and Γ is the Gamma function.

For non-relativistic betas ($Q \ll m_e c^2$), this expression can be approximated by:[18]

$$F(Z,T) \approx \frac{2\pi\eta}{1 - e^{-2\pi\eta}}.$$

Other approximations can be found in the literature.[19][20]

9.13.2   Kurie plot

A **Kurie plot** (also known as a **Fermi–Kurie plot**) is a graph used in studying beta decay developed by Franz N. D. Kurie, in which the square root of the number of beta particles whose momenta (or energy) lie within a certain narrow range, divided by the Fermi function, is plotted against beta-particle energy.[21][22] It is a straight line for allowed transitions and some forbidden transitions, in accord with the Fermi beta-decay theory. The energy-axis (x-axis) intercept of a Kurie plot corresponds to the maximum energy imparted to the electron/positron (the decay's Q-value). With a Kurie plot one can find the limit on the effective mass of a neutrino.[23]

9.14   History

9.14.1   Discovery and characterization of β⁻ decay

Radioactivity was discovered in 1896 by Henri Becquerel in uranium, and subsequently observed by Marie and Pierre Curie in thorium and in the new elements polonium and radium. In 1899, Ernest Rutherford separated radioactive emissions into two types: alpha and beta (now beta minus), based on penetration of objects and ability to cause ionization. Alpha rays could be stopped by thin sheets of paper or aluminium, whereas beta rays could penetrate several millimetres of aluminium. (In 1900, Paul Villard identified a still more penetrating type of radiation, which Rutherford identified as a fundamentally new type in 1903, and termed gamma rays).

In 1900, Becquerel measured the mass-to-charge ratio (m/e) for beta particles by the method of J.J. Thomson used to study cathode rays and identify the electron. He found that m/e for a beta particle is the same as for Thomson's electron, and therefore suggested that the beta particle is in fact an electron.

In 1901, Rutherford and Frederick Soddy showed that alpha and beta radioactivity involves the transmutation of atoms into atoms of other chemical elements. In 1913, after the products of more radioactive decays were known, Soddy and Kazimierz Fajans independently proposed their radioactive displacement law, which states that beta (i.e., β−) emission from one element produces another element one place to the right in the periodic table, while alpha emission produces an element two places to the left.

9.14.2 Neutrinos in beta decay

Historically, the study of beta decay provided the first physical evidence of the neutrino. Measurements of the beta particle (electron) kinetic energy spectrum in 1911 by Lise Meitner and Otto Hahn and in 1913 by Jean Danysz showed multiple lines on a diffuse background, offering the first hint of a continuous spectrum.[24] In 1914, James Chadwick used a magnetic spectrometer with one of Hans Geiger's new counters to make a more accurate measurement and showed that the spectrum was continuous.[24][25] This was in apparent contradiction to the law of conservation of energy, since if beta decay were simply electron emission as assumed at the time, then the energy of the emitted electron should equal the energy difference between the initial and final nuclear states and lead to a narrow energy distribution, as observed for both alpha and gamma decay.[26] For beta decay, however, the observed broad continuous spectrum suggested that energy is lost in the beta decay process.

In 1920–1927, Charles Drummond Ellis (along with James Chadwick and colleagues) further established that the beta decay spectrum is continuous, ending all controversies. It also had an effective upper bound in energy, which was a severe blow to Bohr's suggestion that conservation of energy might be true only in a statistical sense, and might be violated in any given decay. Now the problem of how to account for the variability of energy in known beta decay products, as well as for conservation of momentum and angular momentum in the process, became acute.

A second problem related to the conservation of angular momentum. Molecular band spectra showed that the nuclear spin of nitrogen-14 is 1 (i.e. equal to the reduced Planck constant), and more generally that the spin is integral for nuclei of even mass number and half-integral for nuclei of odd mass number, as later explained by the proton-neutron model of the nucleus.[26] Beta decay leaves the mass number unchanged, so that the change of nuclear spin must be an integer. However the electron spin is 1/2, so that angular momentum would not be conserved if beta decay were simply electron emission.

In a famous letter written in 1930, Wolfgang Pauli suggested that, in addition to electrons and protons, atomic nuclei also contained an extremely light neutral particle, which he called the neutron. He suggested that this "neutron" was also emitted during beta decay (thus accounting for the known missing energy, momentum, and angular momentum) and had simply not yet been observed. In 1931, Enrico Fermi renamed Pauli's "neutron" to neutrino and, in 1934, he published a very successful model of beta decay in which neutrinos were produced. The neutrino interaction with matter was so weak that detecting it proved a severe experimental challenge, which was finally met in 1956 in the Cowan–Reines neutrino experiment.[27] However, the properties of neutrinos were (with a few minor modifications) as predicted by Pauli and Fermi.

9.14.3 Discovery of other types of beta decay

In 1934, Frédéric and Irène Joliot-Curie bombarded aluminium with alpha particles to effect the nuclear reaction 4
2He + 27
13Al → 30
15P + 1
0n, and observed that the product isotope 30
15P emits a positron identical to those found in cosmic rays by Carl David Anderson in 1932. This was the first example of β+ decay (positron emission), which they termed artificial radioactivity since 30
15P is a short-lived nuclide which does not exist in nature.

The theory of electron capture was first discussed by Gian-Carlo Wick in a 1934 paper, and then developed by Hideki Yukawa and others. K-electron capture was first observed in 1937 by Luis Alvarez, in the nuclide ^{48}V.[28][29][30] Alvarez went on to study electron capture in ^{67}Ga and other nuclides.[28][31][32]

9.14.4 Non-conservation of parity

In 1956, Chien-Shiung Wu and coworkers proved in the Wu experiment that parity is not conserved in beta decay.[33][34] This surprising fact had been postulated shortly before in an article by Tsung-Dao Lee and Chen Ning Yang.[35]

9.15   See also

- Double beta decay

- Electron capture

- Neutrino

- Alpha decay

- Betavoltaics

- Particle radiation

- Radionuclide

- Tritium illumination, a form of fluorescent lighting powered by beta decay

- Pandemonium effect

- Total absorption spectroscopy

9.16   References

- Tuli, J. K. (2011). *Nuclear Wallet Cards* (PDF) (8th ed.). Brookhaven National Laboratory.

[1] Konya, J.; Nagy, N. M. (2012). *Nuclear and Radiochemistry*. Elsevier. pp. 74–75. ISBN 978-0-12-391487-3.

[2] Basdevant, Jean-Louis; Rich, James; Spiro, Michael (2005). *Fundamentals in Nuclear Physics: From Nuclear Structure to Cosmology*. Springer. ISBN 978-0387016726.

[3] Zuber, Kai (2011). *Neutrino Physics* (2 ed.). CRC Press. p. 466. ISBN 9781420064711.

[4] Loveland, Walter D. (2005). *Modern Nuclear Chemistry*. Wiley. p. 232. ISBN 0471115320.

[5] Tatjana Jevremovic (21 April 2009). *Nuclear Principles in Engineering*. Springer Science & Business Media. p. 201. ISBN 978-0-387-85608-7.

[6] H. Frauenfelder, R. Bobone, E. Von Goeler, N. Levine, H. R. Lewis, R. N. Peacock, A. Rossi and G. DePasquali, Physical Review 106 (1957) 386

[7] E. J. Konopinski and M. E. Rose, *The Theory of nuclear Beta Decay*, in: *Alpha-, Beta- and Gamma-Ray Spectroscopy*, ed. by Kai Siegbahn, Vol. 2, North-Holland Publishing Company, Amsterdam, 1966

[8] Kenneth S. Krane (5 November 1987). *Introductory Nuclear Physics*. Wiley. ISBN 978-0-471-80553-3.

[9] "Interactive Chart of Nuclides". National Nuclear Data Center, Brookhaven National Laboratory. Retrieved 2014-09-18.

[10] "WWW Table of Radioactive Isotopes, Copper 64". *LBNL Isotopes Project*. Lawrence Berkeley National Laboratory. Retrieved 2014-09-18.

[11] "WWW Table of Radioactive Isotopes, Potassium 40". *LBNL Isotopes Project*. Lawrence Berkeley National Laboratory. Retrieved 2014-09-18.

[12] S.M. Bilenky (October 5, 2010). "Neutrinoless double beta-decay". *Physics of Particles and Nuclei* **41** (5). arXiv:1001.1946. Bibcode:2010PPN....41..690B. doi:10.1134/S1063779610050035.

[13] An Overview Of Neutron Decay J. Byrne in Quark-Mixing, CKM Unitarity (H.Abele and D.Mund, 2002), see p.XV

[14] Jung,M.et al. (1992). "First observation of bound-state β–decay". *Physical Review Letters* **69**(15):2164–2167. Bibcode:1992J. doi:10.1103/PhysRevLett.69.2164. PMID 10046415.

[15] Bosch, F. et al. (1996). "Observation of bound-state beta minus decay of fully ionized ^{187}Re: ^{187}Re–^{187}Os Cosmochronometry". *Physical Review Letters* **77** (26): 5190–5193. Bibcode:1996PhRvL..77.5190B. doi:10.1103/PhysRevLett.77.5190. PMID 10062738.

[16] Nave, C. R. "Energy and Momentum Spectra for Beta Decay". *HyperPhysics*. Retrieved 2013-03-09.

[17] Fermi, E. (1934). "Versuch einer Theorie der β-Strahlen. I". *Zeitschrift für Physik* **88** (3–4): 161–177. Bibcode:1934ZPhy...8. doi:10.1007/BF01351864.

[18] Mott, N. F.; Massey, H. S. W. (1933). *The Theory of Atomic Collisions*. Clarendon Press. LCCN 34001940.

[19] Venkataramaiah, P.; Gopala, K.; Basavaraju, A.; Suryanarayana, S. S.; Sanjeeviah, H. (1985). "A simple relation for the Fermi function". *Journal of Physics G* **11** (3): 359–364. Bibcode:1985JPhG...11..359V. doi:10.1088/0305-4616/11/3/014.

[20] Schenter, G. K.; Vogel, P. (1983). "A simple approximation of the fermi function in nuclear beta decay". *Nuclear Science and Engineering* **83** (3): 393–396. OSTI 5307377.

[21] Kurie, F. N. D.; Richardson, J. R.; Paxton, H. C. (1936). "The Radiations Emitted from Artificially Produced Radioactive Substances. I. The Upper Limits and Shapes of the β-Ray Spectra from Several Elements". *Physical Review* **49** (5): 368–381. Bibcode:1936PhRv...49..368K. doi:10.1103/PhysRev.49.368.

[22] Kurie, F. N. D. (1948). "On the Use of the Kurie Plot". *Physical Review* **73** (10): 1207. Bibcode:1948PhRv...73.1207K. doi:10.1103/PhysRev.73.1207.

[23] Rodejohann, Werner (2012). "Neutrinoless double beta decay and neutrino physics". arXiv:1206.2560v2.

[24] Jensen, Carsten (2000). *Controversy and Consensus: Nuclear Beta Decay 1911-1934*. Birkhäuser Verlag. ISBN 3-7643-5313-9.

[25] Chadwick, James (1914). "Intensitätsverteilung im magnetischen Spektren der β-Strahlen von Radium B + C". *Verhandlungen der Deutschen Physikalischen Gesellschaft* (in German) (Deutsche Physikalische Gesellschaft) **16**: 383–391.

[26] Brown, Laurie M. (1978). "The idea of the neutrino". *Physics Today* **31** (9): 23–8. Bibcode:1978PhT....31i..23B. doi:10.1063/1.

[27] C. L Cowan Jr., F. Reines, F. B. Harrison, H. W. Kruse, A. D McGuire (July 20, 1956). "Detection of the Free Neutrino: a Confirmation". *Science* **124** (3212): 103–4. Bibcode:1956Sci...124..103C. doi:10.1126/science.124.3212.103. PMID 17796274.

[28] Segré, E. (1987). "K-Electron Capture by Nuclei". In Trower, P. W. *Discovering Alvarez: Selected Works of Luis W. Alvarez*. University of Chicago Press. pp. 11–12. ISBN 978-0-226-81304-2.

[29] "The Nobel Prize in Physics 1968: Luis Alvarez". The Nobel Foundation. Retrieved 2009-10-07.

[30] Alvarez, L. W. (1937). "Nuclear K Electron Capture". *Physical Review* **52** (2): 134–135. Bibcode:1937PhRv...52..134A. doi:10.1103/PhysRev.52.134.

[31] Alvarez, L. W. (1938). "Electron Capture and Internal Conversion in Gallium 67". *Physical Review* **53** (7): 606. Bibcode:1938PhRv. doi:10.1103/PhysRev.53.606.

[32] Alvarez, L. W. (1938). "The Capture of Orbital Electrons by Nuclei". *Physical Review* **54** (7): 486–497. Bibcode:1938PhRv...54. doi:10.1103/PhysRev.54.486.

[33] C. S. Wu; E. Ambler; R. W. Hayward; D. D. Hoppes; R. P. Hudson (1957). "Experimental Test of Parity Conservation in Beta Decay". *Physical Review* **105**: 1413–1415. Bibcode:1957PhRv..105.1413W. doi:10.1103/PhysRev.105.1413.

[34] http://blogs.scientificamerican.com/guest-blog/2013/10/15/channeling-ada-lovelace-chien-shiung-wu-courageous-hero-of/

[35] T. D. Lee, C. N. Yang (1956). "Question of Parity Conservation in Weak Interactions". *Physical Review* **104**: 254–258. Bibcode:1956PhRv..104..254L. doi:10.1103/PhysRev.104.254.

9.17   Bibliography

- Sin-Itiro Tomonaga (1997). *The Story of Spin*. University of Chicago Press.

9.18 External links

-  **The Live Chart of Nuclides - IAEA** with filter on decay type

- Definition of Beta Disintegration (Decay) at Science Dictionary

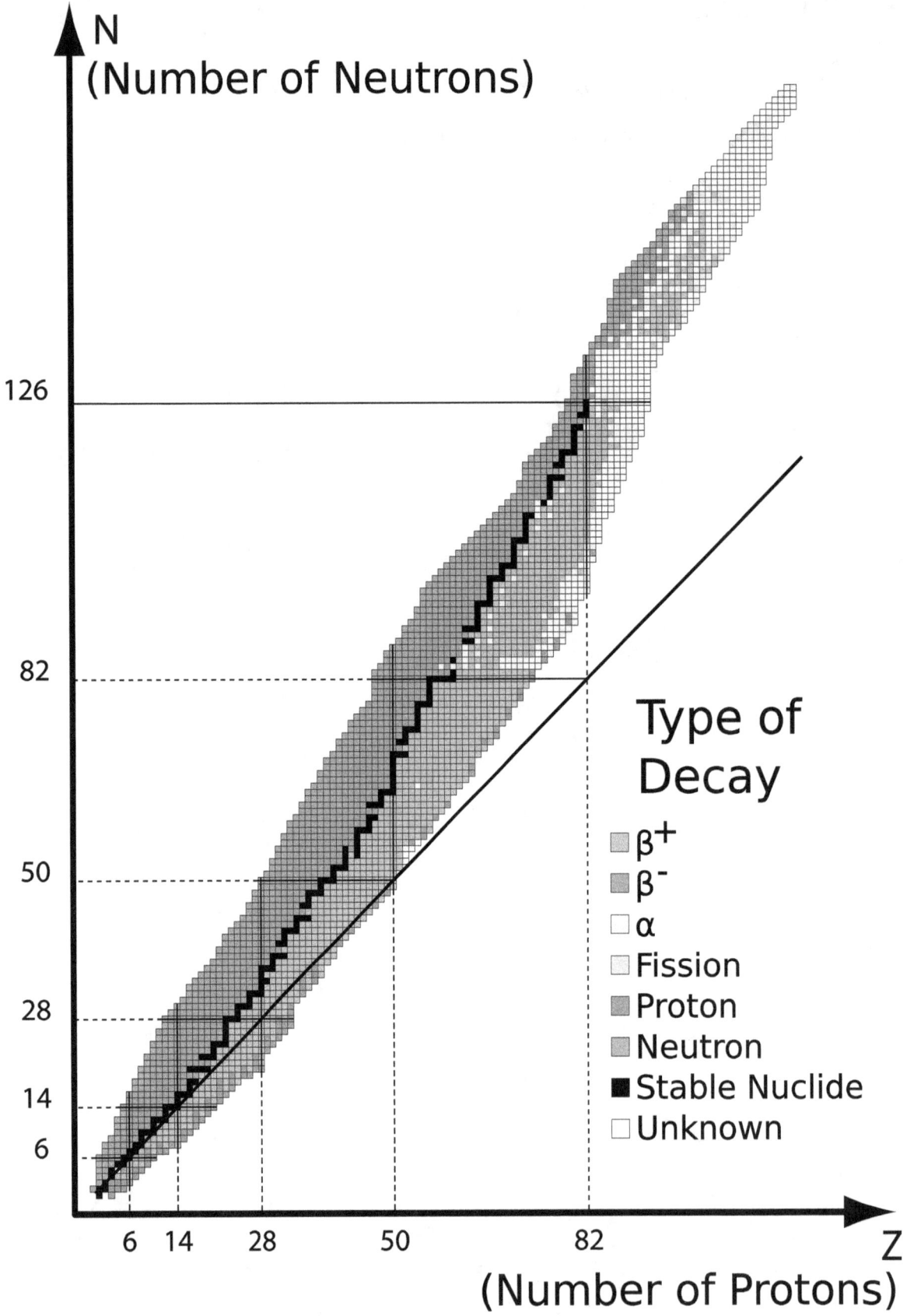

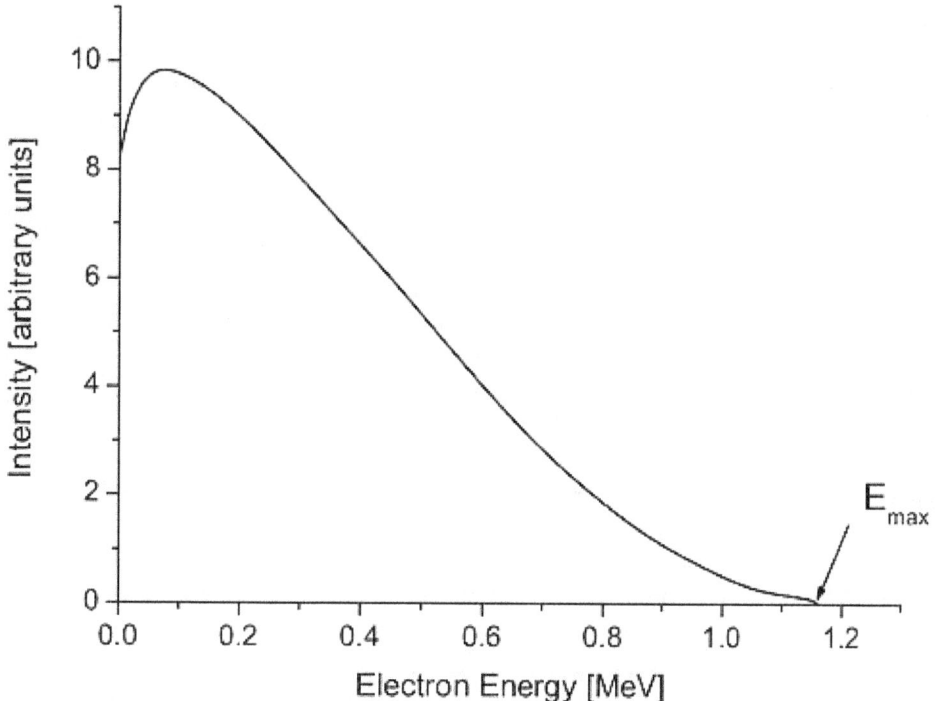

Beta spectrum of ^{210}Bi. $E_{max} = Q = 1.16$ MeV is the maximum energy

Chapter 10

Electron capture

This article is about the radioactive decay mode. For the fragmentation method used in mass spectrometry, see Electron capture ionization. For the detector used in gas chromatography, see Electron-capture dissociation.

Electron capture is a process in which the proton-rich nucleus of an electrically neutral atom absorbs an inner atomic electron, thereby changing a nuclear proton to a neutron and simultaneously causing the emission of an electron neutrino. The atom, now in an excited state, then transitions to its ground state. An outer electron replaces the electron that was captured and an X-ray photon is emitted. Electron capture sometimes results in the Auger effect, where an electron is ejected from the atom and a positive ion results. Sometimes, a gamma ray is emitted because the nucleus is also temporarily in an excited state. Following electron capture, the atomic number is reduced by one, but there is no change in atomic mass. Electron capture is an example of weak interaction, one of the four fundamental forces.

Electron capture is the primary decay mode for isotopes with a relative superabundance of protons in the nucleus, but with insufficient energy difference between the isotope and its prospective daughter (the isobar with one less positive charge) for the nuclide to decay by emitting a positron. Electron capture is an alternate decay mode for radioactive isotopes with insufficient energy to decay by positron emission. It is sometimes called **inverse beta decay**, though this term can also refer to the interaction of an electron antineutrino with a proton.[1]

If the energy difference between the parent atom and the daughter atom is less than 1.022 MeV, positron emission is forbidden as not enough decay energy is available to allow it, and thus electron capture is the sole decay mode. For example, rubidium-83 (37 protons, 46 neutrons) will decay to krypton-83 (36 protons, 47 neutrons) solely by electron capture (the energy difference, or decay energy, is about 0.9 MeV).

A free proton cannot normally be changed to a free neutron by this process; the proton and neutron must be part of a larger nucleus. In the process of electron capture, one of the orbital electrons, usually from the K or L electron shell (**K-electron capture**, also **K-capture**, or **L-electron capture**, **L-capture**), is captured by a proton in the nucleus, forming a neutron and emitting an electron neutrino.

Since a proton is changed to a neutron during electron capture, the number of neutrons in the nucleus increases by 1, the number of protons decreases by 1, and the atomic mass number remains unchanged. By changing the number of protons, electron capture transforms the nuclide into a new element. The atom, although still neutral in charge, now exists in an excited state with the inner shell missing an electron. An outer shell electron eventually makes a transition to replace the missing inner electron and thereby moves into a lower energy state. During this process, that electron will emit an X-ray photon (a type of electromagnetic radiation) and other electrons may also be emitted (see Auger electrons). Often the nucleus will be in an excited state also, and will emit a gamma ray as it transitions to the ground state energy of the new nuclide.

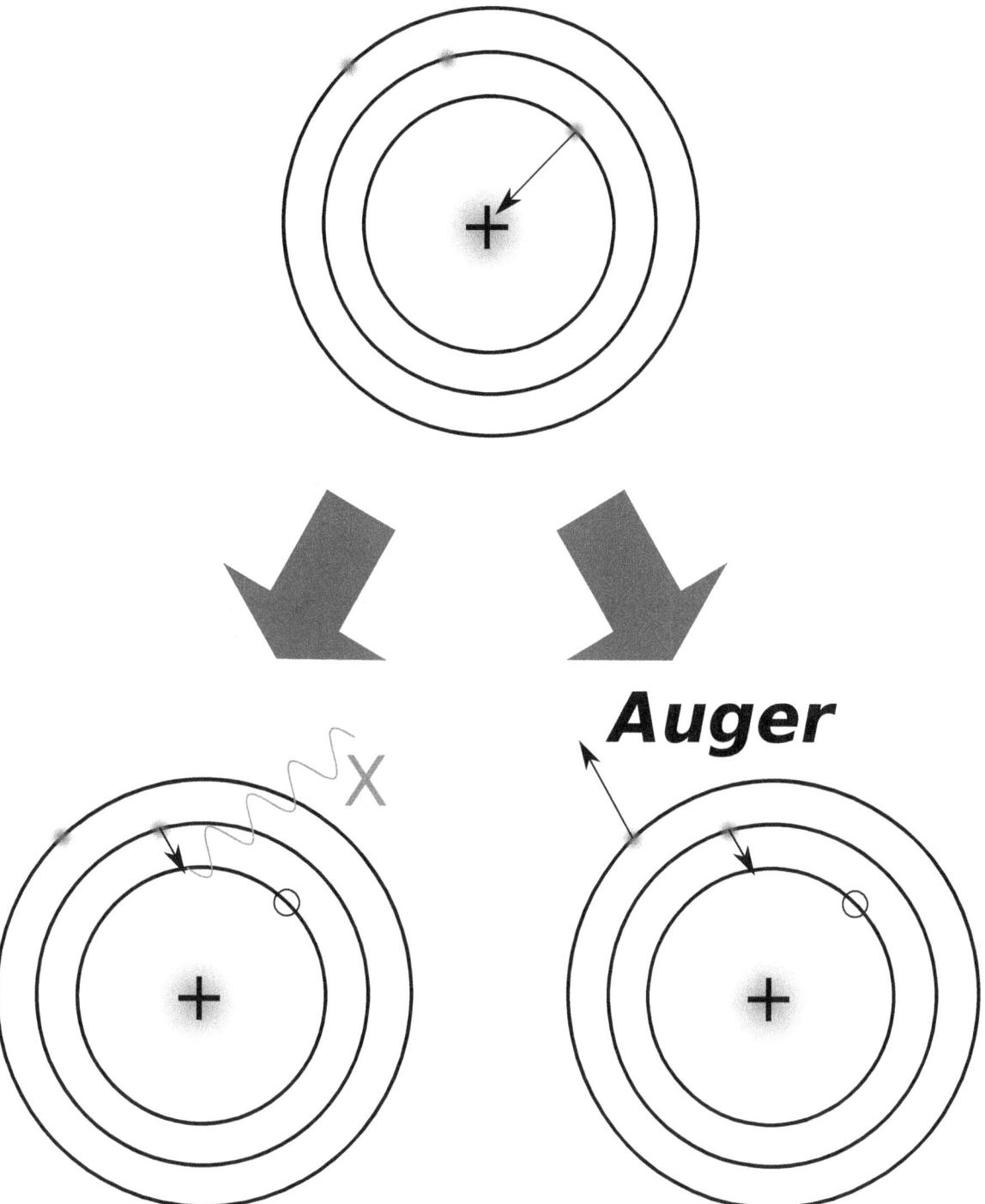

Scheme of two types of electron capture. Top*: The nucleus absorbs an electron.* Lower left*: An outer electron replaces the "missing" electron. An x-ray, equal in energy to the difference between the two electron shells, is emitted.* Lower right*: In the Auger effect, the energy released when the outer electron replaces the inner electron is transferred to an outer electron. The outer electron is ejected from the atom, leaving a positive ion.*

10.1 History

The theory of electron capture was first discussed by Gian-Carlo Wick in a 1934 paper, and then developed by Hideki Yukawa and others. K-electron capture was first observed by Luis Alvarez, in vanadium-48. He reported it in a 1937 paper in *Physical Review*.[2][3][4] Alvarez went on to study electron capture in gallium-67 and other nuclides.[2][5][6]

10.2 Reaction details

The electron that is captured is one of the atom's own electrons, and not a new, incoming electron, as might be suggested by the way the above reactions are written. Radioactive isotopes that decay by pure electron capture can be inhibited from radioactive decay if they are fully ionized ("stripped" is sometimes used to describe such ions). It is hypothesized that such elements, if formed by the r-process in exploding supernovae, are ejected fully ionized and so do not undergo radioactive decay as long as they do not encounter electrons in outer space. Anomalies in elemental distributions are thought to be partly a result of this effect on electron capture. Inverse decays can also be induced by full ionisation; for instance, ^{163}Ho decays into ^{163}Dy by electron capture; however, a fully ionised ^{163}Dy decays into a bound state of ^{163}Ho by the process of bound-state β^- decay.[7]

Chemical bonds can also affect the rate of electron capture to a small degree (in general, less than 1%) depending on the proximity of electrons to the nucleus. For example in ^{7}Be, a difference of 0.9% has been observed between half-lives in metallic and insulating environments.[8] This relatively large effect is due to the fact that beryllium is a small atom whose valence electrons are close to the nucleus.

Around the elements in the middle of the periodic table, isotopes that are lighter than stable isotopes of the same element tend to decay through electron capture, while isotopes heavier than the stable ones decay by electron emission. Electron capture happens most often in the heavier neutron-deficient elements where the mass change is smallest and positron emission isn't always possible. When the loss of mass in a nuclear reaction is greater than zero but less than 2m[0-1e-], the process cannot occur by positron emission but is spontaneous for electron capture.

10.3 Common examples

Some common radioisotopes that decay by electron capture include:

For a full list, see the table of nuclides.

10.4 References

[1] "The Reines-Cowan Experiments: Detecting the Poltergeist" (PDF). *Los Alamos National Laboratory* **25**: 3. 1997.

[2] Luis W. Alvarez, W. Peter Trower (1987). "Chapter 3: K-Electron Capture by Nuclei (with the commentary of Emilio Segré)" In *Discovering Alvarez: selected works of Luis W. Alvarez, with commentary by his students and colleagues.* University of Chicago Press, pp. 11–12, ISBN 978-0-226-81304-2.

[3] "Luis Alvarez, The Nobel Prize in Physics 1968", biography, nobelprize.org. Accessed October 7, 2009.

[4] Alvarez, Luis W. (1937). "Nuclear K Electron Capture". *Physical Review* **52**: 134–135. Bibcode:1937PhRv...52..134A. doi:10.1103/PhysRev.52.134.

[5] Alvarez, Luis W. (1937). "Electron Capture and Internal Conversion in Gallium 67". *Physical Review* **53**:606. Bibcode:1938PhRv. doi:10.1103/PhysRev.53.606.

[6] Alvarez, Luis W. (1938). "The Capture of Orbital Electrons by Nuclei". *Physical Review* **54**:486–497. Bibcode:1938PhRv...54. doi:10.1103/PhysRev.54.486.

[7] Fritz Bosch (1995). "Manipulation of Nuclear Lifetimes in Storage Rings" (PDF). *Physica Scripta* **T59**: 221–229.

[8] B. Wang et al. (2006). "Change of the [7]Be electron capture half-life in metallic environments". *The European Physical Journal A* **28**: 375–377. (subscription required)

10.5   External links

-  **The LIVEChart of Nuclides - IAEA** with filter on electron capture

Chapter 11

Neutron electric dipole moment

"NEDM" redirects here. For the Sussex experiment, see Sussex/RAL/ILL neutron EDM experiment.

The **neutron electric dipole moment (nEDM)** is a measure for the distribution of positive and negative charge inside the neutron. A finite electric dipole moment can only exist if the centers of the negative and positive charge distribution inside the particle do not coincide. So far, no neutron EDM has been found. The current best upper limit amounts to $|d_n| < 2.9 \times 10^{-26} \, e \cdot \mathrm{cm}$.[1]

11.1 Theory

A permanent electric dipole moment of a fundamental particle violates both parity (P) and time reversal symmetry (T). This can be understood by examining the neutron with its magnetic dipole moment and hypothetical electric dipole moment. Under time reversal, the magnetic dipole moment changes its direction, whereas the electric dipole moment stays unchanged. Under parity, the electric dipole moment changes its direction but not the magnetic dipole moment. As the resulting system under P and T is not symmetric with respect to the initial system, these symmetries are violated in the case of the existence of an EDM. Having also CPT symmetry, the combined symmetry CP is violated as well.

11.1.1 Standard Model prediction

As it is depicted above, in order to generate a finite nEDM one needs processes that violate CP symmetry. CP violation has been observed in weak interactions and is included in the Standard Model of particle physics via the CP-violating phase in the CKM matrix. However, the amount of CP violation is very small and therefore also the contribution to the nEDM: $|d_n| \sim 10^{-31} \, e \cdot \mathrm{cm}$.[2]

11.1.2 Matter–antimatter asymmetry

Main article: Baryogenesis

From the asymmetry between matter and antimatter in the universe, one suspects that there must be a sizeable amount of CP-violation. Measuring a neutron electric dipole moment at a much higher level than predicted by the Standard Model would therefore directly confirm this suspicion and improve our understanding of CP-violating processes.

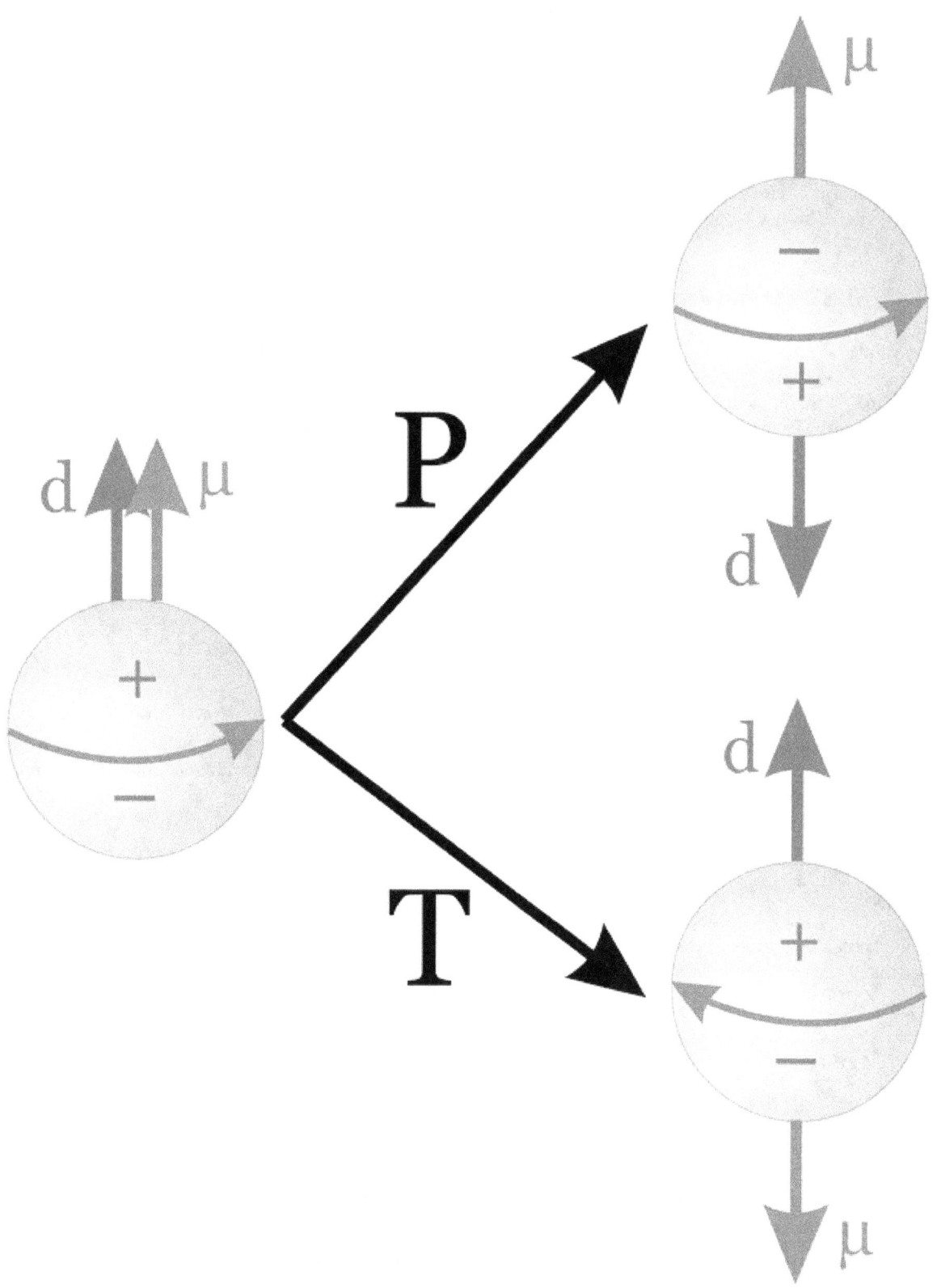

Parity (P) and time reversal (T) violation due to an electric dipole moment

11.1.3 Strong CP problem

Main article: CP-violation

As the neutron is built up of quarks, it is also susceptible to CP violation stemming from strong interactions. Quantum chromodynamics – the theoretical description of the strong force – naturally includes a term which breaks CP-symmetry. The strength of this term is characterized by the angle θ. The current limit on the nEDM constrains this angle to be less than 10^{-10} rad. This fine-tuning of the θ-angle, which is naturally expected to be of order 1, is the strong CP problem.

11.1.4 SUSY CP problem

Supersymmetric extensions to the Standard Model, such as the Minimal Supersymmetric Standard Model, generally lead to a large CP-violation. Typical predictions for the neutron EDM arising from the theory range between 10^{-25} e·cm and 10^{-28} e·cm.[3][4] As in the case of the strong interaction, the limit on the neutron EDM is already constraining the CP violating phases. The fine-tuning is, however, not as severe yet.

11.2 Experimental technique

In order to extract the neutron EDM, one measures the Larmor precession of the neutron spin in the presence of parallel and antiparallel magnetic and electric fields. The precession frequency for each of the two cases is given by

$$h\nu = 2\mu_{\mathrm{n}}B \pm 2d_{\mathrm{n}}E$$

the addition or subtraction of the frequencies stemming from the precession of the magnetic moment around the magnetic field and the precession of the electric dipole moment around the electric field. From the difference of those two frequencies one readily obtains a measure of the neutron EDM:

$$d_{\mathrm{n}} = \frac{h\,\Delta\nu}{4E}$$

The biggest challenge of the experiment (and at the same time the source of the biggest systematic false effects) is to ensure that the magnetic field does not change during these two measurements.

11.3 History

The first experiments searching for the electric dipole moment of the neutron used beams of thermal (and later cold) neutrons to conduct the measurement. It started with the experiment by Smith, Purcell and Ramsey in 1951 (and published in 1957) obtaining a limit of $|d_{\mathrm{n}}| < 5\times10^{-20}$ e·cm.[5] Beams of neutrons were used until 1977 for nEDM experiments. At this point, systematic effects related to the high velocities of the neutrons in the beam became insurmountable. The final limit obtained with a neutron beam amounts to $|d_{\mathrm{n}}| < 3\times10^{-24}$ e·cm.[6]

After that, experiments with ultracold neutrons took over. It started in 1980 with an experiment at the Leningrad Nuclear Physics Institute obtaining a limit of $|d_{\mathrm{n}}| < 1.6\times10^{-24}$ e·cm.[7] This experiment and especially the experiment starting in 1984 at the Institut Laue-Langevin pushed the limit down by another two orders of magnitude yielding the above quoted best upper limit in 2006.

During these 50 years of experiments, six orders of magnitude have been covered thereby putting stringent constraints on theoretical models.[8]

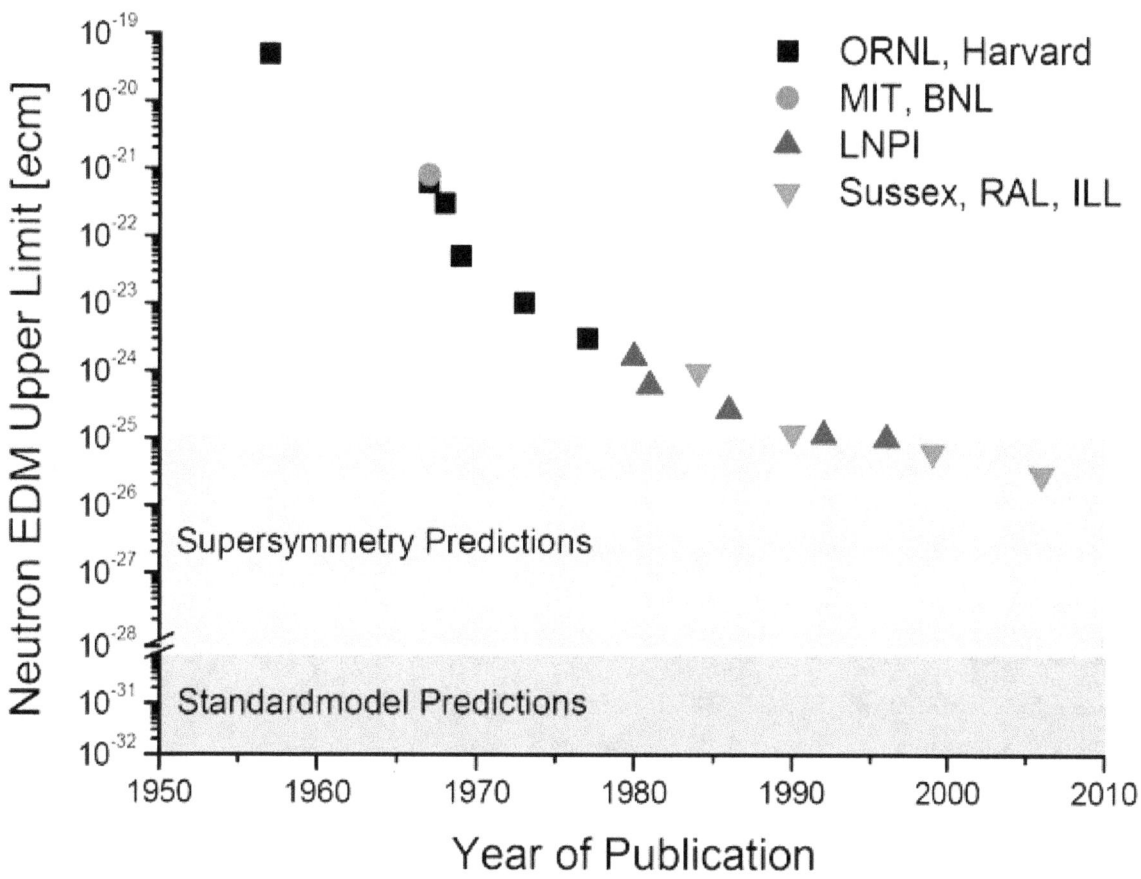

Measured upper limits of the neutron EDM. Given are also the predictions stemming from Supersymmetry and the Standard Model

11.4 Current experiments

Currently, there are at least five experiments aiming at improving the current limit (or measuring for the first time) on the neutron EDM with a sensitivity down to 10^{-28} e·cm over the next 10 years, thereby covering the range of prediction coming from supersymmetric extensions to the Standard Model.

- nEDM experiment running (n2EDM under construction) at the UCN source at the Paul Scherrer Institute[9]

- UCN nEDM experiment under construction at TRIUMF[10]

- Cryogenic neutron EDM experiment being set up at the Institut Laue-Langevin[11]

- nEDM experiment being envisaged at the Spallation Neutron Source[12]

- nEDM experiment being built at the Institut Laue-Langevin[13]

- nEDM experiment being built at the Forschungsreaktor München II[14]

11.5 References

[1] Baker, C. A. et al. (2006). "Improved Experimental Limit on the Electric Dipole Moment of the Neutron". *Physical Review Letters* **97** (13): 131801. arXiv:hep-ex/0602020. Bibcode:2006PhRvL..97m1801B. doi:10.1103/PhysRevLett.97.131801. PMID 17026025.

[2] Dar, S. (2000). "The Neutron EDM in the SM : A Review". arXiv:hep-ph/0008248 [hep-ph].

[3] Abel, S.; Khalil, S.; Lebedev, O. (2001). "EDM constraints in supersymmetric theories". *Nuclear Physics B* **606**: 151–182. arXiv:hep-ph/0103320. Bibcode:2001NuPhB.606..151A. doi:10.1016/S0550-3213(01)00233-4.

[4] Pospelov, M.; Ritz, A. (2005). "Electric dipole moments as probes of new physics". *Annals of Physics* **318**: 119–169. arXiv:hep-ph/0504231. Bibcode:2005AnPhy.318..119P. doi:10.1016/j.aop.2005.04.002.

[5] Smith, J. H.; Purcell, E. M.; Ramsey, N. F. (1957). "Experimental Limit to the Electric Dipole Moment of the Neutron". *Physical Review* **108**: 120–122. Bibcode:1957PhRv..108..120S. doi:10.1103/PhysRev.108.120.

[6] Dress, W. B. et al. (1977). "Search for an electric dipole moment of the neutron". *Physical Review D* **15**: 9–21. Bibcode:1977PhRvD. doi:10.1103/PhysRevD.15.9.

[7] Altarev, I. S. et al. (1980). "A search for the electric dipole moment of the neutron using ultracold neutrons". *Nuclear Physics A* **341** (2): 269–283. Bibcode:1980NuPhA.341..269A. doi:10.1016/0375-9474(80)90313-9.

[8] Ramsey, N.F. (1982). "Electric-Dipole Moments of Particles". *Ann. Rev. Nucl. Part. Sci.* **32**: 211–233. Bibcode:1982ARNPS1R. doi:10.1146/annurev.ns.32.120182.001235.

[9] nEDM Collaboration at PSI Website: http://nedm.web.psi.ch

[10] TRIUMF Ultracold Neutron Source

[11] hepwww.rl.ac.uk Cryogenic EDM

[12] p25ext.lanl.gov on EDM

[13] nrd.pnpi.spb.ru Neutron EDM page

[14] nEDM experiment FRM-II

Chapter 12

Neutron magnetic moment

The **neutron magnetic moment** is the intrinsic magnetic dipole moment of the neutron, symbol μ_n. Protons and neutrons, both nucleons, comprise the nucleus of atoms, and both nucleons behave as small magnets whose strengths are measured by their magnetic moments. The neutron interacts with normal matter primarily through the nuclear force and through its magnetic moment. The neutron's magnetic moment is exploited to probe the atomic structure of materials using scattering methods and to manipulate the properties of neutron beams in particle accelerators. The neutron was determined to have a magnetic moment by indirect methods in the mid 1930s. Luis Alvarez and Felix Bloch made the first accurate, direct measurement of the neutron's magnetic moment in 1940. The existence of the neutron's magnetic moment indicates the neutron is not an elementary particle. For an elementary particle to have an intrinsic magnetic moment, it must have both spin and electric charge. The neutron has spin $1/2$ $\hbar$, but it has no net charge. The existence of the neutron's magnetic moment was puzzling and defied a correct explanation until the quark model for particles was developed in the 1960s. The neutron is composed of three quarks, and the magnetic moments of these elementary particles combine to give the neutron its magnetic moment.

12.1 Description

The best available measurement for the value of the magnetic moment of the neutron is $\mu_n = -1.91304272(45)$ μN.[1] Here μN is the nuclear magneton, a physical constant and standard unit for the magnetic moments of nuclear components. In SI units, $\mu_n = -9.6623647(23) \times 10^{-27}$ J·T^{-1}. A magnetic moment is a vector quantity, and the direction of the neutron's magnetic moment is defined by its spin. The torque on the neutron resulting from an external magnetic field is towards aligning the neutron's spin vector opposite to the magnetic field vector.

The nuclear magneton is the spin magnetic moment of a Dirac particle, a charged, spin $1/2$ elementary particle, with a proton's mass m_p. In SI units, the nuclear magneton is

$$\mu_N = \frac{e\hbar}{2m_p},$$

where e is the elementary charge and $\hbar$ is the reduced Planck constant.[2] The magnetic moment of this particle is parallel to its spin. Since the neutron has no charge, it should have no magnetic moment by this expression. The non-zero magnetic moment of the neutron indicates that it is not an elementary particle.[3] The sign of the neutron's magnetic moment is that of a negatively charged particle. Similarly, the fact that the magnetic moment of the proton, $\mu_p = 2.793$ μN, is not equal to 1 μN indicates that it too is not an elementary particle.[2] Protons and neutrons are composed of quarks, and the magnetic moments of the quarks can be used to compute the magnetic moments of the nucleons.

Although the neutron interacts with normal matter primarily through either nuclear or magnetic forces, the magnetic interactions are about seven orders of magnitude weaker than the nuclear interactions. The influence of the neutron's magnetic moment is therefore only apparent for low energy, or slow, neutrons. Because the value for the magnetic

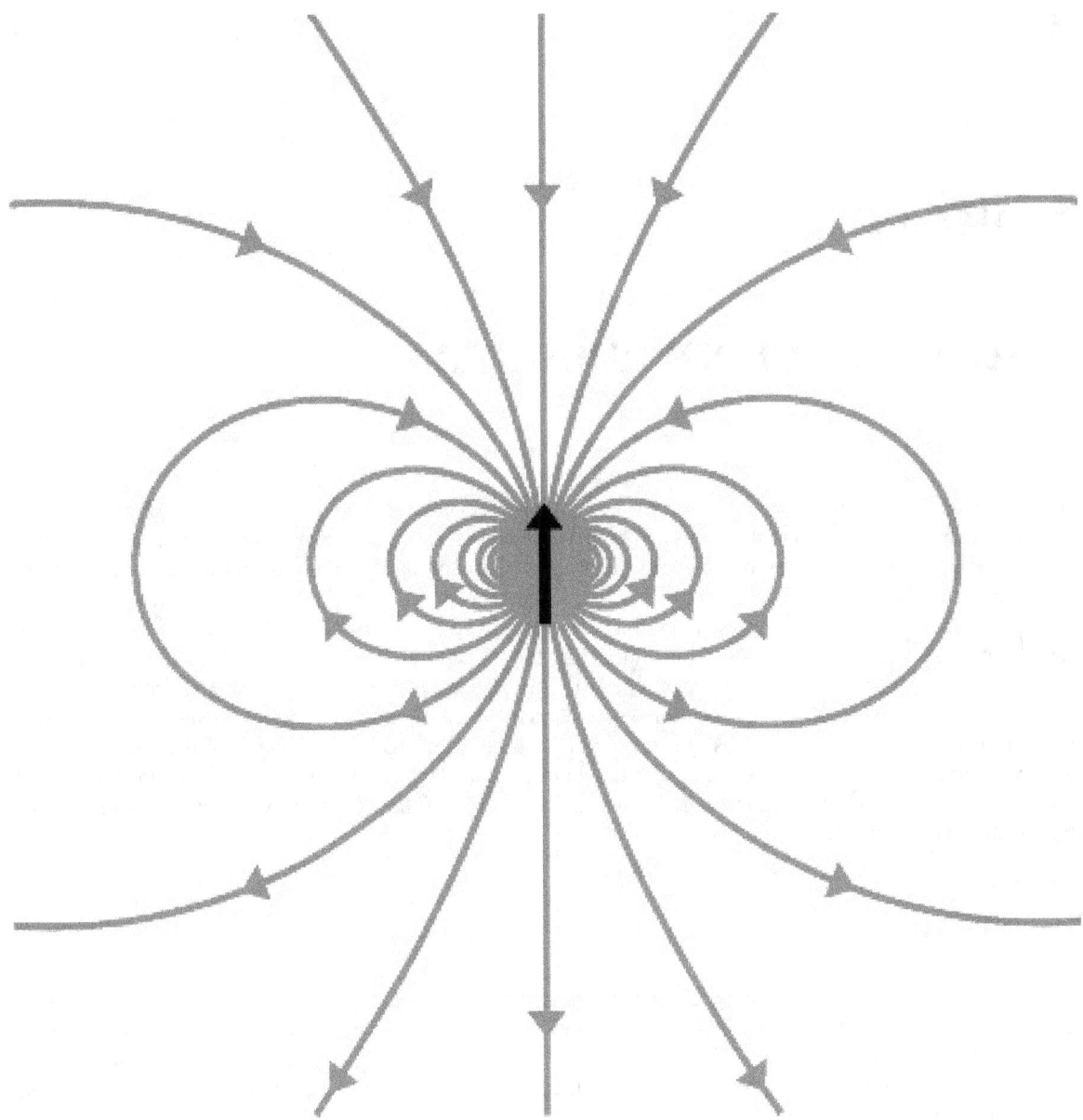

Schematic diagram depicting the spin of the neutron as the black arrow and magnetic field lines associated with the neutron's negative magnetic moment. The spin of the neutron is upward in this diagram, but the magnetic field lines at the center of the dipole are downward.

moment is inversely proportional to particle mass, the nuclear magneton is about 1/2000 as large as the Bohr magneton. The magnetic moment of the electron is therefore about 1000 times larger than that of the neutron.[4]

The magnetic moment of the antineutron has the same magnitude as, but has the opposite sign, that of the neutron.[5]

12.2 Measurement

See also: Discovery of the neutron

Soon after the neutron was discovered in 1932, indirect evidence suggested the neutron had an unexpected non-zero value for its magnetic moment. Attempts to measure the neutron's magnetic moment originated with the discovery by Otto Stern in 1933 in Hamburg that the proton had an anomalously large magnetic moment.[6][7] The proton's magnetic moment had been determined by measuring the deflection of a beam of molecular hydrogen by a magnetic field.[8] Stern won the Nobel Prize in 1943 for this discovery.[9]

By 1934 groups led by Stern, now in Pittsburgh, and I. I. Rabi in New York had independently measured the magnetic moments of the proton and deuteron.[10][11][12] The measured values for these particles were only in rough agreement between the groups, but the Rabi group confirmed the earlier Stern measurements that the magnetic moment for the proton was unexpectedly large.[13][14] Since a deuteron is composed of a proton and a neutron with aligned spins, the neutron's magnetic moment could be inferred by subtracting the deuteron and proton magnetic moments. The resulting value was not zero and had sign opposite to that of the proton. Values for the magnetic moment of the neutron were also determined by R. Bacher[15] at Ann Arbor (1933) and I.Y. Tamm and S.A. Altshuler[16] (1934) in the Soviet Union from studies of the hyperfine structure of atomic spectra. Although Tamm and Altshuler's estimate had the correct sign and order of magnitude ($\mu_n = -0.5~\mu N$), the result was met with skepticism.[13][17] By the late 1930s, accurate values for the magnetic moment of the neutron had been deduced by the Rabi group using measurements employing newly developed nuclear magnetic resonance techniques.[14] The large value for the proton's magnetic moment and the inferred negative value for the neutron's magnetic moment were unexpected and could not be explained.[13] The anomalous values for the magnetic moments of the nucleons would remain a puzzle until the quark model was developed in the 1960s.

The refinement and evolution of the Rabi measurements led to the discovery in 1939 that the deuteron also possessed an electric quadrupole moment.[14][18] This electrical property of the deuteron had been interfering with the measurements by the Rabi group. The discovery meant that the physical shape of the deuteron was not symmetric, which provided valuable insight into the nature of the nuclear force binding nucleons. Rabi was awarded the Nobel Prize in 1944 for his resonance method for recording the magnetic properties of atomic nuclei.[19]

The value for the neutron's magnetic moment was first directly measured by Luis Alvarez and Felix Bloch at Berkeley, California in 1940,[20] using an extension of the magnetic resonance methods developed by Rabi. Alvarez and Bloch determined the magnetic moment of the neutron to be $\mu_n = -1.93(2)~\mu N$. By directly measuring the magnetic moment of free neutrons, or individual neutrons free of the nucleus, Alvarez and Bloch resolved all doubts and ambiguities about this anomalous property of neutrons.[21]

12.3   Neutron *g*-factor and gyromagnetic ratio

The magnetic moment of a nucleon is sometimes expressed in terms of its *g*-factor, a dimensionless scalar. The convention defining the *g*-factor for composite particles, such as the neutron or proton, is

$$\mu = \frac{g\mu_N}{\hbar} I$$

where μ is the intrinsic magnetic moment, I is the spin angular momentum, and g is the effective *g*-factor.[22] While the g-factor is dimensionless, for composite particles it is defined relative to the natural unit of the nuclear magneton. For the neutron, I is 1/2 $\hbar$, so the neutron's *g*-factor, symbol g_n, is $-3.82608545(90)$.[23]

The gyromagnetic ratio, symbol γ, of a particle or system is the ratio of its magnetic moment to its spin angular momentum, or

$$\mu = \gamma I$$

For nucleons, the ratio is conventionally written in terms of the proton mass and charge, by the formula

$$\gamma = \frac{g\mu_N}{\hbar} = g\frac{e}{2m_p}$$

The neutron's gyromagnetic ratio, symbol γ_n, is $-1.83247179(43)\times10^8~s^{-1}\cdot T^{-1}$.[24] The gyromagnetic ratio is also the ratio between the observed angular frequency of Larmor precession (in rad s^{-1}) and the strength of the magnetic field in nuclear magnetic resonance applications,[25] such as in MRI imaging. For this reason, the value of γ_n is often given in units of MHz/T. The quantity $\gamma_n/2\pi$ ("gamma bar") is therefore convenient, which has the value −29.1646943(69) MHz·T^{-1}.[26]

12.4 Physical significance

When a neutron is put into a magnetic field produced by an external source, it is subject to a torque tending to orient its magnetic moment parallel to the field (hence its spin antiparallel to the field).[27] Like any magnet, the amount of this torque is proportional both to the magnetic moment and the external magnetic field. Since the neutron has spin angular momentum, this torque will cause the neutron to precess with a well-defined frequency, called the Larmor frequency. It is this phenomenon that enables the measurement of nuclear properties through nuclear magnetic resonance. The Larmor frequency can be determined by the product of the gyromagnetic ratio with the magnetic field strength. Since the sign of γ_n is negative, the neutron's spin angular momentum precesses counterclockwise about the direction of the external magnetic field.[28]

The magnetic moment of the neutron has been exploited to probe the properties of matter using scattering or diffraction techniques. These methods provide information that is complementary to X-ray spectroscopy. In particular, the magnetic moment of the neutron is used to determine magnetic properties of materials at length scales of 1–100 Å using cold or thermal neutrons.[29] Bertram Brockhouse and Clifford Shull won the Nobel Prize in physics in 1994 for developing these scattering techniques.[30]

Since neutrons are neutral particles, they do not have to overcome Coulomb repulsion as they approach charged targets, as experienced by protons or alpha particles. Neutrons can deeply penetrate matter. On the other hand, without an electric charge, neutron beams cannot be controlled by the conventional electromagnetic methods employed for particle accelerators. The magnetic moment of the neutron allows some control of neutrons using magnetic fields, however,[31][32] including the formation of polarized neutron beams.

Since an atomic nucleus consists of a bound state of protons and neutrons, the magnetic moments of the nucleons contribute to the nuclear magnetic moment, or the magnetic moment for the nucleus as a whole. The nuclear magnetic moment also includes contributions from the orbital motion of the nucleons. The deuteron has the simplest example of a nuclear magnetic moment, with measured value 0.857 μN. This value is within 3% of the sum of the moments of the proton and neutron, which gives 0.879 μN. In this calculation, the spins of the nucleons are aligned, but their magnetic moments offset because of the neutron's negative magnetic moment.[33]

12.5 Anomalous magnetic moments and meson physics

The anomalous values for the magnetic moments of the nucleons presented a theoretical quandary for the 30 years from the time of their discovery in the early 1930s to the development of the quark model in the early 1960s. Considerable theoretical efforts were expended in trying to understand the origins of these magnetic moments, but the failures of these theories were glaring.[34] Much of the theoretical focus was on developing a nuclear-force equivalence to the remarkably successful theory explaining the small anomalous magnetic moment of the electron.

In quantum electrodynamics (QED), the anomalous magnetic moment of a particle stems from the small contributions of quantum mechanics to the magnetic moment of that particle.[35] The g-factor for a "Dirac" magnetic moment is predicted to be $g = -2$ for a negatively charged, spin 1/2 particle. For particles such as the electron, this "classical" result differs from the observed value by a small fraction of a percent; the difference compared to the classical value is the anomalous magnetic moment. The actual g-factor for the electron is measured to be −2.00231930436153(53).[36] QED results from the mediation of the electromagnetic force by photons. The physical picture is that the *effective* magnetic moment of the electron results from the contributions of the "bare" electron, which is the Dirac particle, and the cloud of "virtual," short-lived electron–positron pairs and photons that surround this particle as a consequence of QED. The small effects of these quantum mechanical fluctuations can be theoretically computed using Feynman diagrams with loops.[37]

The one-loop contribution to the anomalous magnetic moment of the electron, corresponding to the first order and largest correction in QED, is found by calculating the vertex function shown in the diagram on the right. The calculation was discovered by Julian Schwinger in 1948.[35][38] Computed to fourth order, the QED prediction for the electron's anomalous magnetic moment agrees with the experimentally measured value to more than 10 significant figures, making the magnetic moment of the electron one of the most accurately verified predictions in the history of physics.[35]

Compared to the electron, the anomalous magnetic moments of the nucleons are enormous.[3] The g-factor for the proton is 5.6, and the chargeless neutron should have no magnetic moment at all. Note, however, that the anomalous magnetic moments of the nucleons, that is, their magnetic moments with the expected Dirac particle magnetic moments subtracted, are roughly equal but of opposite sign: $\mu_p - 1.00\,\mu N = +1.79\,\mu N$, $\mu_n - 0.00\,\mu N = -1.91\,\mu N$.[39]

The Yukawa interaction for nucleons was discovered in the mid-1930s, and this nuclear force is mediated by pion mesons.[40] In parallel with the theory for the electron, the hypothesis was that higher-order loops involving nucleons and pions may generate the anomalous magnetic moments of the nucleons.[2] The physical picture was that the *effective* magnetic moment of the neutron arose from the combined contributions of the "bare" neutron, which is zero, and the cloud of "virtual" pions and photons that surround this particle as a consequence of the nuclear and electromagnetic forces.[41] The Feynman diagram at right is roughly the first order diagram, with the role of the virtual particles played by pions. As noted by Abraham Pais, "between late 1948 and the middle of 1949 at least six papers appeared reporting on second order calculations of nucleon moments."[34] These theories were also, as noted by Pais, "a flop" – they gave results that grossly disagreed with observation. Nevertheless, serious efforts continued along these lines for the next couple of decades, to little success.[2][41][42] These theoretical approaches were incorrect because the nucleons are composite particles with their magnetic moments arising from their elementary components, quarks.

12.6   Quark model for nucleon magnetic moments

In the quark model for hadrons, the neutron is composed of one up quark (charge $+2/3\ e$) and two down quarks (charge $-1/3\ e$).[43] The magnetic moment of the neutron can be modeled as a sum of the magnetic moments of the constituent quarks,[44] although this simple model belies the complexities of the Standard Model of particle physics.[45] The calculation assumes that the quarks behave like pointlike Dirac particles, each having their own magnetic moment, as computed using an expression similar to the one above for the nuclear magneton:

$$\mu_q = \frac{e_q \hbar}{2m_q},$$

where the q-subscripted variables refer to quark magnetic moment, charge, or mass. Simplistically, the magnetic moment of the neutron can be viewed as resulting from the vector sum of the three quark magnetic moments, plus the orbital magnetic moments caused by the movement of the three charged quarks within the neutron.

In one of the early successes of the Standard Model (SU(6) theory), in 1964 Mirza A. B. Beg, Benjamin W. Lee, and Abraham Pais theoretically calculated the ratio of proton to neutron magnetic moments to be $-3/2$, which agrees with the experimental value to within 3%.[46][47][48] The measured value for this ratio is $-1.45989806(34)$.[49] A contradiction of the quantum mechanical basis of this calculation with the Pauli exclusion principle, led to the discovery of the color charge for quarks by Oscar W. Greenberg in 1964.[46]

From the nonrelativistic, quantum mechanical wavefunction for baryons composed of three quarks, a straightforward calculation gives fairly accurate estimates for the magnetic moments of neutrons, protons, and other baryons.[44] For a neutron, the end result of this calculation is that the magnetic moment of the neutron is given by $\mu_n = 4/3\ \mu_d - 1/3\ \mu_u$, where μ_d and μ_u are the magnetic moments for the down and up quarks, respectively. This result combines the intrinsic magnetic moments of the quarks with their orbital magnetic moments, and assumes the three quarks are in a particular, dominant quantum state.

The results of this calculation are encouraging, but the masses of the up or down quarks were assumed to be 1/3 the mass of a nucleon.[44] The masses of the quarks are actually only about 1% that of a nucleon.[45] The discrepancy stems from the complexity of the Standard Model for nucleons, where most of their mass originates in the gluon fields, virtual particles, and their associated energy that are essential aspects of the strong force.[45][50] Furthermore, the complex system of quarks and gluons that constitute a neutron requires a relativistic treatment.[51] Nucleon magnetic moments have been successfully computed from first principles, requiring significant computing resources.[52][53]

12.7 See also

- Bohr magneton

- Electron magnetic moment

- Proton magnetic moment

- Nuclear magnetic moment

- Anomalous magnetic moment

- Neutron diffraction

- Neutron triple-axis spectrometry

- LARMOR neutron microscope

- Antineutron

12.8 References

[1] Beringer, J.; et al. (Particle Data Group) (2012). "Review of Particle Physics, 2013 partial update" (PDF). *Phys. Rev. D* **86**: 010001. Bibcode:2012PhRvD..86a0001B. doi:10.1103/PhysRevD.86.010001. Retrieved May 8, 2015.

[2] Bjorken, J.D.; Drell, S.D. (1964). *Relativistic Quantum Mechanics*. New York: McGraw-Hill. pp. 241–246. ISBN 0070054932.

[3] Hausser, O. (1981). "Nuclear Moments". In Lerner, R. G.; Trigg, G. L. *Encyclopedia of Physics*. Reading, Massachusetts: Addison-Wesley Publishing Company. pp. 679–680. ISBN 0201043130.

[4] "CODATA values of the fundamental constants". *NIST*. Retrieved May 8, 2015.

[5] Schreckenbach, K. (2013). "Physics of the Neutron". In Stock, R. *Encyclopedia of Nuclear Physics and its Applications*. Weinheim, Germany: Wiley-VCH Verlag GmbH & Co. pp. 321–354. ISBN 978-3-527-40742-2.

[6] Frisch, R.; Stern, O. (1933). "Über die magnetische Ablenkung von Wasserstoffmolekülen und das magnetische Moment des Protons. I / Magnetic Deviation of Hydrogen Molecules and the Magnetic Moment of the Proton. I.". *Z. Phys.* **85**: 4–16. Bibcode:1933ZPhy...85....4F. doi:10.1007/bf01330773. Retrieved May 9, 2015.

[7] Esterman, I.; Stern, O. (1933). "Über die magnetische Ablenkung von Wasserstoffmolekülen und das magnetische Moment des Protons. II / Magnetic Deviation of Hydrogen Molecules and the Magnetic Moment of the Proton. I.". *Z. Phys.* **85**: 17–24. Bibcode:1933ZPhy...85...17E. doi:10.1007/bf01330774. Retrieved May 9, 2015.

[8] Toennies, J. P.; Schmidt-Bocking, H.; Friedrich, B.; Lower, J. C. A. "Otto Stern (1888–1969): The founding father of experimental atomic physics". arXiv.org. pp. 1–39. Retrieved May 9, 2015.

[9] "The Nobel Prize in Physics 1943". Nobel Foundation. Retrieved 2015-01-30.

[10] Esterman,I.;Stern,O. (1934)."Magnetic moment of the deuton".*Physical Review***45**:761(A109).Bibcode:1934PhRv...45. doi:10.1103/PhysRev.45.739. Retrieved May 9, 2015.

[11] Rabi, I.I.; Kellogg, J.M.; Zacharias, J.R. (1934). "The magnetic moment of the proton". *Physical Review* **46**: 157–163. Bibcode:1934PhRv...46..157R. doi:10.1103/physrev.46.157. Retrieved May 9, 2015.

[12] Rabi, I.I.; Kellogg, J.M.; Zacharias, J.R. (1934). "The magnetic moment of the deuton". *Physical Review* **46**: 163–165. Bibcode:1934PhRv...46..163R. doi:10.1103/physrev.46.163. Retrieved May 9, 2015.

[13] Breit, G.; Rabi, I.I. (1934). "On the interpretation of present values of nuclear moments". *Physical Review* **46**: 230–231. Bibcode:1934PhRv...46..230B. doi:10.1103/physrev.46.230. Retrieved May 9, 2015.

[14] John S. Rigden (1987). *Rabi, Scientist and Citizen*. New York: Basic Books, Inc. pp. 99–114. ISBN 9780674004351. Retrieved May 9, 2015.

[15]Bacher,R.F. (1933)."Note on the Magnetic Moment of the Nitrogen Nucleus".*Physical Review***43**:1001–1002.Bibcode:B. doi:10.1103/physrev.43.1001. Retrieved 2015-02-10.

[16] Tamm, I.Y.; Altshuler, S.A. (1934). "Magnetic Moment of the Neutron". *Doklady Akad. Nauk SSSR* **8**: 455. Retrieved 2015-01-30.

[17] Sergei Vonsovsky (1975). *Magnetism of Elementary Particles.* Moscow: Mir Publishers. pp. 73–75.

[18] Kellogg, J.M.; Rabi, I.I.; Ramsey, N.F.; Zacharias, J.R. (1939). "An electrical quadrupole moment of the deuteron". *Physical Review* **55**: 318–319. Bibcode:1939PhRv...55..318K. doi:10.1103/physrev.55.318. Retrieved May 9, 2015.

[19] "The Nobel Prize in Physics 1944". Nobel Foundation. Retrieved 2015-01-25.

[20] Alvarez, L. W; Bloch, F. (1940). "A quantitative determination of the neutron magnetic moment in absolute nuclear magnetons". *Physical Review* **57**: 111–122. Bibcode:1940PhRv...57..111A. doi:10.1103/physrev.57.111.

[21] Ramsey, Norman F. (1987). "Chapter 5: The Neutron Magnetic Moment". In Trower, W. Peter. *Discovering Alvarez: Selected Works of Luis W. Alvarez with Commentary by His Students and Colleagues.* University of Chicago Press. pp. 30–32. ISBN 978-0226813042. Retrieved May 9, 2015.

[22] Povh, B.; Rith, K.; Scholz, C.; Zetsche, F. (2002). *Particles and Nuclei: An Introduction to the Physical Concepts.* Berlin: Springer-Verlag. pp. 74–75,259–260. ISBN 978-3-540-43823-6. Retrieved May 10, 2015.

[23] "CODATA values of the fundamental constants". *NIST.* Retrieved May 8, 2015.

[24] "CODATA values of the fundamental constants". *NIST.* Retrieved May 8, 2015.

[25] Jacobsen, Neil E. (2007). *NMR spectroscopy explained.* Hoboken, New Jersey: Wiley-Interscience. ISBN 9780471730965. Retrieved May 8, 2015.

[26] "CODATA values of the fundamental constants". *NIST.* Retrieved May 8, 2015.

[27] B. D. Cullity, C. D. Graham (2008). *Introduction to Magnetic Materials* (2 ed.). Hoboken, New Jersey: Wiley-IEEE Press. p. 103. ISBN 0-471-47741-9. Retrieved May 8, 2015.

[28] M. H. Levitt (2001). *Spin dynamics: basics of nuclear magnetic resonance.* West Sussex, England: John Wiley & Sons. pp. 25–30. ISBN 0-471-48921-2.

[29] S.W. Lovesey (1986). *Theory of Neutron Scattering from Condensed Matter Volume 1: Nuclear Scattering.* Oxford: Clarendon Press. pp. 1–30. ISBN 0198520298.

[30] "The Nobel Prize in Physics 1994". Nobel Foundation. Retrieved 2015-01-25.

[31] Oku, T.; Suzuki, J. et al. (2007). "Highly polarized cold neutron beam obtained by using a quadrupole magnet". *Physica B* **397**: 188–191. Bibcode:2007PhyB..397..188O. doi:10.1016/j.physb.2007.02.055.

[32] Arimoto, Y.; Geltenbort, S. et al. (2012). "Demonstration of focusing by a neutron accelerator". *Physical Review A* **86**: 023843. Bibcode:2012PhRvA..86b3843A. doi:10.1103/PhysRevA.86.023843. Retrieved May 9, 2015.

[33] Semat, Henry (1972). *Introduction to Atomic and Nuclear Physics: 5th edition.* London: Holt, Rinehart and Winston. p. 556. ISBN 978-1-4615-9701-8. Retrieved May 8, 2015.

[34] Pais, Abraham (1986). *Inward Bound.* Oxford: Oxford University Press. p. 299. ISBN 0198519974.

[35] See section 6.3 in Peskin, M. E.; Schroeder, D. V. (1995). *An Introduction to Quantum Field Theory.* Reading, Massachusetts: Perseus Books. pp. 175–198. ISBN 978-0201503975.

[36] "CODATA values of the fundamental constants". *NIST.* Retrieved May 11, 2015.

[37] Aoyama, T.; Hayakawa, M.; Kinoshita, T.; Nio, M. (2008). "Revised value of the eighth-order QED contribution to the anomalous magnetic moment of the electron". *Physical Review D* **77** (5): 053012. arXiv:0712.2607. Bibcode:2008PhRvD..77e3012A. doi:10.1103/PhysRevD.77.053012.

[38] Schwinger, J. (1948). "On Quantum-Electrodynamics and the Magnetic Moment of the Electron". *Physical Review* **73** (4): 416–417. Bibcode:1948PhRv...73..416S. doi:10.1103/PhysRev.73.416.

[39] See chapter 1, section 6 in deShalit, A.; Feschbach, H. (1974). *Theoretical Nuclear Physics Volume I: Nuclear Structure*. New York: John Wiley and Sons. p. 31. ISBN 0471203858.

[40] Brown, L.M.; Rechenberg, H. (1996). *The Origin of the Concept of Nuclear Forces*. Bristol and Philadelphia: Institute of Physics Publishing. pp. 95–312. ISBN 0750303735.

[41] Drell, S.; Zachariasen, F. (1961). *Electromagnetic Structure of Nucleons*. New York: Oxford University Press. pp. 1–130.

[42] Drell, S.; Pagels, H.R. (1965). "Anomalous Magnetic Moment of the Electron, Muon, and Nucleon". *Physical Review* **140**: B397–B407. Bibcode:1965PhRv..140..397D. doi:10.1103/PhysRev.140.B397. Retrieved May 10, 2015.

[43] Gell, Y.; Lichtenberg, D. B. (1969). "Quark model and the magnetic moments of proton and neutron". *Il Nuovo Cimento A*. Series 10 **61**: 27–40. Bibcode:1969NCimA..61...27G. doi:10.1007/BF02760010.

[44] Perkins, Donald H. (1982), *Introduction to High Energy Physics*, Addison Wesley, Reading, Massachusetts, pp. 201–202, ISBN 0-201-05757-3

[45] Cho, Adiran (2 April 2010). "Mass of the Common Quark Finally Nailed Down". *http://news.sciencemag.org"*. *American Association for the Advancement of Science*. Retrieved 27 September 2014.

[46] Greenberg, O. W. (2009), "Color charge degree of freedom in particle physics", *Compendium of Quantum Physics*, Springer Berlin Heidelberg, pp. 109–111, doi:10.1007/978-3-540-70626-7_32

[47] Beg, M.A.B.; Lee, B.W.; Pais, A. (1964). "SU(6) and electromagnetic interactions". *Physical Review Letters* **13**: 514–517, erratum 650. Bibcode:1964PhRvL..13..514B. doi:10.1103/physrevlett.13.514.

[48] Sakita, B. (1964). "Electromagnetic properties of baryons in the supermultiplet scheme of elementary particles". *Physical Review Letters* **13**: 643–646. Bibcode:1964PhRvL..13..643S. doi:10.1103/physrevlett.13.643.

[49] Mohr, P.J.; Taylor, B.N. and Newell, D.B. (2011), "The 2010 CODATA Recommended Values of the Fundamental Physical Constants" (Web Version 6.0). The database was developed by J. Baker, M. Douma, and S. Kotochigova. (2011-06-02). National Institute of Standards and Technology, Gaithersburg, Maryland 20899. Retrieved May 9, 2015.

[50] Wilczek, F. (2003). "The Origin of Mass" (PDF). *MIT Physics Annual*: 24–35. Retrieved May 8, 2015.

[51] Ji, Xiangdong (1995). "A QCD Analysis of the Mass Structure of the Nucleon". *Phys.Rev.Lett.* **74**: 1071–1074. arXiv:hep-ph/9410274. Bibcode:1995PhRvL..74.1071J. doi:10.1103/PhysRevLett.74.1071. Retrieved May 8, 2015.

[52] Martinelli, G.; Parisi, G.; Petronzio, R.; Rapuano, F. (1982). "The proton and neutron magnetic moments in lattice QCD". *Physics Letters B* **116**: 434–436. Bibcode:1982PhLB..116..434M. doi:10.1016/0370-2693(82)90162-9. Retrieved May 8, 2015.

[53] Kincade, Kathy (2 February 2015). "Pinpointing the magnetic moments of nuclear matter". *http://phys.org"*. *Phys.org*. Retrieved *May 8, 2015.*

12.9 Bibliography

- S.W. Lovesey (1986). Theory of Neutron Scattering from Condensed Matter. Oxford University Press. ISBN 0198520298.

- Donald H. Perkins (1982). Introduction to High Energy Physics. Reading, Massachusetts: Addison Wesley, ISBN 0-201-05757-3.

- John S. Rigden (1987). Rabi, Scientist and Citizen. New York: Basic Books, Inc., ISBN 0-465-06792-1.

- Sergei Vonsovsky (1975). Magnetism of Elementary Particles. Moscow: Mir Publishers.

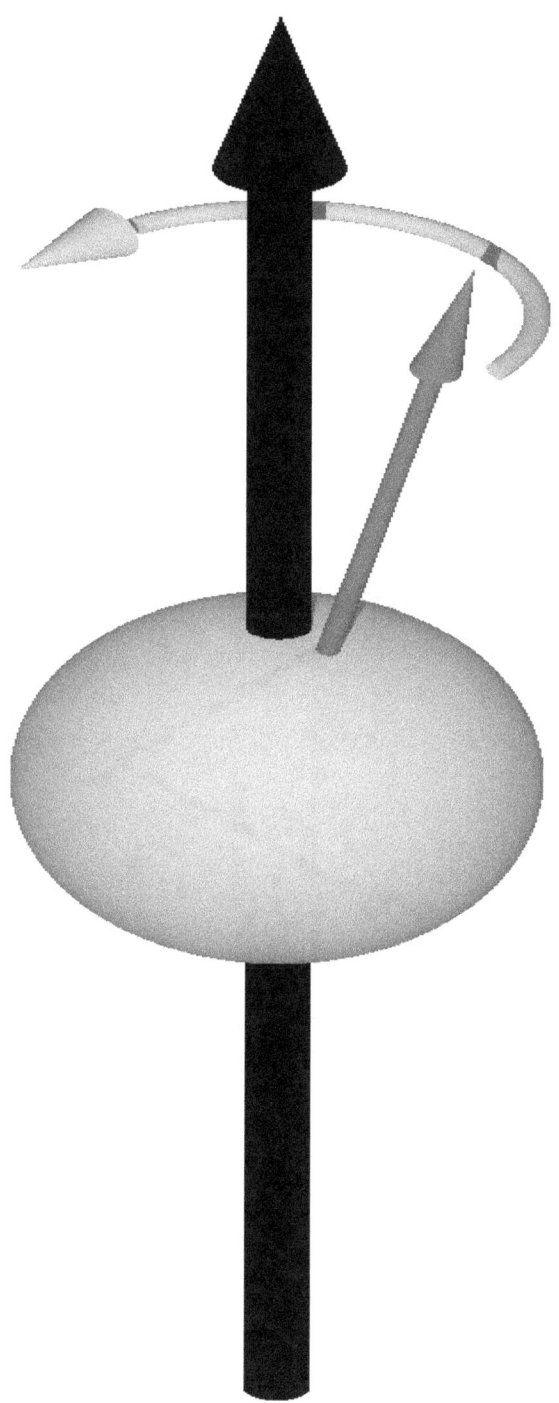

Direction of Larmor precession for a neutron. The central arrow denotes the magnetic field, the small red arrow the spin of the neutron.

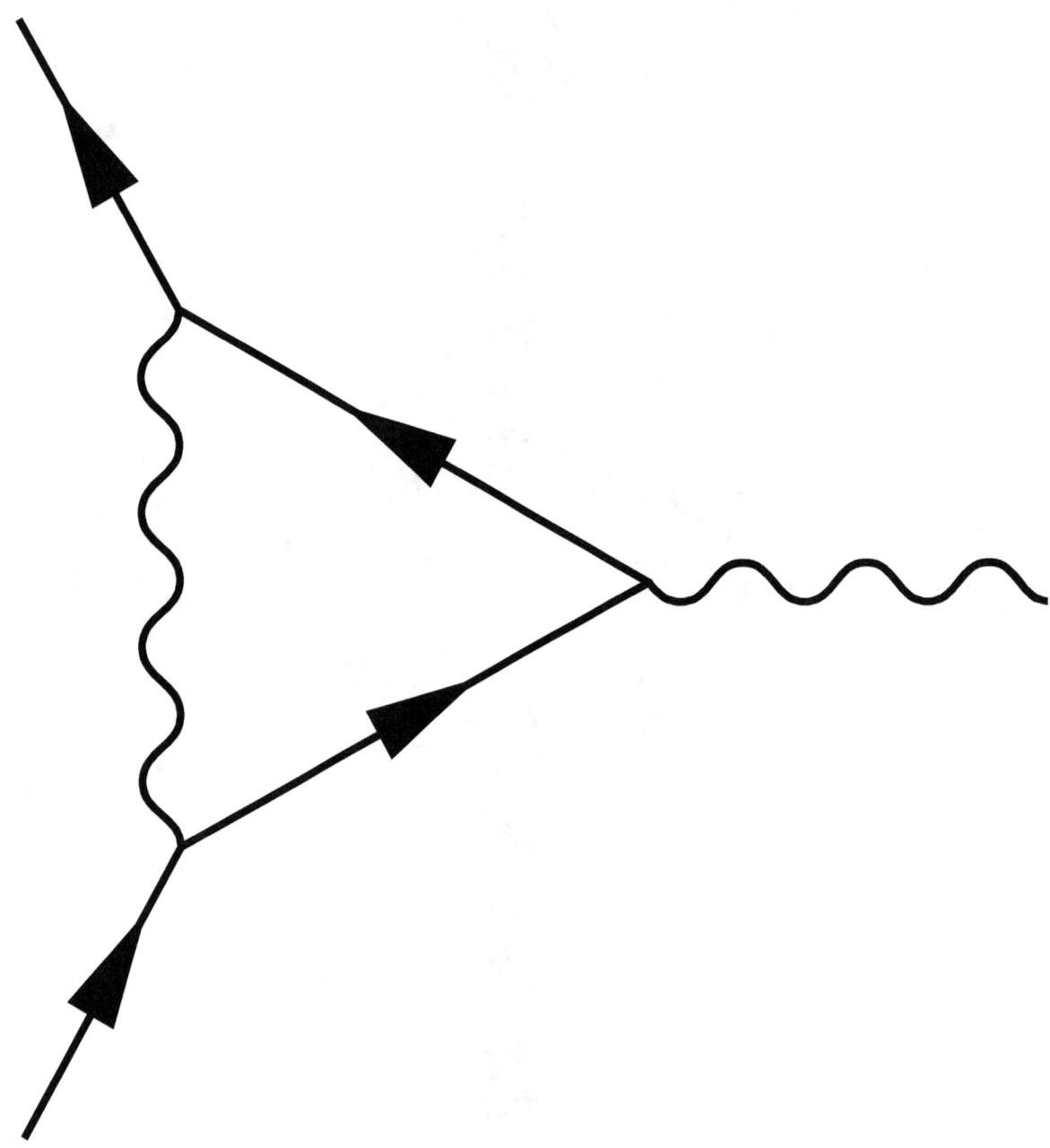

One-loop correction to the fermion's magnetic dipole moment. The solid lines at top and bottom represent the fermion (electron or nucleon), the wavey lines represent the photon mediating the electromagnetic force. The middle solid lines represent a virtual pair of particles, electron and positron for QED, pions for the nuclear force.

Chapter 13

Antineutron

The **antineutron** is the antiparticle of the neutron with symbol n. It differs from the neutron only in that some of its properties have equal magnitude but opposite sign. It has the same mass as the neutron, and no net electric charge, but has opposite baryon number (+1 for neutron, −1 for the antineutron). This is because the antineutron is composed of antiquarks, while neutrons are composed of quarks. In particular, the antineutron consists of one up antiquark and two down antiquarks.

Since the antineutron is electrically neutral, it cannot easily be observed directly. Instead, the products of its annihilation with ordinary matter are observed. In theory, a free antineutron should decay into an antiproton, a positron and a neutrino in a process analogous to the beta decay of free neutrons. There are theoretical proposals that neutron–antineutron oscillations exist, a process which would occur only if there is an undiscovered physical process that violates baryon number conservation.[1][2][3]

The antineutron was discovered in proton–proton collisions at the Bevatron (Lawrence Berkeley National Laboratory) by Bruce Cork in 1956, one year after the antiproton was discovered.

13.1 Magnetic moment

The magnetic moment of the antineutron is the opposite of that of the neutron.[4] It is +1.91 μN for the antineutron but −1.91 μN for the neutron (relative to the direction of the spin). Here μN is the nuclear magneton.

13.2 See also

- Antimatter

- Neutron magnetic moment

- List of particles

13.3 References

[1] R. N. Mohapatra (2009). "Neutron-Anti-Neutron Oscillation: Theory and Phenomenology". *Journal of Physics G* **36** (10): 104006. arXiv:0902.0834. Bibcode:2009JPhG...36j4006M. doi:10.1088/0954-3899/36/10/104006.

[2] C. Giunti, M. Laveder (19 August 2010). "Neutron Oscillations". *Neutrino Unbound*. Istituto Nazionale di Fisica Nucleare. Retrieved 2010-08-19.

[3] Y. A. Kamyshkov (16 January 2002). "Neutron → Antineutron Oscillations" (PDF). *NNN 2002 Workshop on "Large Detectors for Proton Decay, Supernovae and Atmospheric Neutrinos and Low Energy Neutrinos from High Intensity Beams" at CERN.* Retrieved 2010-08-19.

[4] Lorenzon, Wolfgang (6 April 2007). "Physics 390: Homework set #7 Solutions" (PDF). *Modern Physics, Physics 390, Winter 2007.* Retrieved 2009-12-22.

13.4 External links

- LBL Particle Data Group: summary tables

- suppression of neutron-antineutron oscillation

- Elementary particles: includes information about antineutron discovery (archived link)

- "Is Antineutron the Same as Neutron?" explains how the antineutron differs from the regular neutron despite having the same, that is zero, charge

Chapter 14

Neutronium

Neutronium (sometimes shortened to **neutrium**[1]) is a proposed name for a substance composed purely of neutrons. The word was coined by scientist Andreas von Antropoff in 1926 (before the discovery of the neutron) for the conjectured "element of atomic number zero" that he placed at the head of the periodic table.[2][3] However, the meaning of the term has changed over time, and from the last half of the 20th century onward it has been also used legitimately to refer to extremely dense substances resembling the neutron-degenerate matter theorized to exist in the cores of neutron stars; henceforth "*degenerate* neutronium" will refer to this. Science fiction and popular literature frequently use the term "neutronium" to refer to a highly dense phase of matter composed primarily of neutrons.

14.1 Neutronium and neutron stars

Main article: Neutron star

Neutronium is used in popular literature to refer to the material present in the cores of neutron stars (stars which are too massive to be supported by electron degeneracy pressure and which collapse into a denser phase of matter). This term is very rarely used in scientific literature, for three reasons:

- There are multiple definitions for the term "neutronium".

- There is considerable uncertainty over the composition of the material in the cores of neutron stars (it could be neutron-degenerate matter, strange matter, quark matter, or a variant or combination of the above).

- The properties of neutron star material should depend on depth due to changing pressure (see below), and no sharp boundary between the crust (consisting primarily of atomic nuclei) and almost protonless inner layer is expected to exist.

When neutron star core material is presumed to consist mostly of free neutrons, it is typically referred to as neutron-degenerate matter in scientific literature.[4]

14.2 Neutronium and the periodic table

The term "neutronium" was coined in 1926 by Andreas von Antropoff for a conjectured form of matter made up of neutrons with no protons or electrons, which he placed as the chemical element of atomic number zero at the head of his new version of the periodic table. It was subsequently placed in the middle of several spiral representations of the periodic system for classifying the chemical elements, such as those of Charles Janet (1928), E. I. Emerson (1944), John D. Clark (1950) and in Philip Stewart's Chemical Galaxy (2005).

Although the term is not used in the scientific literature either for a condensed form of matter, or as an element, there have been reports that, besides the free neutron, there may exist two bound forms of neutrons without protons.[5] If neutronium were considered to be an element, then these neutron clusters could be considered to be the isotopes of that element. However, these reports have not been further substantiated.

- Mononeutron: An isolated neutron undergoes beta decay with a mean lifetime of approximately 15 minutes (half-life of approximately 10 minutes), becoming a proton (the nucleus of hydrogen), an electron and an antineutrino.

- Dineutron: The dineutron, containing two neutrons was unambiguously observed in the decay of beryllium-16, in 2012 by researchers at Michigan State University.[6][7] It is not a bound particle, but had been proposed as an extremely short-lived state produced by nuclear reactions involving tritium. It has been suggested to have a transitory existence in nuclear reactions produced by helions that result in the formation of a proton and a nucleus having the same atomic number as the target nucleus but a mass number two units greater. There had been evidence of dineutron emission from neutron-rich isotopes such as beryllium−16 where mononeutron decay would result in a less stable isotope. The dineutron hypothesis had been used in nuclear reactions with exotic nuclei for a long time.[8] Several applications of the dineutron in nuclear reactions can be found in review papers.[9] Its existence has been proven to be relevant for nuclear structure of exotic nuclei.[10] A system made up of only two neutrons is not bound, though the attraction between them is very nearly enough to make them so.[11] This has some consequences on nucleosynthesis and the abundance of the chemical elements.[9][12]

- Trineutron: A trineutron state consisting of three bound neutrons has not been detected, and is not expected to exist even for a short time.

- Tetraneutron: A tetraneutron is a hypothetical particle consisting of four bound neutrons. Reports of its existence have not been replicated. If confirmed, it would require revision of current nuclear models.[13][14]

- Pentaneutron: Calculations indicate that the hypothetical pentaneutron state, consisting of a cluster of five neutrons, would not be bound.[15]

Although not called "neutronium", the National Nuclear Data Center's *Nuclear Wallet Cards* lists as its first "isotope" an "element" with the symbol **n** and atomic number $Z = 0$ and mass number $A = 1$. This isotope is described as decaying to element **H** with a half life of 10.24±0.02 min.

14.3 Properties

See also: Neutron star § Properties

Due to beta (β^-) decay of mononeutron and extreme instability of aforementioned heavier "isotopes", degenerate neutronium is not expected to be stable under ordinary pressures. Free neutrons decay with a half-life of 10 minutes, 11 seconds. A teaspoon of degenerate neutronium gas would have a mass of two billion tonnes, and if moved to standard temperature and pressure, would emit 57 billion joules of β^- decay energy in the first half-life (average of 95 MW of power).[16] This energy may be absorbed as the neutronium gas expands. Though, in the presence of atomic matter compressed to the state of electron degeneracy, the β^- decay may be inhibited due to Pauli exclusion principle, thus making free neutrons stable. Also, elevated pressures should make neutrons degenerate themselves. Compared to ordinary elements, neutronium should be more compressible due to the absence of electrically charged protons and electrons. This makes neutronium more energetically favorable than (positive-Z) atomic nuclei and leads to their conversion to (degenerate) neutronium through electron capture, a process which is believed to occur in stellar cores in the final seconds of the lifetime of massive stars, where it is facilitated by cooling via ν
e emission. As a result, degenerate neutronium can have a density of 4×10^{17} kg/m^3,[17] roughly 13 magnitudes denser than the densest known ordinary substances. It was theorized that extreme pressures may deform the neutrons into a cubic symmetry, allowing tighter packing of neutrons,[18] or cause a strange matter formation.

14.4   In fiction

The term "neutronium" has been popular in science fiction since at least the middle of the 20th century. It typically refers to an extremely dense, incredibly strong form of matter. While presumably inspired by the concept of neutron-degenerate matter in the cores of neutron stars, the material used in fiction bears at most only a superficial resemblance, usually depicted as an extremely strong solid under Earth-like conditions, or possessing exotic properties such as the ability to manipulate time and space. In contrast, all proposed forms of neutron star core material are fluids and are extremely unstable at pressures lower than that found in stellar cores. According to one analysis, a neutron star with a mass below about 0.2 solar masses will explode.[19]

Noteworthy appearances of neutronium in fiction include the following:

- In Hal Clement's short story *Proof* (1942), neutronium is the only form of solid matter known to Solarians, the inhabitants of the Sun's interior.

- In Vladimir Savchenko's *Black Stars* (1960), neutronium is a mechanically and thermally indestructible substance. It is also used to make antimatter, which leads to an annihilation accident.

- In *Doctor Who* (1963), neutronium is a substance which can shield spaces from time-shear when used as shielding in time-vessels.

- In Larry Niven's *Known Space* fictional universe (1964), neutronium is actual neutron star core material, but it is stable in smaller quantities.

- In the *Star Trek* universe, neutronium is an extremely hard and durable substance, often used as armor, which conventional weapons cannot penetrate or even dent. The substance is referred to in the storyline dialogue of "The Doomsday Machine", "A Piece of the Action", "Evolution", "Relics", "To the Death", "What You Leave Behind", "Phage", "Prey", and "Think Tank".

- In Peter F. Hamilton's novel *The Neutronium Alchemist* (1997), neutronium is created by the "aggressive" setting off of a superweapon.

- In the *Stargate* universe, neutronium is a substance which is the basis of the technology of the advanced Asgard race, as well as a primary component of human-form Replicators.

- In Greg Bear's *The Forge of God* (1987), alien aggressors inject two high-mass weapons made of neutronium and antineutronium into the Earth which orbit the Earth's core until they meet and annihilate, destroying the planet.

- Action Comics #376 (May, 1969), "The Only Way to Kill Superman"[20] has Superman flying into a white dwarf star to grab a couple of handfuls of neutronium to fashion earplugs that will protect his ears from a hypersonic blast. In the panel on page seven, he states that each handful weighs a million tons.

- In the SF webcomic *Schlock Mercenary,* neutronium is used as a fuel.

14.5   See also

- Compact star

14.6   References

[1] "Neutrium: The Most Neutral Hypothetical State of Matter Ever". *io9.com*. 2012. Retrieved 2013-02-11.

[2] von Antropoff, A. (1926). "Eine neue Form des periodischen Systems der Elementen" (pdf). *Zeitschrift für Angewandte Chemie* **39** (23): 722–725. doi:10.1002/ange.19260392303.

[3] Stewart, P. J. (2007). "A century on from Dmitrii Mendeleev: Tables and spirals, noble gases and Nobel prizes". *Foundations of Chemistry* **9** (3): 235–245. doi:10.1007/s10698-007-9038-x.

[4] Angelo, J. A. (2006). *Encyclopedia of Space and Astronomy*. Infobase Publishing. p. 178. ISBN 978-0-8160-5330-8.

[5] Timofeyuk, N.K. (2003). "Do multineutrons exist?". *Journal of Physics G* **29**(2):L9.arXiv:nucl-th/0301020.Bibcode:2003JPhGT. doi:10.1088/0954-3899/29/2/102.

[6] Schirber, M. (2012). "Nuclei Emit Paired-up Neutrons". *Physics* **5**: 30. Bibcode:2012PhyOJ...5...30S. doi:10.1103/Physics.5.30.

[7] Spyrou, A.; Kohley, Z.; Baumann, T.; Bazin, D. et al. (2012). "First Observation of Ground State Dineutron Decay: ^{16}Be". *Physical Review Letters* **108** (10): 102501. Bibcode:2012PhRvL.108j2501S. doi:10.1103/PhysRevLett.108.102501. PMID 22463404.

[8] Bertulani, C. A.; Baur, G. (1986). "Coincidence Cross-sections for the Dissociation of Light Ions in High-energy Collisions" (pdf). *Nuclear Physics A* **480** (3–4): 615–628. Bibcode:1988NuPhA.480..615B. doi:10.1016/0375-9474(88)90467-8.

[9] Bertulani, C. A.; Canto, L. F.; Hussein, M. S. (1993). "The Structure And Reactions Of Neutron-Rich Nuclei" (pdf). *Physics Reports* **226** (6): 281–376. Bibcode:1993PhR...226..281B. doi:10.1016/0370-1573(93)90128-Z.

[10] Hagino, K.; Sagawa, H.; Nakamura, T.; Shimoura, S. (2009). "Two-particle correlations in continuum dipole transitions in Borromean nuclei".*Physical Review C***80**(3):1301.arXiv:0904.4775.Bibcode:2009PhRvC..80c1301H.doi:10.1103/PhysRevC.

[11] MacDonald, J.; Mullan, D. J. (2009). "Big Bang Nucleosynthesis: The Strong Nuclear Force meets the Weak Anthropic Principle". *Physical Review D* **80** (4): 3507. arXiv:0904.1807. Bibcode:2009PhRvD..80d3507M. doi:10.1103/PhysRevD.80.043507.

[12] Kneller, J. P.; McLaughlin, G. C. (2004). "The Effect of Bound Dineutrons upon BBN". *Physical Review D* **70** (4): 3512. arXiv:astro-ph/0312388. Bibcode:2004PhRvD..70d3512K. doi:10.1103/PhysRevD.70.043512.

[13] Bertulani, C. A.; Zelevinsky, V. (2002). "Is the tetraneutron a bound dineutron-dineutron molecule?". *Journal of Physics G* **29** (10): 2431. arXiv:nucl-th/0212060. Bibcode:2003JPhG...29.2431B. doi:10.1088/0954-3899/29/10/309.

[14] Timofeyuk, N. K. (2002). "On the existence of a bound tetraneutron". arXiv:nucl-th/0203003 [nucl-th].

[15] Bevelacqua,J.J. (1981). "Particle stability of the pentaneutron".*Physics Letters B***102**(2–3):79–80.Bibcode:1981PhLB..102B. doi:10.1016/0370-2693(81)91033-9.

[16] "Neutrinos give neutron stars a chill". *Ars Technica OpenForum*. Retrieved 4 December 2013.

[17] Zarkonnen (2002). "Neutronium". *Everything2.com*. Retrieved 2013-02-11.

[18] Felipe J. Llanes-Estrada; Gaspar Moreno Navarro (2011). "Cubic neutrons". arXiv:1108.1859v1 [nucl-th].

[19] K. Sumiyoshi; S. Yamada; H. Suzuki; W. Hillebrandt (21 Jul 1997). "The fate of a neutron star just below the minimum mass: does it explode?". *Max-Planck-Institut für Astrophysik, Germany; RIKEN, U. Tokyo, and KEK, Japan*. arXiv:astro-ph/9707230. Given this assumption... the minimum possible mass of a neutron star is 0.189 (solar masses)

[20] Action Comics #376 (May, 1969), "The Only Way to Kill Superman"

14.7 Further reading

- Glendenning, N. K. (2000). *Compact Stars: Nuclear Physics, Particle Physics, and General Relativity* (2nd ed.). Springer. ISBN 978-0-387-98977-8.

Chapter 15

Tetraneutron

A **tetraneutron** is a hypothesised stable cluster of four neutrons. The existence of this cluster of particles is not supported by current models of nuclear forces.[1] There is some empirical evidence suggesting that this particle does exist, based on an experiment by Francisco-Miguel Marqués and co-workers at the Ganil accelerator in Caen using a novel detection method in observations of the disintegration of beryllium and lithium nuclei.[2] However, subsequent attempts to replicate this observation have failed.

15.1 Marqués' experiment

As with many particle accelerator experiments, Marques' team fired atomic nuclei at carbon targets and observed the "spray" of particles from the resulting collisions. In this case the experiment involved firing beryllium-14, beryllium-15 and lithium-11 nuclei at a small carbon target, the most successful being beryllium-14. This isotope of beryllium has a nuclear halo that consists of four clustered neutrons; this allows it to be easily separated intact in the high-speed collision with the carbon target. Their approach to the production and detection of bound neutron clusters was new and novel.[2] Current nuclear models suggest that four separate neutrons should result when beryllium-10 is produced, but the single signal detected in the production of beryllium-10 suggested a multineutron cluster in the breakup products; most likely a beryllium-10 nucleus and four neutrons fused together into a tetraneutron.

15.2 Since Marqués' experiment

A later analysis of the detection method used in the Marques' experiment suggested that at least part of the original analysis was flawed,[3] and attempts to reproduce these observations with different methods have not successfully detected any neutron clusters.[4] If, however, the existence of stable tetraneutrons were ever independently confirmed, considerable adjustments would have to be made to current nuclear models. Bertulani and Zelevinsky[5] proposed that, if it existed, the tetraneutron could be formed by a bound state of two dineutron molecules. However, attempts to model interactions that might give rise to multineutron clusters have failed,[6][7][8] and it "does not seem possible to change modern nuclear Hamiltonians to bind a tetraneutron without destroying many other successful predictions of those Hamiltonians. This means that, should a recent experimental claim of a bound tetraneutron be confirmed, our understanding of nuclear forces will have to be significantly changed."[9]

15.3 See also

- Neutron

- Neutronium

15.4 Notes

[1] Cierjacks, S. et al. (1965). "Further Evidence for the Nonexistence of Particle-Stable Tetraneutrons". *Physical Review* **137** (2B): 345–346. Bibcode:1965PhRv..137..345C. doi:10.1103/PhysRev.137.B345.

[2] Marqués, F. M. et al. (2002). "Detection of neutron clusters". *Physical Review C* **65** (4): 044006. arXiv:nucl-ex/0111001. Bibcode:2002PhRvC..65d4006M. doi:10.1103/PhysRevC.65.044006.

[3] Sherrill, B. M.; Bertulani, C. A (2004). "Proton-tetraneutron elastic scattering". *Physical Review C* **69** (2): 027601. arXiv:nucl-th/0312110. Bibcode:2004PhRvC..69b7601S. doi:10.1103/PhysRevC.69.027601.

[4] Aleksandrov, D. V. et al. (2005). "Search for Resonances in the Three- and Four-Neutron Systems in the ^{7}Li (^{7}Li, ^{11}C) $3n$ and ^{7}Li (^{7}Li, ^{10}C) $4n$ Reactions". *JETP Letters* **81** (2): 43–46. Bibcode:2005JETPL..81...43A. doi:10.1134/1.1887912.

[5] Bertulani, C. A.; Zelevinsky, V. G. (2003). "Tetraneutron as a dineutron-dineutron molecule". *Journal of Physics G* **29** (10): 2431–2437. arXiv:nucl-th/0212060. Bibcode:2003JPhG...29.2431B. doi:10.1088/0954-3899/29/10/309.

[6] Lazauskas, R.; Carbonell, J. (2005). "Three-neutron resonance trajectories for realistic interaction models". *Physical Review C* **71** (4): 044004. arXiv:nucl-th/0502037v2. Bibcode:2005PhRvC..71d4004L. doi:10.1103/PhysRevC.71.044004.

[7] Arai, K. (2003). "Resonance states of ^{5}H and ^{5}Be in a microscopic three-cluster model". *Physical Review C* **68** (3): 034303. Bibcode:2003PhRvC..68c4303A. doi:10.1103/PhysRevC.68.034303.

[8] Hemmdan, A.; Glöckle, W.; Kamada, H. (2002). "Indications for the nonexistence of three-neutron resonances near the physical region".*Physical Review C*66(3):054001.arXiv:nucl-th/0208007.Bibcode:2002PhRvC..66e4001H.doi:10.1103/PhysRevC.66.

[9] Pieper, S. C. (2003). "Can Modern Nuclear Hamiltonians Tolerate a Bound Tetraneutron?". *Physical Review Letters* **90** (25): 252501. arXiv:nucl-th/0302048. Bibcode:2003PhRvL..90y2501P. doi:10.1103/PhysRevLett.90.252501.

15.5 External links

- Announcement of possible tetraneutron observations

- Announcement of possible tetraneutron observations (French)

Chapter 16

Neutron star

For the story by Larry Niven, see Neutron Star (short story).

A **neutron star** is a type of compact star that can result from the gravitational collapse of a massive star after a supernova. Neutron stars are the densest and smallest stars known to exist in the universe; with a radius of only about 12–13 km (7 mi), they can have a mass of about twice that of the Sun.

Neutron stars are composed almost entirely of neutrons, which are subatomic particles without net electrical charge and with slightly larger mass than protons. Neutron stars are very hot and are supported against further collapse by quantum degeneracy pressure due to the phenomenon described by the Pauli exclusion principle, which states that no two neutrons (or any other fermionic particles) can occupy the same place and quantum state simultaneously.

A neutron star has a mass of at least 1.1 and perhaps up to 3 solar masses ($M\odot$),[1][2] though the highest observed mass is 2.01 $M\odot$ Neutron stars typically have a surface temperature around ~6×10^5 K.[3][4][5][6][lower-alpha 1] Neutron stars have overall densities of 3.7×10^{17} to 5.9×10^{17} kg/m^3 (2.6×10^{14} to 4.1×10^{14} times the density of the Sun),[lower-alpha 2] which is comparable to the approximate density of an atomic nucleus of 3×10^{17} kg/m^3.[7] The neutron star's density varies from below 1×10^9 kg/m^3 in the crust—increasing with depth—to above 6×10^{17} or 8×10^{17} kg/m^3 deeper inside (denser than an atomic nucleus).[8] A normal-sized matchbox containing neutron-star material would have a mass of approximately 5 trillion tons or ~1000 km^3 of Earth rock.

In general, compact stars of less than 1.39 $M\odot$ (the Chandrasekhar limit) are white dwarfs, whereas compact stars with a mass between 1.4 $M\odot$ and 3 $M\odot$ (the Tolman–Oppenheimer–Volkoff limit) should be neutron stars. The maximum observed mass of neutron stars is about 2 $M\odot$. The smallest observed mass of a stellar black hole is about 5 $M\odot$, though compact stars with more than 10 $M\odot$ will overcome the neutron degeneracy pressure and gravitational collapse will usually occur to produce a black hole.[9] Between 3 $M\odot$ and 5 $M\odot$, hypothetical intermediate-mass stars such as quark stars and electroweak stars have been proposed, but none have been shown to exist. The equations of state of matter at such high densities are not precisely known because of the theoretical and empirical difficulties.

Some neutron stars rotate very rapidly (up to 716 times a second,[10][11] or approximately 43,000 revolutions per minute) and emit beams of electromagnetic radiation as pulsars. Indeed, the discovery of pulsars in 1967 first suggested that neutron stars exist. Gamma-ray bursts may be produced from rapidly rotating, high-mass stars that collapse to form a neutron star, or from the merger of binary neutron stars. There are thought to be on the order of 10^8 neutron stars in the galaxy, but they can only be easily detected in certain instances, such as if they are a pulsar or part of a binary system. Non-rotating and non-accreting neutron stars are virtually undetectable; however, the Hubble Space Telescope has observed one thermally radiating neutron star, called RX J185635-3754.

16.1   Formation

Any main sequence star with an initial mass of above 8 $M\odot$ has the potential to become a neutron star. As the star evolves away from the main sequence, subsequent nuclear burning produces an iron-rich core. When all nuclear fuel in the core has been exhausted, the core must be supported by degeneracy pressure alone. Further deposits of material

Radiation from the pulsar PSR B1509-58, a rapidly spinning neutron star, makes nearby gas glow in X-rays (gold, from Chandra) and illuminates the rest of the nebula, here seen in infrared (blue and red, from WISE).

from shell burning cause the core to exceed the Chandrasekhar limit. Electron-degeneracy pressure is overcome and the core collapses further, sending temperatures soaring to over 5×10^9 K. At these temperatures, photodisintegration (the breaking up of iron nuclei into alpha particles by high- energy gamma rays) occurs. As the temperature climbs even higher, electrons and protons combine to form neutrons, releasing a flood of neutrinos. When densities reach nuclear density of 4×10^{17} kg/m^3, neutron degeneracy pressure halts the contraction. The infalling outer atmosphere of the star is flung outwards, becoming a Type II or Type Ib supernova. The remnant left is a neutron star. If it has a mass greater than about 5 $M\odot$, it collapses further to become a black hole. Other neutron stars are formed within close binaries.

As the core of a massive star is compressed during a Type II, Type Ib or Type Ic supernova, and collapses into a neutron star, it retains most of its angular momentum. Because it has only a tiny fraction of its parent's radius (and therefore its moment of inertia is sharply reduced), a neutron star is formed with very high rotation speed, and then gradually slows down. Neutron stars are known that have rotation periods from about 1.4 ms to 30 s. The neutron star's density also gives it very high surface gravity, with typical values ranging from 10^{12} to 10^{13} m/s^2 (more than 10^{11} times of that of Earth).[6]

One measure of such immense gravity is the fact that neutron stars have an escape velocity ranging from 100,000 km/s to 150,000 km/s, that is, from a third to half the speed of light. Matter falling onto the surface of a neutron star would be accelerated to tremendous speed by the star's gravity. The force of impact would likely destroy the object's component atoms, rendering all its matter identical, in most respects, to the rest of the star.

16.2   Properties

Gravitational light deflection at a neutron star. Due to relativistic light deflection more than half of the surface is visible (each chequered patch here represents 30 degrees by 30 degrees).[12] In natural units, the mass of the depicted star is 1 and its radius 4, or twice its Schwarzschild radius.[12]

The gravitational field at the star's surface is about 2×10^{11} times stronger than on Earth. Such a strong gravitational field acts as a gravitational lens and bends the radiation emitted by the star such that parts of the normally invisible rear surface become visible.[12] If the radius of the neutron star is $3GM/c^2$ or less, then the photons may be trapped in an orbit, thus making the whole surface of that neutron star visible, along with destabilizing orbits at that and less than that of the radius. A fraction of the mass of a star that collapses to form a neutron star is released in the supernova explosion from which it forms (from the law of mass-energy equivalence, $E = mc^2$). The energy comes from the gravitational binding energy of a neutron star.

Neutron star relativistic equations of state provided by Jim Lattimer include a graph of radius vs. mass for various models.[13] The most likely radii for a given neutron star mass are bracketed by models AP4 (smallest radius) and MS2 (largest radius). BE is the ratio of gravitational binding energy mass equivalent to observed neutron star gravitational mass of "M" kilograms with radius "R" meters,[14]

$$BE = \frac{0.60\,\beta}{1-\frac{\beta}{2}} \quad \beta = GM/Rc^2$$

Given current values

$$G = 6.6742 \times 10^{-11}\,\mathrm{m^3kg^{-1}s^{-2}}\ [15]$$

$$c^2 = 8.98755 \times 10^{16}\,\mathrm{m^2s^{-2}}$$

$$M_{solar} = 1.98844 \times 10^{30}\,\mathrm{kg}$$

and star masses "M" commonly reported as multiples of one solar mass,

$$M_x = \frac{M}{M_\odot}$$

then the relativistic fractional binding energy of a neutron star is

$$BE = \frac{885.975\,M_x}{R - 738.313\,M_x}$$

A 2 $M\odot$ neutron star would not be more compact than 10,970 meters radius (AP4 model). Its mass fraction gravitational binding energy would then be 0.187, −18.7% (exothermic). This is not near 0.6/2 = 0.3, −30%.

A neutron star is so dense that one teaspoon (5 milliliters) of its material would have a mass over 5.5×10^{12} kg (that is 1100 tonnes per 1 nanolitre), about 900 times the mass of the Great Pyramid of Giza.[lower-alpha 3] Hence, the gravitational force of a typical neutron star is such that if an object were to fall from a height of one meter, it would only take one microsecond to hit the surface of the neutron star, and would do so at around 2000 kilometers per second, or 7.2 million kilometers per hour.[16]

The temperature inside a newly formed neutron star is from around 10^{11} to 10^{12} kelvin.[8] However, the huge number of neutrinos it emits carry away so much energy that the temperature falls within a few years to around 10^6 kelvin.[8] Even at 1 million kelvin, most of the light generated by a neutron star is in X-rays.

The pressure increases from 3×10^{33} to 1.6×10^{35} Pa from the inner crust to the center.[17]

The equation of state for a neutron star is still not known. It is assumed that it differs significantly from that of a white dwarf, whose equation of state is that of a degenerate gas that can be described in close agreement with special relativity. However, with a neutron star the increased effects of general relativity can no longer be ignored. Several equations of state have been proposed (FPS, UU, APR, L, SLy, and others) and current research is still attempting to constrain the theories to make predictions of neutron star matter.[6][18] This means that the relation between density and mass is not fully known, and this causes uncertainties in radius estimates. For example, a 1.5 $M\odot$ neutron star could have a radius of 10.7, 11.1, 12.1 or 15.1 kilometres (for EOS FPS, UU, APR or L respectively).[18]

16.3 Structure

Current understanding of the structure of neutron stars is defined by existing mathematical models, but it might be possible to infer through studies of neutron-star oscillations. Similar to asteroseismology for ordinary stars, the inner structure might be derived by analyzing observed frequency spectra of stellar oscillations.[6]

Current models indicate that matter at the surface of a neutron star is composed of ordinary atomic nuclei crushed into a solid lattice with a sea of electrons flowing through the gaps between them. It is possible that the nuclei at the surface are iron, due to iron's high binding energy per nucleon.[19] It is also possible that heavy elements, such as iron, simply sink beneath the surface, leaving only light nuclei like helium and hydrogen.[19] If the surface temperature exceeds 10^6 kelvin (as in the case of a young pulsar), the surface should be fluid instead of the solid phase observed in cooler neutron stars (temperature <10^6 kelvin).[19]

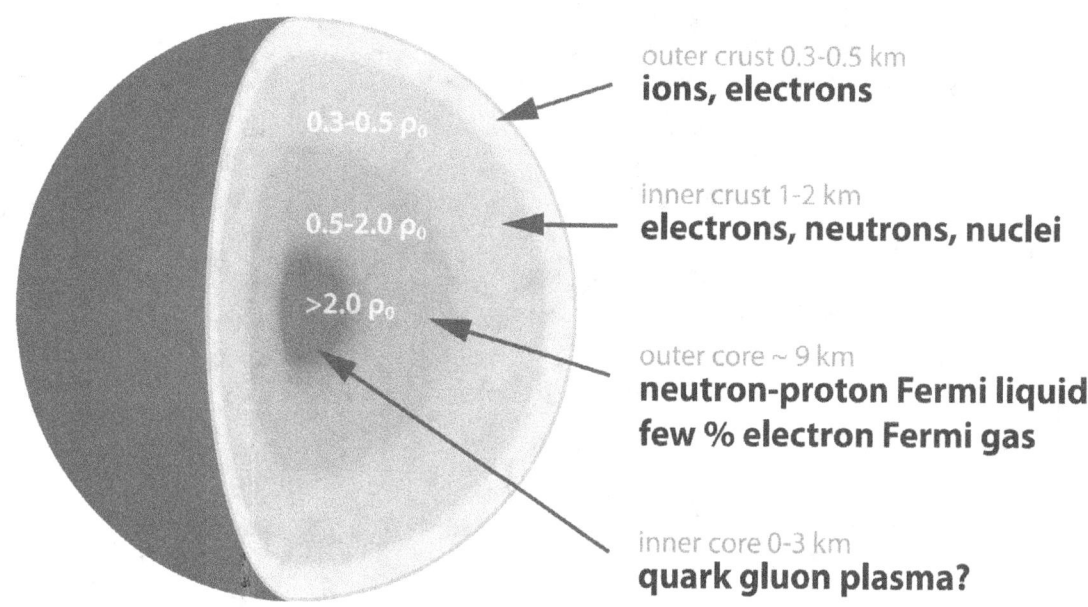

Cross-section of neutron star. Densities are in terms of ρ_0 the saturation nuclear matter density, where nucleons begin to touch.

The "atmosphere" of the star is hypothesized to be at most several micrometers thick, and its dynamic is fully controlled by the star's magnetic field. Below the atmosphere one encounters a solid "crust". This crust is extremely hard and very smooth (with maximum surface irregularities of ~5 mm), because of the extreme gravitational field.[20] The expected hierarchy of phases of nuclear matter in the inner crust has been characterized as nuclear pasta.[21]

Proceeding inward, one encounters nuclei with ever increasing numbers of neutrons; such nuclei would decay quickly on Earth, but are kept stable by tremendous pressures. As this process continues at increasing depths, neutron drip becomes overwhelming, and the concentration of free neutrons increases rapidly. In this region, there are nuclei, free electrons, and free neutrons. The nuclei become increasingly small (gravity and pressure overwhelming the strong force) until the core is reached, by definition the point where they disappear altogether.

The composition of the superdense matter in the core remains uncertain. One model describes the core as superfluid neutron-degenerate matter (mostly neutrons, with some protons and electrons). More exotic forms of matter are possible, including degenerate strange matter (containing strange quarks in addition to up and down quarks), matter containing high-energy pions and kaons in addition to neutrons,[6] or ultra-dense quark-degenerate matter.

16.4   History of discoveries

In 1934, Walter Baade and Fritz Zwicky proposed the existence of the neutron star,[22][lower-alpha 4] only a year after the discovery of the neutron by Sir James Chadwick.[25] In seeking an explanation for the origin of a supernova, they tentatively proposed that in supernova explosions ordinary stars are turned into stars that consist of extremely closely packed neutrons that they called neutron stars. Baade and Zwicky correctly proposed at that time that the release of the gravitational binding energy of the neutron stars powers the supernova: "In the supernova process, mass in bulk is annihilated". Neutron stars were thought to be too faint to be detectable and little work was done on them until November 1967, when Franco Pacini (1939–2012) pointed out that if the neutron stars were spinning and had large magnetic fields, then electromagnetic waves would be emitted. Unbeknown to him, radio astronomer Antony Hewish and his research assistant Jocelyn Bell at Cambridge were shortly to detect radio pulses from stars that are now believed to be highly magnetized, rapidly spinning neutron stars, known as pulsars.

Isolated Neutron Star RX J185635-3754   HST • WFPC2
PRC97-32 • ST ScI OPO • September 25, 1997
F. Walter (State University of New York at Stony Brook) and NASA

The first direct observation of a neutron star in visible light. The neutron star is RX J185635-3754.

In 1965, Antony Hewish and Samuel Okoye discovered "an unusual source of high radio brightness temperature in the Crab Nebula".[26] This source turned out to be the Crab Pulsar that resulted from the great supernova of 1054.

In 1967, Iosif Shklovsky examined the X-ray and optical observations of Scorpius X-1 and correctly concluded that the radiation comes from a neutron star at the stage of accretion.[27]

In 1967, Jocelyn Bell and Antony Hewish discovered regular radio pulses from CP 1919. This pulsar was later interpreted as an isolated, rotating neutron star. The energy source of the pulsar is the rotational energy of the neutron star. The majority of known neutron stars (about 2000, as of 2010) have been discovered as pulsars, emitting regular radio pulses.

In 1971, Riccardo Giacconi, Herbert Gursky, Ed Kellogg, R. Levinson, E. Schreier, and H. Tananbaum discovered 4.8

second pulsations in an X-ray source in the constellation Centaurus, Cen X-3. They interpreted this as resulting from a rotating hot neutron star. The energy source is gravitational and results from a rain of gas falling onto the surface of the neutron star from a companion star or the interstellar medium.

In 1974, Antony Hewish was awarded the Nobel Prize in Physics "for his decisive role in the discovery of pulsars" without Jocelyn Bell who shared in the discovery.

In 1974, Joseph Taylor and Russell Hulse discovered the first binary pulsar, PSR B1913+16, which consists of two neutron stars (one seen as a pulsar) orbiting around their center of mass. Einstein's general theory of relativity predicts that massive objects in short binary orbits should emit gravitational waves, and thus that their orbit should decay with time. This was indeed observed, precisely as general relativity predicts, and in 1993, Taylor and Hulse were awarded the Nobel Prize in Physics for this discovery.

In 1982, Don Backer and colleagues discovered the first millisecond pulsar, PSR B1937+21. This objects spins 642 times per second, a value that placed fundamental constraints on the mass and radius of neutron stars. Many millisecond pulsars were later discovered, but PSR B1937+21 remained the fastest-spinning known pulsar for 24 years, until PSR J1748-2446ad (which spins more than 700 times a second) was discovered.

In 2003, Marta Burgay and colleagues discovered the first double neutron star system where both components are detectable as pulsars, PSR J0737-3039. The discovery of this system allows a total of 5 different tests of general relativity, some of these with unprecedented precision.

In 2010, Paul Demorest and colleagues measured the mass of the millisecond pulsar PSR J1614–2230 to be 1.97±0.04 $M\odot$, using Shapiro delay.[28] This was substantially higher than any previously measured neutron star mass (1.67 $M\odot$, see PSR J1903+0327), and places strong constraints on the interior composition of neutron stars.

In 2013, John Antoniadis and colleagues measured the mass of PSR J0348+0432 to be 2.01±0.04 $M\odot$, using white dwarf spectroscopy.[29] This confirmed the existence of such massive stars using a different method. Furthermore, this allowed, for the first time, a test of general relativity using such a massive neutron star.

16.5   Rotation

Neutron stars rotate extremely rapidly after their formation due to the conservation of angular momentum; like spinning ice skaters pulling in their arms, the slow rotation of the original star's core speeds up as it shrinks. A newborn neutron star can rotate several times a second; sometimes, the neutron star absorbs orbiting matter from a companion star, increasing the rotation to several hundred times per second, reshaping the neutron star into an oblate spheroid.

Over time, neutron stars slow down (spin down) because their rotating magnetic fields radiate energy; older neutron stars may take several seconds for each revolution.

The rate at which a neutron star slows its rotation is usually constant and very small: the observed rates of decline are between 10^{-10} and 10^{-21} seconds for each rotation. Therefore, for a typical slow down rate of 10^{-15} seconds per rotation, a neutron star now rotating in 1 second will rotate in 1.000003 seconds after a century, or 1.03 seconds after 1 million years.

Sometimes a neutron star will *spin up* or undergo a *glitch*, a sudden small increase of its rotation speed. Glitches are thought to be the effect of a starquake — as the rotation of the star slows down, the shape becomes more spherical. Due to the stiffness of the "neutron" crust, this happens as discrete events when the crust ruptures, similar to tectonic earthquakes. After the starquake, the star will have a smaller equatorial radius, and because angular momentum is conserved, rotational speed increases. Recent work, however, suggests that a starquake would not release sufficient energy for a neutron star glitch; it has been suggested that glitches may instead be caused by transitions of vortices in the superfluid core of the star from one metastable energy state to a lower one.[30]

Neutron stars have been observed to "pulse" radio and x-ray emissions believed to be caused by particle acceleration near the magnetic poles, which need not be aligned with the rotation axis of the star. Through mechanisms not yet entirely understood, these particles produce coherent beams of radio emission. External viewers see these beams as pulses of radiation whenever the magnetic pole sweeps past the line of sight. The pulses come at the same rate as the rotation of the neutron star, and thus, appear periodic. Neutron stars that emit such pulses are called pulsars.

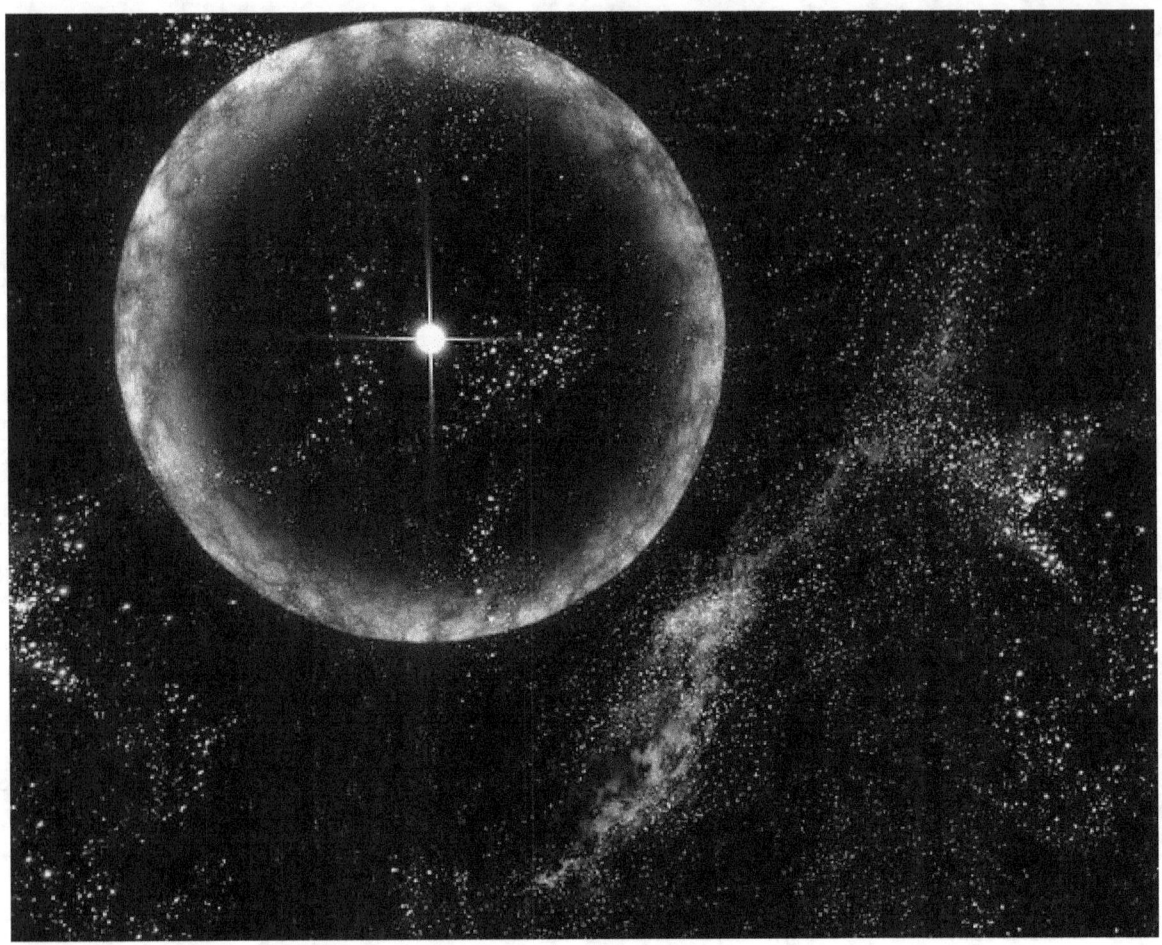

NASA artist's conception of a "starquake", or "stellar quake".

The most rapidly rotating neutron star currently known, PSR J1748-2446ad, rotates at 716 rotations per second.[31] A recent paper reported the detection of an X-ray burst oscillation (an indirect measure of spin) at 1122 Hz from the neutron star XTE J1739-285.[32] However, at present, this signal has only been seen once, and should be regarded as tentative until confirmed in another burst from this star.

16.6 Population and distances

At present, there are about 2000 known neutron stars in the Milky Way and the Magellanic Clouds, the majority of which have been detected as radio pulsars. Neutron stars are mostly concentrated along the disk of the Milky Way although the spread perpendicular to the disk is large because the supernova explosion process can impart high speeds (400 km/s) to the newly formed neutron star.

Some of the closest neutron stars are RX J1856.5-3754 about 400 light years away and PSR J0108-1431 at about 424 light years.[33] RX J1856.5-3754 is a member of a close group of neutron stars called The Magnificent Seven. Another nearby neutron star that was detected transiting the backdrop of the constellation Ursa Minor has been nicknamed Calvera by its Canadian and American discoverers, after the villain in the 1960 film *The Magnificent Seven*. This rapidly moving object was discovered using the ROSAT/Bright Source Catalog.

16.7   Binary neutron stars

About 5% of all known neutron stars are members of a binary system. The formation and evolution scenario of binary neutron stars is a rather exotic and complicated process.[34] The companion stars may be either ordinary stars, white dwarfs or other neutron stars. According to modern theories of binary evolution it is expected that neutron stars also exist in binary systems with black hole companions. Such binaries are expected to be prime sources for emitting gravitational waves. Neutron stars in binary systems often emit X-rays, which is caused by the heating of material (gas) accreted from the companion star. Material from the outer layers of a (bloated) companion star is sucked towards the neutron star as a result of its very strong gravitational field. As a result of this process binary neutron stars may also coalesce into black holes if the accretion of mass takes place under extreme conditions.[35] It has been proposed that coalescence of binaries consisting of two neutron stars may be responsible for producing short gamma-ray bursts. Such events may also be responsible for producing all chemical elements beyond iron,[36] as opposed to the supernova nucleosynthesis theory.

Circinus X-1: X-ray light rings from a binary neutron star (24 June 2015; Chandra X-ray Observatory).

16.8 Subtypes

- Neutron star

  - Protoneutron star (PNS), theorized.[37]

  - Radio-quiet neutron stars

  - Radio loud neutron star
    - Single pulsars–general term for neutron stars that emit directed pulses of radiation towards us at regular intervals (due to their strong magnetic fields).
      - Rotation-powered pulsar *("radio pulsar")*
        - Magnetar–a neutron star with an extremely strong magnetic field (1000 times more than a regular neutron star), and long rotation periods (5 to 12 seconds).
          - Soft gamma repeater (SGR)
          - Anomalous X-ray pulsar (AXP)
    - Binary pulsars
      - Low-mass X-ray binaries (LMXB)
      - Intermediate-mass X-ray binaries (IMXB)
      - High-mass X-ray binaries (HMXB)
      - Accretion-powered pulsar *("X-ray pulsar")*
        - X-ray burster–a neutron star with a low mass binary companion from which matter is accreted resulting in irregular bursts of energy from the surface of the neutron star.
        - Millisecond pulsar (MSP) *("recycled pulsar")*
          - Sub-millisecond pulsar[38]

  - Exotic star

    - Quark star–currently a hypothetical type of neutron star composed of quark matter, or strange matter. As of 2008, there are three candidates.

    - Electroweak star–currently a hypothetical type of extremely heavy neutron star, in which the quarks are converted to leptons through the electroweak force, but the gravitational collapse of the star is prevented by radiation pressure. As of 2010, there is no evidence for their existence.

    - Preon star–currently a hypothetical type of neutron star composed of preon matter. As of 2008, there is no evidence for the existence of preons.

16.9 Giant nucleus

A neutron star has some of the properties of an atomic nucleus, including density (within an order of magnitude) and being composed of nucleons. In popular scientific writing, neutron stars are therefore sometimes described as giant nuclei. However, in other respects, neutron stars and atomic nuclei are quite different. In particular, a nucleus is held together by the strong interaction, whereas a neutron star is held together by gravity, and thus the density and structure of neutron stars is more variable. It is generally more useful to consider such objects as stars.

16.10 Examples of neutron stars

- PSR J0108-1431 – closest neutron star

- LGM-1 – the first recognized radio-pulsar

- PSR B1257+12 – the first neutron star discovered with planets (a millisecond pulsar)

- SWIFT J1756.9-2508 – a millisecond pulsar with a stellar-type companion with planetary range mass (below brown dwarf)

- PSR B1509-58 source of the "Hand of God" photo shot by the Chandra X-ray Observatory.

- PSR J0348+0432 – the most massive neutron star with a well-constrained mass, $2.01 \pm 0.04 M_\odot$.

16.11   See also

- Dragon's Egg

- Neutronium

- Preon-degenerate matter

- Rotating radio transients

16.12   Gallery

- Play media

  Video - Neutron stars contain 500,000 Earth-masses in 25 km (16 mi) dia. sphere.

- Play media

  Video - Neutron stars colliding (animation).

- Play media

  Video - Neutron star collision.

16.13   Notes

[1] A neutron star's density increases as its mass increases, and its radius decreases non-linearly. (NASA mass radius graph)

[2] 3.7×10^{17} kg/m^3 derives from mass 2.68×10^{30} kg / volume of star of radius 12 km; 5.9×10^{17} kg m^{-3} derives from mass 4.2×10^{30} kg per volume of star radius 11.9 km

[3] The average density of material in a neutron star of radius 10 km is 1.1×10^{12} kg cm^{-3}. Therefore, 5 ml of such material is 5.5×10^{12} kg, or 5 500 000 000 metric tons. This is about 15 times the total mass of the human world population. Alternatively, 5 ml from a neutron star of radius 20 km radius (average density 8.35×10^{10} kg cm^{-3}) has a mass of about 400 million metric tons, or about the mass of all humans.

[4] Even before the discovery of neutron, in 1931, neutron stars were *anticipated* by Lev Landau, who wrote about stars where "atomic nuclei come in close contact, forming one gigantic nucleus"[23]). However, the widespread opinion that Landau *predicted* neutron stars proves to be wrong.[24]

16.14 References

[1] Özel, Feryal; Psaltis, Dimitrios; Narayan, Ramesh; Santos Villarreal, Antonio (September 2012). "On the Mass Distribution and Birth Masses of Neutron Stars". *The Astrophysical Journal* **757** (1): 13. arXiv:1201.1006. Bibcode:2012ApJ...757...55O. doi:10.1088/0004-637X/757/1/55. Retrieved 14 May 2015.

[2] Chamel, N.; Haensel, P.; Zdunik, J.L.; Fantina, A.F. (19 November 2013). "On the Maximum Mass of Neutron Stars" (PDF). *International Journal of Modern Physics* **1** (28): 1330018. arXiv:1307.3995. Bibcode:2013IJMPE..2230018C. doi:10.1142/S0218301313 Retrieved 14 May 2015.

[3] Bulent Kiziltan (2011). *Reassessing the Fundamentals: On the Evolution, Ages and Masses of Neutron Stars.* Universal-Publishers. ISBN 1-61233-765-1.

[4] Neutron star mass measurements

[5] "Nasa Ask an Astrophysist: Maximum Mass of a Neutron Star".

[6] Paweł Haensel; A Y Potekhin; D G Yakovlev (2007). *Neutron Stars.* Springer. ISBN 0-387-33543-9.

[7] "Calculating a Neutron Star's Density". Retrieved 2006-03-11. NB 3×10^{17} kg/m^3 is 3×10^{14} g/cm^3

[8] "Introduction to neutron stars". Retrieved 2007-11-11.

[9] , a 10 *M*⊙ star will collapse into a black hole.

[10] Hessels, Jason; Ransom, Scott M.; Stairs, Ingrid H.; Freire, Paulo C. C. et al. (2006). "A Radio Pulsar Spinning at 716 Hz". *Science* **311** (5769): 1901–1904. arXiv:astro-ph/0601337. Bibcode:2006Sci...311.1901H. doi:10.1126/science.1123430. PMID 16410486.

[11] Naeye, Robert (2006-01-13). "Spinning Pulsar Smashes Record". *Sky & Telescope.* Retrieved 2008-01-18.

[12] Zahn, Corvin (1990-10-09). "Tempolimit Lichtgeschwindigkeit" (in German). Retrieved 2009-10-09. Durch die gravitative Lichtablenkung ist mehr als die Hälfte der Oberfläche sichtbar. Masse des Neutronensterns: 1, Radius des Neutronensterns: 4, ... dimensionslosen Einheiten (c, G = 1)

[13] Neutron Star Masses and Radii, p. 9/20, bottom

[14] J. M. Lattimer and M. Prakash, "Neutron Star Structure and the Equation of State" Astrophysical J. 550(1) 426 (2001); http://arxiv.org/abs/astro-ph/0002232

[15] Measurement of Newton's Constant Using a Torsion Balance with Angular Acceleration Feedback , Phys. Rev. Lett. 85(14) 2869 (2000)

[16] Miscellaneous Facts

[17] Neutron degeneracy pressure (Archive). Physics Forums. Retrieved on 2011-10-09.

[18] NASA. Neutron Star Equation of State Science Retrieved 2011-09-26 Archived February 20, 2013 at the Wayback Machine

[19] V. S. Beskin (1999). "*Radiopulsars*". УФН. Т.169, №11, p.1173-1174

[20] neutron star

[21] Pons, José A.; Viganò, Daniele; Rea, Nanda (2013). "Too much "pasta" for pulsars to spin down". *Nature Physics* **9** (7): 431–434. arXiv:1304.6546. Bibcode:2013NatPh...9..431P. doi:10.1038/nphys2640.

[22] Baade, Walter & Zwicky, Fritz (1934). "Remarks on Super-Novae and Cosmic Rays". *Phys. Rev.* **46** (1): 76–77. Bibcode:1934PhRv...46.. doi:10.1103/PhysRev.46.76.2.

[23] Landau L.D. (1932). "On the theory of stars". *Phys. Z. Sowjetunion* **1**: 285–288.

[24] P. Haensel, A. Y. Potekhin, & D. G. Yakovlev (2007). *Neutron Stars 1: Equation of State and Structure* (New York: Springer), page 2 http://adsabs.harvard.edu/abs/2007ASSL..326.....H

[25] Chadwick, James (1932). "On the possible existence of a neutron". *Nature* **129** (3252): 312. Bibcode:1932Natur.129Q.312C. doi:10.1038/129312a0.

[26] Hewish, A. & Okoye, S. E. (1965). "Evidence of an unusual source of high radio brightness temperature in the Crab Nebula". *Nature* **207** (4992): 59–60. Bibcode:1965Natur.207...59H. doi:10.1038/207059a0.

[27] Shklovsky, I.S. (April 1967). "On the Nature of the Source of X-Ray Emission of SCO XR-1". *Astrophys. J.* **148** (1): L1–L4. Bibcode:1967ApJ...148L...1S. doi:10.1086/180001.

[28] Demorest, PB; Pennucci, T; Ransom, SM; Roberts, MS et al. (2010). "A two-solar-mass neutron star measured using Shapiro delay". *Nature* **467** (7319): 1081–1083. arXiv:1010.5788. Bibcode:2010Natur.467.1081D. doi:10.1038/nature09466. PMID 20981094.

[29] Antoniadis, J (2012). "A Massive Pulsar in a Compact Relativistic Binary". *Science* **340** (6131): 1233232. arXiv:1304.6875. Bibcode:2013Sci...340..448A2010. doi:10.1126/science.1233232.

[30] Alpar, M Ali (January 1, 1998). "Pulsars, glitches and superfluids". Physicsworld.com.

[31] [astro-ph/0601337] A Radio Pulsar Spinning at 716 Hz

[32] University of Chicago Press – Millisecond Variability from XTE J1739285 – 10.1086/513270

[33] Posselt, B.; Neuhäuser, R.; Haberl, F. (March 2009). "Searching for substellar companions of young isolated neutron stars". *Astronomy and Astrophysics* **496** (2): 533–545. arXiv:0811.0398. Bibcode:2009A&A...496..533P. doi:10.1051/0004-6361/200810156.

[34] Tauris & van den Heuvel (2006), in Compact Stellar X-ray Sources. Eds. Lewin and van der Klis, Cambridge University Press http://adsabs.harvard.edu/abs/2006csxs.book..623T

[35] Compact Stellar X-ray Sources (2006). Eds. Lewin and van der Klis, Cambridge University

[36] Urry, Meg (July 20, 2013). "Gold comes from stars". CNN.

[37] Neutrino-Driven Protoneutron Star Winds, Todd A. Thompson.

[38] Nakamura, T. (1989). "Binary Sub-Millisecond Pulsar and Rotating Core Collapse Model for SN1987A". *Progress of Theoretical Physics* **81** (5): 1006–1020. Bibcode:1989PThPh..81.1006N. doi:10.1143/PTP.81.1006.

- "ASTROPHYSICS: ON OBSERVED PULSARS". *scienceweek.com*. Retrieved 6 August 2004.

- Norman K. Glendenning; R. Kippenhahn; I. Appenzeller; G. Borner et al. (2000). *Compact Stars* (2nd ed.).

- Kaaret; Prieskorn; in 't Zand; Brandt et al. (2006). "Evidence for 1122 Hz X-Ray Burst Oscillations from the Neutron-Star X-Ray Transient XTE J1739-285". *The Astrophysical Journal* **657** (2): L97. arXiv:astro-ph/0611716. Bibcode:2007ApJ...657L..97K. doi:10.1086/513270.

16.15   External links

- Introduction to neutron stars

- Neutron Stars for Undergraduates and its Errata

- NASA on pulsars

- "NASA Sees Hidden Structure Of Neutron Star In Starquake". SpaceDaily.com. April 26, 2006

- "Mysterious X-ray sources may be lone neutron stars". *New Scientist.*

- "Massive neutron star rules out exotic matter". *New Scientist.* According to a new analysis, exotic states of matter such as free quarks or BECs do not arise inside neutron stars.

- "Neutron star clocked at mind-boggling velocity". *New Scientist.* A neutron star has been clocked traveling at more than 1500 kilometers per second.

Chapter 17

Neutron detection

Neutron detection is the effective detection of neutrons entering a well-positioned detector. There are two key aspects to effective neutron detection: hardware and software. Detection hardware refers to the kind of neutron detector used (the most common today is the scintillation detector) and to the electronics used in the detection setup. Further, the hardware setup also defines key experimental parameters, such as source-detector distance, solid angle and detector shielding. Detection software consists of analysis tools that perform tasks such as graphical analysis to measure the number and energies of neutrons striking the detector.

17.1 Basic physics of neutron detection

17.1.1 Signatures by which a neutron may be detected

Atomic and subatomic particles are detected by the signature they produce through interaction with their surroundings. The interactions result from the particles' fundamental characteristics.

- Charge: Neutrons are neutral particles and do not ionize directly; hence they are harder than charged particles to detect directly. Further, their paths of motion are only weakly affected by electric and magnetic fields.

- Mass: The neutron mass of 1.0086649156(6) u.[1] is not directly detectable, but does influence reactions through which it can be detected.

- Reactions: Neutrons react with a number of materials through elastic scattering producing a recoiling nucleus, inelastic scattering producing an excited nucleus, or absorption with transmutation of the resulting nucleus. Most detection approaches rely on detecting the various reaction products.

- Magnetic moment: Although neutrons have a magnetic moment of $-1.9130427(5)$ μN, techniques for detection of the magnetic moment are too insensitive to use for neutron detection.

- Electric dipole moment: The neutron is predicted to have only a tiny electric dipole moment, which has not yet been detected. Hence it is not a viable detection signature.

- Decay: Outside the nucleus, free neutrons are unstable and have a mean lifetime of 885.7±0.8 s (about 14 minutes, 46 seconds).[1] Free neutrons decay by emission of an electron and an electron antineutrino to become a proton, a process known as beta decay:[2]

    n0 → p+ + e− + ν
    e.

    Although the p+ and e− produced by neutron decay are detectable, the decay rate is too low to serve as the basis for a practical detector system.

17.1.2   Classic neutron detection options

As a result of these properties, detection approaches for neutrons fall into several major categories:[3]

- Absorptive reactions with prompt reactions - low energy neutrons are typically detected indirectly through absorption reactions. Typical absorber materials used have high cross sections for absorption of neutrons and include helium-3, lithium-6, boron-10, and uranium-235. Each of these reacts by emission of high energy ionized particles, the ionization track of which can be detected by a number of means. Commonly used reactions include ^{3}He(n,p) ^{3}H, ^{6}Li(n,α) ^{3}H, ^{10}B(n,α) ^{7}Li and the fission of uranium.[3]

- Activation processes - Neutrons may be detected by reacting with absorbers in a radiative capture, spallation or similar reaction, producing reaction products that then decay at some later time, releasing beta particles or gammas. Selected materials (e.g., indium, gold, rhodium, iron (^{56}Fe(n,p) ^{56}Mn), aluminum (^{27}Al(n,α)^{24}Na), niobium (^{93}Nb(n,2n) ^{92m}Nb), & silicon (^{28}Si(n,p) ^{28}Al)) have extremely large cross sections for the capture of neutrons within a very narrow band of energy. Use of multiple absorber samples allows characterization of the neutron energy spectrum. Activation also allows recreation of an historic neutron exposure (e.g., forensic recreation of neutron exposures during an accidental criticality).[3]

- Elastic scattering reactions (also referred to as proton-recoil) - High energy neutrons are typically detected indirectly through elastic scattering reactions. Neutron collide with the nucleus of atoms in the detector, transferring energy to that nucleus and creating an ion, which is detected. Since the maximum transfer of energy occurs when the mass of the atom with which the neutron collides is comparable to the neutron mass, hydrogenous[4] materials are often the preferred medium for such detectors.[3]

17.2   Types of neutron detectors

17.2.1   Gas proportional detectors

Gas proportional detectors can be adapted to detect neutrons. While neutrons do not typically cause ionization, the addition of a nuclide with high neutron cross-section allows the detector to respond to neutrons. Nuclides commonly used for this purpose are helium-3, lithium-6, boron-10 and uranium-235. Since these materials are most likely to react with thermal neutrons (i.e., neutrons that have slowed to equilibrium with their surroundings), they are typically surrounded by moderating materials.

Further refinements are usually necessary to isolate the neutron signal from the effects of other types of radiation. Since the energy of a thermal neutron is relatively low, charged particle reaction is discrete (i.e., essentially monoenergetic) while other reactions such as gamma reactions will span a broad energy range, it is possible to discriminate among the sources.

As a class, gas ionization detectors measure the number (count rate), and not the energy of neutrons.

^{3}He gas-filled proportional detectors

An isotope of Helium, ^{3}He provides for an effective neutron detector material because ^{3}He reacts by absorbing thermal neutrons, producing a ^{1}H and ^{3}H ion. Its sensitivity to gamma rays is negligible, providing a very useful neutron detector. Unfortunately the supply of ^{3}He is limited to production as a byproduct from the decay of tritium (which has a 12.3 year half-life); tritium is produced either as part of weapons programs as a booster for nuclear weapons or as a byproduct of reactor operation.

BF$_3$ gas-filled proportional detectors

As elemental boron is not gaseous, neutron detectors containing boron may alternately use boron trifluoride (BF$_3$) enriched to 96% boron-10 (natural boron is 20% ^{10}B, 80% ^{11}B).[5] It should be noted that boron trifluoride is highly toxic.

Boron lined proportional detectors

Alternately, boron-lined gas-filled proportional counters react similarly to BF_3 gas-filled proportional detectors, with the exception that the walls are coated with ^{10}B. In this design, since the reaction takes place on the surface, only one of the two particles will escape into the proportional counter.

17.2.2   Scintillation neutron detectors

Scintillation neutron detectors include liquid organic scintillators,[6] crystals,[7][8] plastics, glass[9] and scintillation fibers.[10]

Neutron-sensitive scintillating glass fiber detectors

Scintillating ^6Li glass for neutron detection was first reported in the scientific literature in 1957[11] and key advances were made in the 1960s and 1970s.[12][13] Scintillating fiber was demonstrated by Atkinson M. *et al.* in 1987[14] and major advances were made in the late 1980s and early 1990s at Pacific Northwest National Laboratory where it was developed as a classified technology.[15][16][17][18][19] It was declassified in 1994 and first licensed by Oxford Instruments in 1997, followed by a transfer to Nucsafe in 1999.[20][21][22] The fiber and fiber detectors are now manufactured and sold commercially by Nucsafe, Inc.[23]

The scintillating glass fibers work by incorporating ^6Li and Ce^{3+} into the glass bulk composition. The ^6Li has a high cross-section for thermal neutron absorption through the $^6Li(n,\alpha)$ reaction. Neutron absorption produces a tritium ion, an alpha particle, and kinetic energy. The alpha particle and triton interact with the glass matrix to produce ionization, which transfers energy to Ce^{3+} ions and results in the emission of photons with wavelength 390 nm - 600 nm as the excited state Ce^{3+} ions return to the ground state. The event results in a flash of light of several thousand photons for each neutron absorbed. A portion of the scintillation light propagates through the glass fiber, which acts as a waveguide. The fibers ends are optically coupled to a pair of photomultiplier tubes (PMTs) to detect photon bursts. The detectors can be used to detect both neutrons and gamma rays, which are typically distinguished using pulse-height discrimination. Substantial effort and progress in reducing fiber detector sensitivity to gamma radiation has been made. Original detectors suffered from false neutrons in a 0.02 mR gamma field. Design, process, and algorithm improvements now enable operation in gamma fields up to 20 mR/h (^{60}Co).

The scintillating fiber detectors have excellent sensitivity, they are rugged, and have fast timing (~60 ns) so that a large dynamic range in counting rates is possible. The detectors have the advantage that they can be formed into any desired shape, and can be made very large or very small for use in a variety of applications.[24] Further, they do not rely on ^3He or any raw material that has limited availability, nor do they contain toxic or regulated materials. Their performance matches or exceeds that of ^3He tubes for gross neutron counting due to the higher density of neutron absorbing species in the solid glass compared to high-pressure gaseous ^3He.[24] Even though the thermal neutron cross section of ^6Li is low compared to ^3He (940 barns vs. 5330 barns), the atom density of ^6Li in the fiber is fifty times greater, resulting in an advantage in effective capture density ratio of approximately 10:1.

$LiCaAlF_6$

$LiCaAlF_6$ is a neutron sensitive inorganic scintillator crystal which like neutron-sensitive scintillating glass fiber detectors makes use of neutron capture by ^6Li. Unlike scintillating glass fiber detectors however the ^6Li is part of the crystalline structure of the scintillator giving it a naturally high ^6Li density. A doping agent is added to provide the crystal with its scintillating properties, two common doping agents are cesium and europium. Europium doped $LiCaAlF_6$ has the advantage over other materials that the number of optical photons produced per neutron capture is around 30.000 which is 5 times higher than for example in neutron-sensitive scintillating glass fibers.[25] This property makes neutron photon discrimination easier. Due to its high ^6Li density this material is suitable for producing light weight compact neutron detectors, as a result $LiCaAlF_6$ has been used for neutron detection at high altitudes on balloon missions.[26] The long decay time of Europium doped $LiCaAlF_6$ makes it less suitable for measurements in high radiation environments, the cesium doped variant has a shorter decay time but suffers from a lower light-yield.

17.2.3 Semiconductor neutron detectors

Semiconductors have been used for neutron detection.[27]

17.2.4 Neutron activation detectors

Activation samples may be placed in a neutron field to characterize the energy spectrum and intensity of the neutrons. Activation reactions that have differing energy thresholds can be used including ^{56}Fe(n,p) ^{56}Mn, ^{27}Al(n,α)^{24}Na, ^{93}Nb(n,2n) ^{92m}Nb, & ^{28}Si(n,p)^{28}Al.[28]

17.2.5 Fast neutron detectors

Fast neutrons are often detected by first moderating (slowing) them to thermal energies. However, during the slowing-down process the information on the original energy of the neutron, its direction of travel, and the time of emission is lost. For many applications, the detection of "fret" neutrons that retain this information is highly desirable.[29]

Typical fast neutron detectors are liquid scintillators,[30] 4-He based noble gas detectors [31] and plastic detectors. Fast neutron detectors differentiate themselves from one another by their 1.) capability of neutron/gamma discrimination (through pulse shape discrimination) and 2.) sensitivity. The capability to distinguish between neutrons and gammas is excellent in noble gas based 4-He detectors due to their low electron density and excellent pulse shape discrimination property.

Detection of fast neutrons poses a range of special problems. A directional fast-neutron detector has been developed using multiple proton recoils in separated planes of plastic scintillator material. The paths of the recoil nuclei created by neutron collision are recorded; determination of the energy and momentum of two recoil nuclei allow calculation of the direction of travel and energy of the neutron that underwent elastic scattering with them.[32]

17.3 Applications

Neutron detection is used for varying purposes. Each application has different requirements for the detection system.

- Reactor instrumentation: Since reactor power is essentially linearly proportional to the Neutron flux, neutron detectors provide an important measure of power in nuclear power and research reactors. Boiling water reactors may have dozens of neutron detectors, one per fuel assembly. Most neutron detectors used in thermal-spectrum nuclear reactors are optimized to detect thermal neutrons.

- Particle physics: Neutron detection has been proposed as a method of enhancing neutrino detectors.[33]

- Materials science: Elastic and inelastic neutron scattering enables experimentalists to characterize the morphology of materials from scales ranging from ångströms to about one micrometer.

- Radiation safety: Neutron radiation is a hazard associated with neutron sources, space travel, accelerators and nuclear reactors. Neutron detectors used for radiation safety must take into account the relative biological effectiveness (i.e., the way damage caused by neutrons varies with energy).

- Cosmic ray detection: Secondary neutrons are one component of particle showers produced in Earth's atmosphere by cosmic rays. Dedicated ground-level neutron detectors, namely neutron monitors, are employed to monitor variations in cosmic ray flux.

- Special nuclear material detection: Special nuclear materials (SNM) such as uranium-233 and plutonium-239 decay by spontaneous fission, yielding neutrons. Neutrons detectors can be used for monitor for SNM in commerce.

17.4 Experimental neutron detection

Experiments that make use of this science include scattering experiments in which neutrons directed and then scattered from a sample are to be detected. Facilities include the ISIS neutron source at the Rutherford Appleton Laboratory, the Spallation Neutron Source at the Oak Ridge National Laboratory, and the Spallation Neutron Source (SINQ) at the Paul Scherrer Institute, in which the neutrons are produced by spallation reaction, and the traditional research reactor facilities in which neutrons are produced during fission of uranium isotopes. Noteworthy among the various neutron detection experiments is the trademark experiment of the European Muon Collaboration, first performed at CERN and now termed the "EMC experiment." The same experiment is performed today with more sophisticated equipment to obtain more definite results related to the original EMC effect.

17.4.1 Challenges in neutron detection in an experimental environment

Neutron detection in an experimental environment is not an easy science. The major challenges faced by modern-day neutron detection include background noise, high detection rates, neutron neutrality, and low neutron energies.

Background noise

The main components of background noise in neutron detection are high-energy photons, which aren't easily eliminated by physical barriers. The other sources of noise, such as alpha and beta particles, can be eliminated by various shielding materials, such as lead, plastic, thermo-coal, etc. Thus, photons cause major interference in neutron detection, since it is uncertain if neutrons or photons are being detected by the neutron detector. Both register similar energies after scattering into the detector from the target or ambient light, and are thus hard to distinguish. Coincidence detection can also be used to discriminate real neutron events from photons and other radiation.

High detection rates

If the detector lies in a region of high beam activity, it is hit continuously by neutrons and background noise at overwhelmingly high rates. This obfuscates collected data, since there is extreme overlap in measurement, and separate events are not easily distinguished from each other. Thus, part of the challenge lies in keeping detection rates as low as possible and in designing a detector that can keep up with the high rates to yield coherent data.

Neutrality of neutrons

Neutrons are neutral and thus do not respond to electric fields. This makes it hard to direct their course towards a detector to facilitate detection. Neutrons also do not ionize atoms except by direct collision, so gaseous ionization detectors are ineffective.

Varying behavior with energy

Detectors relying on neutron absorption are generally more sensitive to low-energy thermal neutrons, and are orders of magnitude less sensitive to high-energy neutrons. Scintillation detectors, on the other hand, have trouble registering the impacts of low-energy neutrons.

17.4.2 Experimental setup and method

Figure 1 shows the typical main components of the setup of a neutron detection unit. In principle, the diagram shows the setup as it would be in any modern particle physics lab, but the specifics describe the setup in Jefferson Lab (Newport News, Virginia).

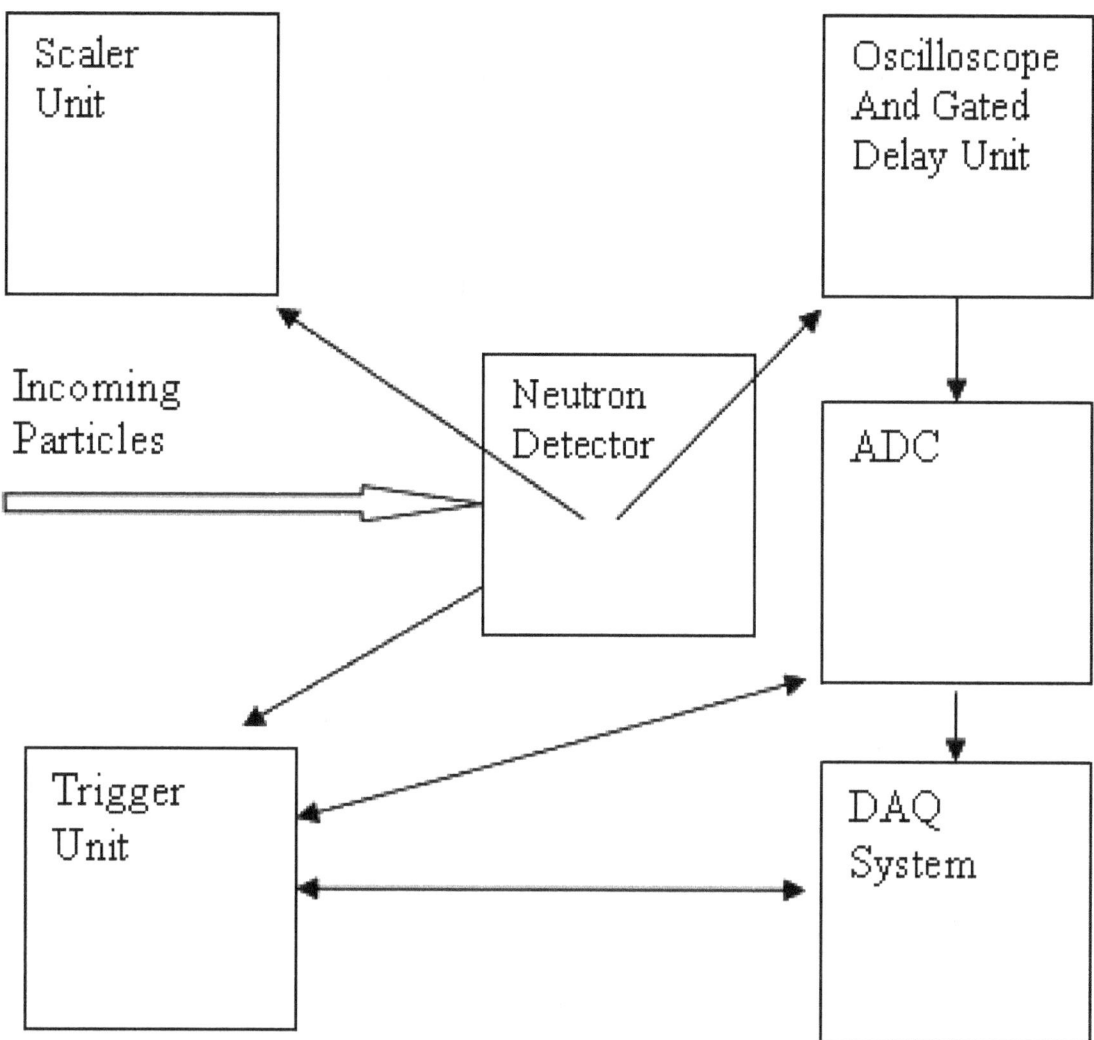

Figure 1: The experimental setup

In this setup, the incoming particles, comprising neutrons and photons, strike the neutron detector; this is typically a scintillation detector consisting of scintillating material, a waveguide, and a photomultiplier tube (PMT), and will be connected to a data acquisition (DAQ) system to register detection details.

The detection signal from the neutron detector is connected to the scaler unit, gated delay unit, trigger unit and the oscilloscope. The scaler unit is merely used to count the number of incoming particles or events. It does so by incrementing its tally of particles every time it detects a surge in the detector signal from the zero-point. There is very little dead time in this unit, implying that no matter how fast particles are coming in, it is very unlikely for this unit to fail to count an event (e.g. incoming particle). The low dead time is due to sophisticated electronics in this unit, which take little time to recover from the relatively easy task of registering a logical high every time an event occurs. The trigger unit coordinates all the electronics of the system and gives a logical high to these units when the whole setup is ready to record an event run.

The oscilloscope registers a current pulse with every event. The pulse is merely the ionization current in the detector caused by this event plotted against time. The total energy of the incident particle can be found by integrating this current pulse with respect to time to yield the total charge deposited at the end of the PMT. This integration is carried out in the analog-digital converter (ADC). The total deposited charge is a direct measure of the energy of the ionizing particle (neutron or photon) entering the neutron detector. This signal integration technique is an established method for measuring ionization in the detector in nuclear physics.[34] The ADC has a higher dead time than the oscilloscope, which has limited memory and needs to transfer events quickly to the ADC. Thus, the ADC samples out approximately one in every 30 events from the oscilloscope for analysis. Since the typical event rate is around 10^6 neutrons every second,[35] this sampling will still accumulate thousands of events every second.

17.4.3 Separating neutrons from photons

The ADC sends its data to a DAQ unit that sorts the data in presentable form for analysis. The key to further analysis lies in the difference between the shape of the photon ionization-current pulse and that of the neutron. The photon pulse is longer at the ends (or "tails") whereas the neutron pulse is well-centered.[35] This fact can be used to identify incoming neutrons and to count the total rate of incoming neutrons. The steps leading to this separation (those that are usually performed at leading national laboratories, Jefferson Lab specifically among them) are gated pulse extraction and plotting-the-difference.

Gated pulse extraction

Ionization current signals are all pulses with a local peak in between. Using a logical AND gate in continuous time (having a stream of "1" and "0" pulses as one input and the current signal as the other), the tail portion of every current pulse signal is extracted. This gated discrimination method is used on a regular basis on liquid scintillators.[36] The gated delay unit is precisely to this end, and makes a delayed copy of the original signal in such a way that its tail section is seen alongside its main section on the oscilloscope screen.

After extracting the tail, the usual current integration is carried out on both the tail section and the complete signal. This yields two ionization values for each event, which are stored in the event table in the DAQ system.

Plotting the difference

In this step lies the crucial point of the analysis: the extracted ionization values are plotted. Specifically, the graph plots energy deposition in the tail against energy deposition in the entire signal for a range of neutron energies. Typically, for a given energy, there are many events with the same tail-energy value. In this case, plotted points are simply made denser with more overlapping dots on the two-dimensional plot, and can thus be used to eyeball the number of events corresponding to each energy-deposition. A considerable random fraction (1/30) of all events is plotted on the graph.

If the tail size extracted is a fixed proportion of the total pulse, then there will be two lines on the plot, having different slopes. The line with the greater slope will correspond to photon events and the line with the lesser slope to neutron events. This is precisely because the photon energy deposition current, plotted against time, leaves a longer "tail" than does the neutron deposition plot, giving the photon tail more proportion of the total energy than neutron tails.

The effectiveness of any detection analysis can be seen by its ability to accurately count and separate the number of neutrons and photons striking the detector. Also, the effectiveness of the second and third steps reveals whether event rates in the experiment are manageable. If clear plots can be obtained in the above steps, allowing for easy neutron-photon separation, the detection can be termed effective and the rates manageable. On the other hand, smudging and indistinguishability of data points will not allow for easy separation of events.

17.4.4 Rate control

Detection rates can be kept low in many ways. Sampling of events can be used to choose only a few events for analysis. If the rates are so high that one event cannot be distinguished from another, physical experimental parameters (shielding, detector-target distance, solid-angle, etc.) can be manipulated to give the lowest rates possible and thus distinguishable events.

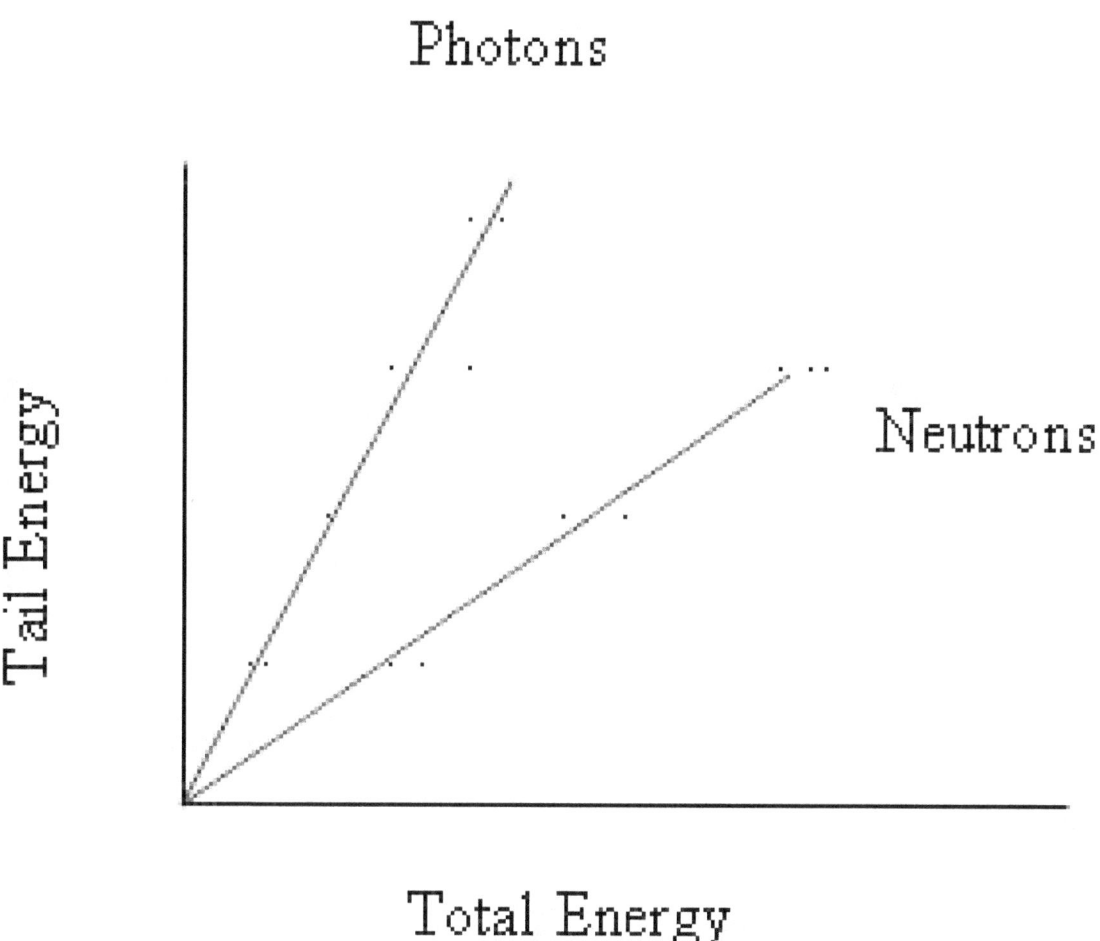

Figure 2: Expected plot of tail energy against energy in the complete pulse plotted for all event energies. Dots represent number densities of events.

17.4.5   Finer detection points

It is important here to observe precisely those variables that matter, since there may be false indicators along the way. For example, ionization currents might get periodic high surges, which do not imply high rates but just high energy depositions for stray events. These surges will be tabulated and viewed with cynicism if unjustifiable, especially since there is so much background noise in the setup.

One might ask how experimenters can be sure that every current pulse in the oscilloscope corresponds to exactly one event. This is true because the pulse lasts about 50 ns, allowing for a maximum of 2×10^7 events every second. This number is much higher than the actual typical rate, which is usually an order of magnitude less, as mentioned above.[35] This means that is it highly unlikely for there to be two particles generating one current pulse. The current pulses last 50 ns each, and start to register the next event after a gap from the previous event.

Although sometimes facilitated by higher incoming neutron energies, neutron detection is generally a difficult task, for all the reasons stated earlier. Thus, better scintillator design is also in the foreground and has been the topic of pursuit ever since the invention of scintillation detectors. Scintillation detectors were invented in 1903 by Crookes but were not very efficient until the PMT (photomultiplier tube) was developed by Curran and Baker in 1944.[34] The PMT gives a reliable and efficient method of detection since it multiplies the detection signal tenfold. Even so, scintillation design has room for improvement as do other options for neutron detection besides scintillation.

17.5   See also

- Bonner sphere – tool for determining neutron energies

- Nested Neutron Spectrometer - A field portable neutron spectrometer based on the Bonner Sphere Principle

- Large Area Neutron Detector

- Neutron Probe

- European Muon Collaboration

- Anger camera - position sensitive neutron detectors are developed using technologies of the Anger camera

- Microchannel plate detector - position sensitive neutron detectors are developed using technologies of the microchannel plate detector

17.6   References

[1] Particle Data Group's Review of Particle Physics 2006

[2] Particle Data Group Summary Data Table on Baryons

[3] Tsoulfanidis, Nicholas (1995). *Measurement and Detection of Radiation* (2nd ed.). Washington, D.C.: Taylor & Francis. pp. 467–501. ISBN 1-56032-317-5.

[4] Materials with a high hydrogen content such as water or plastic

[5] Boron Trifluoride (BF_3) Neutron Detectors

[6] Yousuke, I.; Daiki, S.; Hirohiko, K.; Nobuhiro, S.; Kenji, I. (2000). "Deterioration of pulse-shape discrimination in liquid organic scintillator at high energies". *Nuclear Science Symposium Conference Record, Volume: 1* (IEEE) **1**: 6/219–6/221 vol.1. doi:10.1109/NSSMIC.2000.949173. ISBN 0-7803-6503-8.

[7] Kawaguchi, N.; Yanagida, T.; Yokota, Y.; Watanabe, K.; Kamada, K.; Fukuda, K.; Suyama, T.; Yoshikawa, A. (2009). "Study of crystal growth and scintillation properties as a neutron detector of 2-inch diameter eu doped LiCaAlF6 single crystal". *Nuclear Science Symposium Conference Record (NSS/MIC)* (IEEE): 1493–1495. doi:10.1109/NSSMIC.2009.5402299. ISBN 978-1-4244-3961-4.

[8] Example crystal scintillator based neutron monitor.

[9] Bollinger, L. M.; Thomas, G. E.; Ginther, R. J. (1962). "Neutron Detection With Glass Scintillators". *Nuclear Instruments and Methods* **17**: 97–116. Bibcode:1962NucIM..17...97B. doi:10.1016/0029-554X(62)90178-7.

[10] Miyanaga, N.; Ohba, N.; Fujimoto, K. (1997). "Fiber scintillator/streak camera detector for burn history measurement in inertial confinement fusion experiment". *Review of Scientific Instruments* **68** (1): 621–623. Bibcode:1997RScI...68..621M. doi:10.1063/1.1147667.

[11] Egelstaff, P. A. et al. (1957). "Glass Scintillators For Prompt Detection Of Intermediate Energy Neutrons". *Nuclear Instruments and Methods* **1**: 197–199. Bibcode:1957NucIn...1..197E. doi:10.1016/0369-643x(57)90042-7.

[12] Bollinger, L. M.; Thomas, G. E.; Ginther, R. J. (1962). "Neutron Detection With Glass Scintillators". *Nuclear Instruments and Methods* **17**: 97–116. Bibcode:1962NucIM..17...97B. doi:10.1016/0029-554X(62)90178-7.

[13] Spowart, A. R. (1976). "Neutron Scintillating Glasses .1. Activation By External Charged-Particles And Thermal-Neutrons". *Nuclear Instruments and Methods* **135**: 441–453. Bibcode:1976NucIM.135..441S. doi:10.1016/0029-554X(76)90057-4.

[14] Atkinson, M.; Fent J.; Fisher C. et al. (1987). "Initial Tests Of A High-Resolution Scintillating Fiber (Scifi) Tracker". *Nuclear Instruments & Methods In Physics Research Section A-Accelerators Spectrometers Detectors And Associated Equipment* **254**: 500–514.

[15] Bliss, M.; Brodzinski R. L.; Craig R. A.; Geelhood B. D.; Knopf M. A.; Miley H. S.; Perkins R. W.; Reeder P. L.; Sunberg D. S.; Warner R. A.; Wogman N. A. (1995). "Glass-fiber-based neutron detectors for high- and low-flux environments". *Proc. SPIE* **2551**: 108. doi:10.1117/12.218622.

[16] Abel, K. H.; Arthur R. J.; Bliss M.; Brite D. W. et al. (1993). "Performance and Applications of Scintillating-Glass-Fiber Neutron Sensors". *Proceedings of the SCIFI 93 Workshop on Scintillating Fiber Detectors*: 463–472.

[17] Abel, K. H.; Arthur R. J.; Bliss M.; Brite D. W. et al. (1994). "Scintillating Glass Fiber-Optic Neutron Sensors". *Materials Research Society Symposium Proceedings* **348**: 203–208.

[18] Bliss, M.; Craig R. A.; Reeder P. L. (1994). "The Physics and Structure-property Relationships of Scintillator Materials: Effect of Thermal History and Chemistry on the Light Output of Scintillating Glasses". *Nuclear Instruments and Methods in Physics Research A* **342**: 357–393. Bibcode:1994NIMPA.342..357B. doi:10.1016/0168-9002(94)90263-1.

[19] Bliss, M.; Craig R. A.; Reeder P. L.; Sunberg D. S.; Weber M. J. (1994). "Relationship Between Microstructure and Efficiency of Scintillating Glasses". *Materials Research Society Symposium Proceedings* **348**: 195–202.

[20] Seymour, R.; Crawford, T. et al. (2001). "Portal, freight and vehicle monitor performance using scintillating glass fiber detectors for the detection of plutonium in the Illicit Trafficking Radiation Assessment Program". *Journal of Radioanalytical and Nuclear Chemistry* **248**: 699–705.

[21] Seymour, R. S.; Craig R. A.; Bliss M.; Richardson B.; Hull C. D.; Barnett D. S. (1998). "Performance of a neutron-sensitive scintillating glass-fiber panel for portal, freight and vehicle monitoring". *Proc. SPIE* **3536**: 148–155. doi:10.1117/12.339067.

[22] Seymour, R. S.; Richardson B.; Morichi M.; Bliss M.; Craig R. A.; Sunberg D. S. (2000). "Scintillating-glass-fiber neutron sensors, their application and performance for plutonium detection and monitoring" (PDF). *Journal of Radioanalytical and Nuclear Chemistry* **243**: 387–388.

[23] Nucsafe Inc. website

[24] Van Ginhoven, R. M.; Kouzes R. T.; Stephens D. L. (2009). "Alternative Neutron Detector Technologies for Homeland Security PIET-43741-TM-840 PNNL-18471".

[25] Yanagida, T. et al. (2011). "Europium and Sodium Codoped LiCaAlF$_6$ Scintillator for Neutron Detection". *Applied Physics Express* **4**: 106401. Bibcode:2011APExp...4j6401Y. doi:10.1143/apex.4.106401.

[26] Kole, M. et al. (2013). "A Balloon-borne Measurement of High Latitude Atmospheric Neutrons Using a LiCAF Neutron Detector". *Nuclear Science Symposium Conference Record (NSS/MIC) (IEEE)*.

[27] Mireshghi, A.; Cho, G.; Drewery, J. S.; Hong, W. S.; Jing, T.; Lee, H.; Kaplan, S. N.; Perez-Mendez, V. (1994). "High efficiency neutron sensitive amorphous silicon pixel detectors". *Nuclear Science* (IEEE) **41** (4 , Part: 1–2): 915–921. Bibcode:1994ITNM.doi:10.1109/23.322831.

[28] van Eijk, C. W. E.; de Haas, J. T. M.; Dorenbos, P.; Kramer, K. W.; Gudel, H. U. (2005). "Development of elpasolite and monoclinic thermal neutron scintillators". *Nuclear Science Symposium Conference Record* (IEEE) **1**: 239–243. doi:10.1109/NSSMIC.2005.1596245.ISBN0-7803-9221-3.

[29] Stromswold, D.C.; AJ Peurrung; RR Hansen; PL Reeder (1999). "Direct Fast-Neutron Detection. PNNL-13068, Pacific Northwest National Laboratory, Richland, WA.".

[30] Pozzi, S. A.; J. L. Dolan; E. C. Miller; M. Flaska; S. D. Clarke; A. Enqvist; P. Peerani; M. A. Smith-Nelson; E. Padovani; J. B. Czirr; L. B. Rees (2011). "Evaluation of New and Existing Organic Scintillators for Fast Neutron Detection". *Proceedings of the Institute of Nuclear Materials Management 52nd Annual Meeting on CD-ROM, Palm Desert, California, USA. July 17 – 22*.

[31] Lewis, J.M.; R. P. Kelley; D. Murer; K. A. Jordan. "Fission signal detection using helium-4 gas fast neutron scintillation detectors". *Appl. Phys. Lett. 105, 014102 (2014);* **105**: 014102. Bibcode:2014ApPhL.105a4102L. doi:10.1063/1.4887366.

[32] Vanier, P. E.; Forman, L.; Dioszegi, I.; Salwen, C.; Ghosh, V. J. (2007). "Calibration and testing of a large-area fast-neutron directional detector". *Nuclear Science Symposium Conference Record* (IEEE): 179–184. doi:10.1109/NSSMIC.2007.4436312. ISBN 978-1-4244-0922-8.

[33] John F. Beacom and Mark R. Vagins (2004). "Antineutrino Spectroscopy with Large Water Čerenkov Detectors". *Physical Review Letters* **93** (17): 171101. arXiv:hep-ph/0309300. Bibcode:2004PhRvL..93q1101B. doi:10.1103/PhysRevLett.93.171101.

[34] Leo, W. R. (1994). *Techniques for Nuclear and Particle Physics Experiments.* Springer.

[35] Cerny, J. C., Dolemal, Z., Ivanov, M. P., Kuzmin, E. P., Svejda, J., Wilhelm, I. (2003). "Study of neutron response and n–γ discrimination by charge comparison method for small liquid scintillation detector". *Nuclear Instruments and Methods in Physics Research A* **527** (3): 512–518. arXiv:nucl-ex/0311022. Bibcode:2004NIMPA.527..512C. doi:10.1016/j.nima.2004.03.179.

[36] Jastaniah, S. D., Sellin, P. J. (2003). "Digital techniques for n–γ pulse shape discrimination capture-gated neutron spectroscopy using liquid". *Nuclear Instruments and Methods in Physics Research A* **517**: 202–210. Bibcode:2004NIMPA.517..202J. doi:10.1016/j.nima.2003.08.178.

17.7 Further reading

- Cates, G. D., Day, D., Liyanage, N. (2004). "Neutron Tagged Bound Proton Structure to Probe the Origin of the EMC Effect" (PostScript). Jefferson Lab. Retrieved 2005-06-09.

- Pozzi, S. A., Mullens, J. A., and Mihalczo, J. T. (2003). "Analysis of neutron and photon detection position for the calibration of plastic (BC-420) and liquid (BC-501) scintillators". *Nuclear Instruments and Methods in Physics Research A* **524** (1–3): 92–101. Bibcode:2004NIMPA.524...92P. doi:10.1016/j.nima.2003.12.036.

- Cecil, R. A., Anderson, B. D., Madey, R. (1979). "Improved Predictions of Neutron Detection Efficiency for Hydrocarbon Scintillators from 1 MeV to about 300 MeV". *Nuclear Instruments and Methods in Physics Research* **161** (3): 439–447. Bibcode:1979NucIM.161..439C. doi:10.1016/0029-554X(79)90417-8.

- DOE Fundamentals Handbook on Instrumentation and Control, Volume 2

Chapter 18

Neutron capture

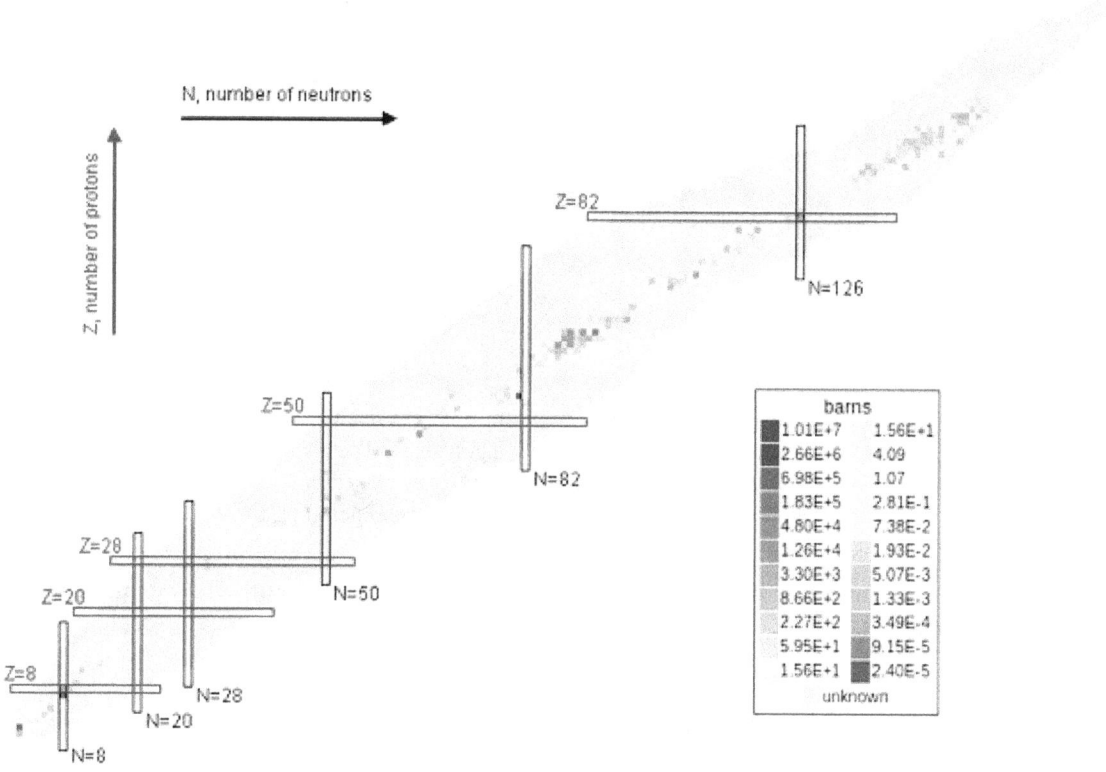

Chart of nuclides showing thermal neutron capture cross section values

Neutron capture is a nuclear reaction in which an atomic nucleus and one or more neutrons collide and merge to form a heavier nucleus.[1] Since neutrons have no electric charge they can enter a nucleus more easily than positively charged protons, which are repelled electrostatically.[1]

Neutron capture plays an important role in the cosmic nucleosynthesis of heavy elements. In stars it can proceed in two ways: as a rapid (r-process) or a slow process (s-process).[1] Nuclei of masses greater than 56 cannot be formed by thermonuclear reactions (i.e. by nuclear fusion), but can be formed by neutron capture.[1]

18.1 Neutron capture at small neutron flux

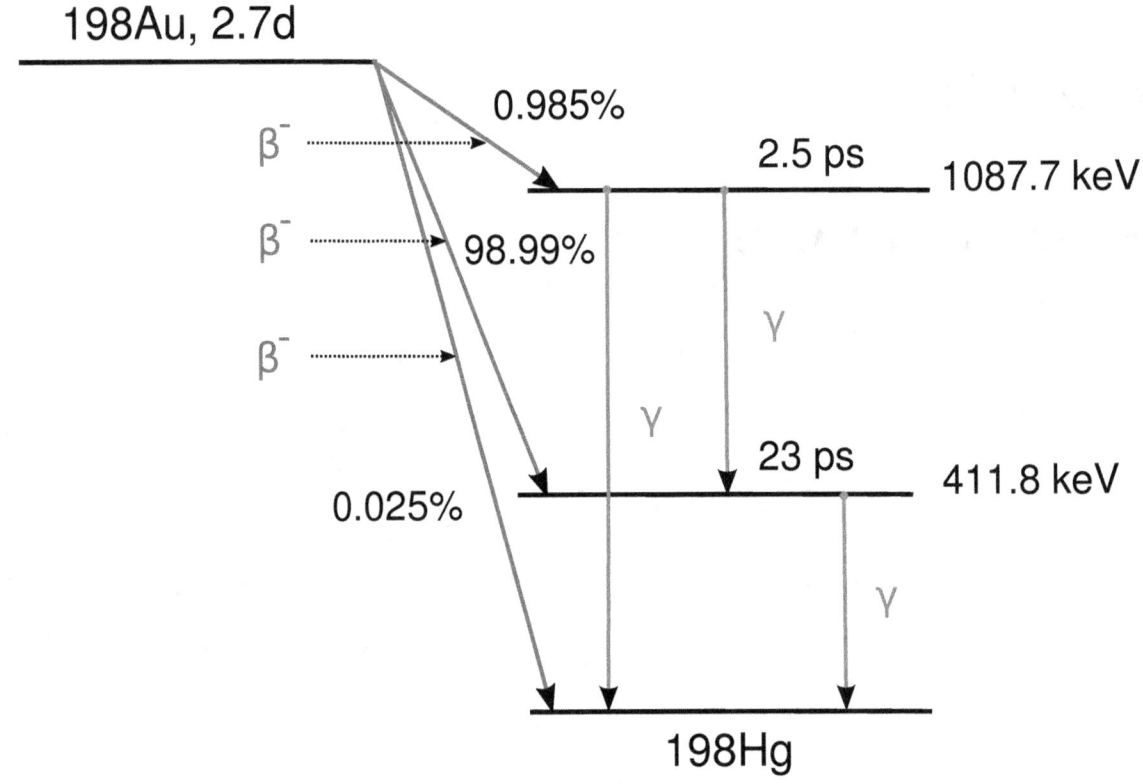

Decay scheme of ^{198}Au

At small neutron flux, as in a nuclear reactor, a single neutron is captured by a nucleus. For example, when natural gold (^{197}Au) is irradiated by neutrons, the isotope ^{198}Au is formed in a highly excited state, and quickly decays to the ground state of ^{198}Au by the emission of γ rays. In this process, the mass number increases by one. This is written as a formula in the form ^{197}Au+n → ^{198}Au+γ, or in short form ^{197}Au(n,γ)^{198}Au. If thermal neutrons are used, the process is called thermal capture.

The isotope ^{198}Au is a beta emitter that decays into the mercury isotope ^{198}Hg. In this process the atomic number rises by one.

18.2 Neutron capture at high neutron flux

The r-process happens inside stars if the neutron flux density is so high that the atomic nucleus has no time to decay via beta emission in between neutron captures. The mass number therefore rises by a large amount while the atomic number (i.e., the element) stays the same. Only afterwards, the highly unstable nuclei decay via many β⁻ decays to stable or unstable nuclei of high atomic number.

18.3 Capture cross section

The absorption neutron cross-section of an isotope of a chemical element is the effective cross sectional area that an atom of that isotope presents to absorption, and is a measure of the probability of neutron capture. It is usually measured in barns (b).

Absorption cross section is often highly dependent on neutron energy. Two of the most commonly specified measures are the cross-section for thermal neutron absorption, and resonance integral which considers the contribution of absorption peaks at certain neutron energies specific to a particular nuclide, usually above the thermal range, but encountered as neutron moderation slows the neutron down from an original high energy.

The thermal energy of the nucleus also has an effect; as temperatures rise, Doppler broadening increases the chance of catching a resonance peak. In particular, the increase in uranium-238's ability to absorb neutrons at higher temperatures (and to do so without fissioning) is a negative feedback mechanism that helps keep nuclear reactors under control.

18.4   Uses

Further information: Neutron activation and Neutron activation analysis

Neutron activation analysis can be used to remotely detect the chemical composition of materials. This is because different elements release different characteristic radiation when they absorb neutrons. This makes it useful in many fields related to mineral exploration and security.

18.5   Neutron absorbers

Main article: Neutron poison
The most important neutron absorber is ^{10}B as ^{10}B$_4$C in control rods, or boric acid as a coolant water additive in

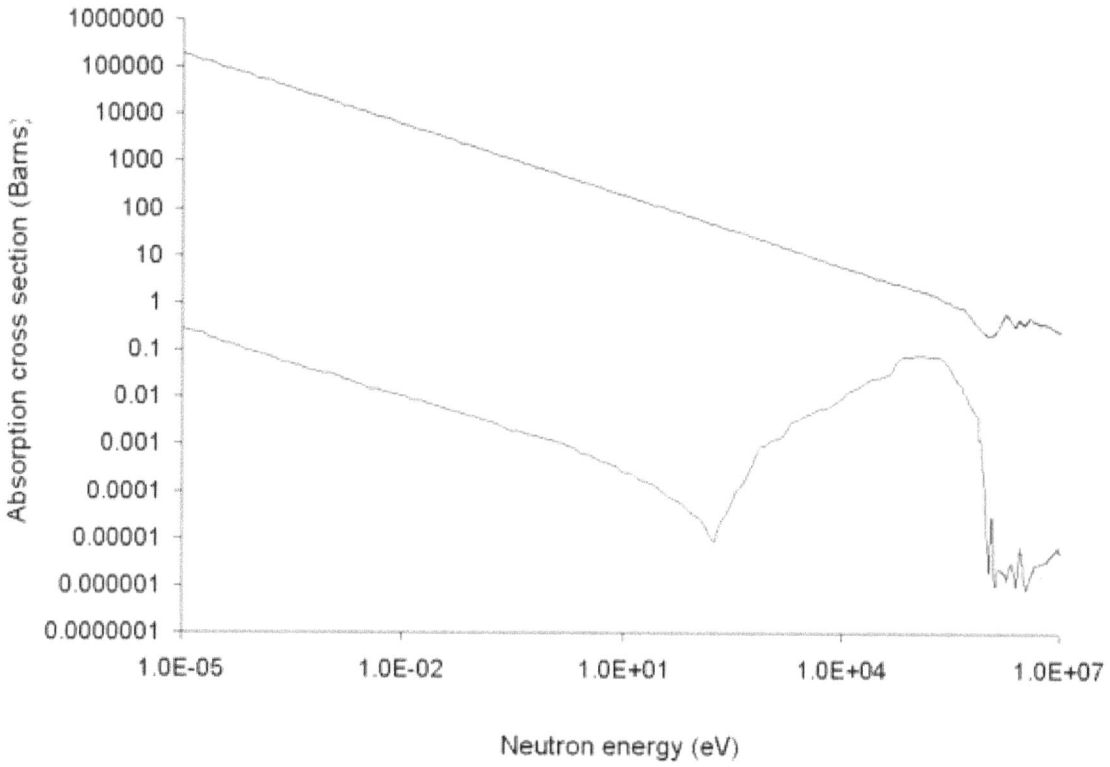

Neutron cross section of boron (top curve is for ^{10}B and bottom curve for ^{11}B)

PWRs. Other important neutron absorbers that are used in nuclear reactors are xenon, cadmium, hafnium, gadolinium, cobalt, samarium, titanium, dysprosium, erbium, europium, molybdenum and ytterbium;[2] all of which usually consist of mixtures of various isotopes—some of which are excellent neutron-absorbers. These also occur in combinations such as Mo_2B_5, hafnium diboride, titanium diboride, dysprosium titanate and gadolinium titanate.

Hafnium, one of the last stable elements to be discovered, presents an interesting case. Even though hafnium is a heavier element, its electron configuration makes it practically identical with the element zirconium, and they are always found in the same ores. However, their nuclear properties are different in a profound way. Hafnium absorbs neutrons avidly (Hf absorbs 600 times more than Zr), and it can be used in reactor control rods, whereas natural zirconium is practically transparent to neutrons. So, zirconium is a very desirable construction material for reactor internal parts, including the metallic cladding of the fuel rods which contain either uranium, plutonium, or mixed oxides of the two elements (MOX fuel).

Hence, it is quite important to be able to separate the zirconium from the hafnium in their naturally-occurring alloy. This can only be done inexpensively by using modern chemical ion-exchange resins.[3] Similar resins are also used in reprocessing nuclear fuel rods, when it is necessary to separate uranium and plutonium, and sometimes thorium.

18.6   See also

- Beta decay

- Induced radioactivity

- List of particles

- Neutron emission

- Radioactive decay

- Rays: α — β — γ — δ — ε

- p-process (proton capture)

18.7   References

[1] Ahmad, Ishfaq; Hans Mes; Jacques Hebert (1966). "Progress of theoretical physics: Resonance in the Nucleus". *Institute of Physics* (Ottawa, Canada: University of Ottawa (Department of Physics)) **3** (3): 556–600.

[2] Prompt Gamma-ray Neutron Activation Analysis. International Atomic Energy Agency

[3] D. Franklin; R. B. Adamson (1 January 1984). *Zirconium in the Nuclear Industry: Sixth International Symposium.* ASTM International. pp. 26–. ISBN 978-0-8031-0270-5. Retrieved 7 October 2012.

18.8   External links

- Thermal Neutron Capture Data

- Thermal Neutron Cross Sections at the International Atomic Energy Agency

Chapter 19

Neutron source

A **neutron source** is any device that emits neutrons, irrespective of the mechanism used to produce the neutrons. Neutron source devices are used in physics, engineering, medicine, nuclear weapons, petroleum exploration, biology, chemistry and nuclear power.

Neutron source variables include the energy of the neutrons emitted by the source, the rate of neutrons emitted by the source, the size of the source, the cost of owning and maintaining the source, and government regulations related to the source.

19.1 Small Sized Devices

Radioisotopes Which Undergo Spontaneous Fission Certain isotopes undergo spontaneous fission with emission of neutrons. The most commonly used spontaneous fission source is the radioactive isotope californium−252. Cf-252 and all other spontaneous fission neutron sources are produced by irradiating uranium or another transuranic element in a nuclear reactor, where neutrons are absorbed in the starting material and its subsequent reaction products, transmuting the starting material into the SF isotope. Cf-252 neutron sources are typically 1/4" to 1/2" in diameter and 1" to 2" in length. When purchased new a typical Cf-252 neutron source emits between 1×10^7 to 1×10^9 neutrons per second but, with a half life of 2.6 years, this neutron output rate drops to half of this original value in 2.6 years. The price of a typical Cf-252 neutron source is from \$15,000 to \$20,000.

Radioisotopes Which Decay With Alpha Particles Packed In A Low-Z Elemental Matrix

Neutrons are produced when alpha particles impinge upon any of several low atomic weight isotopes including isotopes of beryllium, carbon and oxygen. This nuclear reaction can be used to construct a neutron source by intermixing a radioisotope that emits alpha particles such as radium or polonium with a low atomic weight isotope, usually in the form of a mixture of powders of the two materials. Typical emission rates for alpha reaction neutron sources range from 1×10^6 to 1×10^8 neutrons per second. As an example, a representative alpha-beryllium neutron source can be expected to produce approximately 30 neutrons for every one million alpha particles. The useful lifetime for these types of sources is highly variable, depending upon the half life of the radioisotope that emits the alpha particles. The size and cost of these neutron sources are also comparable to spontaneous fission sources. Usual combinations of materials are plutonium-beryllium (PuBe), americium-beryllium (AmBe), or americium-lithium (AmLi).

Radioisotopes Which Decay With High Energy Photons Co-located With Beryllium or Deuterium

Gamma radiation with an energy exceeding the neutron binding energy of a nucleus can eject a neutron. Two examples and their decay products:

- ^{9}Be + >1.7 Mev photon → 1 neutron + 2 ^{4}He

- ^{2}H (deuterium) + >2.26 MeV photon → 1 neutron + ^{1}H

Sealed Tube Neutron Generators Some accelerator-based neutron generators exist that work by inducing fusion between beams of deuterium and/or tritium ions and metal hydride targets which also contain these isotopes.

19.2 Medium Sized Devices

Plasma Focus and Plasma Pinch Devices The plasma focus neutron source (see Plasma focus, not to be confused with the so-called Farnsworth-Hirsch fusor) produces controlled nuclear fusion by creating a dense plasma within which ionized deuterium and/or tritium gas is heated to temperatures sufficient for creating fusion.

Inertial electrostatic confinement This group of devices use an electric field to heat a plasma to fusion conditions and produce neutrons. Various applications from a hobby enthusiast scene up to commercial applications have developed, mostly in the US.

Light Ion Accelerators Traditional particle accelerators with hydrogen (H), deuterium (D), or tritium (T) ion sources may be used to produce neutrons using targets of deuterium, tritium, lithium, beryllium, and other low-Z materials. Typically these accelerators operate with energies in the > 1 MeV range,

High Energy Bremsstrahlung Photoneutron/photofission Systems
Neutrons (so called photoneutrons) are produced when photons above the nuclear binding energy of a substance are incident on that substance, causing it to undergo giant dipole resonance after which it either emits a neutron or undergoes fission. The number of neutrons released by each fission event is dependent on the substance. Typically photons begin to produce neutrons on interaction with normal matter at energies of about 7 to 40 MeV, which means that megavoltage photon radiotherapy facilities may produce neutrons as well, and require special shielding for them. In addition, electrons of energy over about 50 MeV may induce giant dipole resonance in nuclides by a mechanism which is the inverse of internal conversion, and thus produce neutrons by a mechanism similar to that of photoneutrons. [1]

19.3 Large Sized Devices

Nuclear Fission Reactors Nuclear fission which takes place within in a reactor produces very large quantities of neutrons and can be used for a variety of purposes including power generation and experiments.

Nuclear Fusion Systems Nuclear fusion, the combining of the heavy isotopes of hydrogen, also has the potential to produces large quantities of neutrons. Small scale fusion systems exist for research purposes at many universities and laboratories around the world. A small number of large scale nuclear fusion systems also exist including the National Ignition Facility in the USA, JET in the UK, and soon the recently started ITER experiment in France.

High Energy Particle Accelerators A spallation source is a high-flux source in which protons that have been accelerated to high energies hit a target material, prompting the emission of neutrons.

19.4 Neutron flux

For most applications, a higher neutron flux is better (since it reduces the time required to conduct the experiment, acquire the image, etc.). Amateur fusion devices, like the fusor, generate only about 300 000 neutrons per second. Commercial fusor devices can generate on the order of 10^9 neutrons per second, which corresponds to a usable flux of less than 10^5 n/(cm^2 s). Large neutron beamlines around the world achieve much greater flux. Reactor-based sources now produce 10^{15} n/(cm^2 s), and spallation sources generate greater than 10^{17} n/(cm^2 s).

19.5   See also

- Fast neutron

- Nuclear fission

- Neutron generator

- Neutron moderator

- Radioactive decay

- Radioactivity

- Slow neutron

19.6   References

[1] Unknown

19.7   External links

- Portable Neutron Generators

- List of Neutron Sources Worldwide

- Neutronsources.org

- Integrated Infrastructure Initiative for Neutron Scattering and Muon Spectroscopy (NMI3)

Chapter 20

Neutron generator

This article is about generators employing accelerators. For more general sources, see Neutron source.
Neutron generators are neutron source devices which contain compact linear accelerators and that produce neutrons by

Nuclear physicist at the Idaho National Laboratory sets up an experiment using an electronic neutron generator.

fusing isotopes of hydrogen together. The fusion reactions take place in these devices by accelerating either deuterium, tritium, or a mixture of these two isotopes into a metal hydride target which also contains deuterium, tritium or a mixture of these isotopes. Fusion of deuterium atoms (D + D) results in the formation of a He-3 ion and a neutron with a kinetic energy of approximately 2.5 MeV. Fusion of a deuterium and a tritium atom (D + T) results in the formation of a He-4 ion and a neutron with a kinetic energy of approximately 14.1 MeV. Neutron generators have applications in medicine, security, and materials analysis.[1]

Thousands of such small, relatively inexpensive systems have been built over the past five decades.

20.1   Neutron generator theory and operation

Small neutron generators using the deuterium (D, hydrogen-2, ^{2}H) tritium (T, hydrogen-3, ^{3}H) fusion reactions are the most common accelerator based (as opposed to isotopic) neutron sources. In these systems, neutrons are produced by creating ions of deuterium, tritium, or deuterium and tritium and accelerating these into a hydride target loaded with deuterium, tritium, or deuterium and tritium. The DT reaction is used more than the DD reaction because the yield of the DT reaction is 50–100 times higher than that of the DD reaction.

$$D + T \rightarrow n + {}^4He \ E_n = 14.1 \ MeV$$

$$D + D \rightarrow n + {}^3He \ E_n = 2.5 \ MeV$$

Neutrons produced from the DT reaction are emitted isotropically (uniformly in all directions) from the target while neutrons from the DD reaction are slightly peaked in the forward (along the axis of the ion beam) direction. In both cases, the associated He nuclei are emitted in the opposite direction of the neutron.

The gas pressure in the ion source region of the neutron tubes generally ranges between 0.1–0.01 mm Hg. The mean free path of electrons must be shorter than the discharge space to achieve ionization (lower limit for pressure) while the pressure must be kept low enough to avoid formation of discharges at the high extraction voltages applied between the electrodes. The pressure in the accelerating region has however to be much lower, as the mean free path of electrons must be longer to prevent formation of a discharge between the high voltage electrodes.[2]

The ion accelerator usually consists of several electrodes with cylindrical symmetry, acting as electric lenses. The ion beam can be focused to a small spot of the target that way. The accelerators usually have several stages, with voltage between the stages not exceeding 200 kV to prevent field emission.[2]

In comparison with radionuclide neutron sources, neutron tubes can produce much higher neutron fluxes and monochromatic neutron energy spectrums can be obtained. The neutron production rate can also be controlled.[2]

20.2   Sealed neutron tubes

The central part of a neutron generator is the particle accelerator itself, sometimes called a neutron tube. Neutron tubes have several components including an ion source, ion optic elements, and a beam target; all of these are enclosed within a vacuum tight enclosure. High voltage insulation between the ion optical elements of the tube is provided by glass and/or ceramic insulators. The neutron tube is, in turn, enclosed in a metal housing, the accelerator head, which is filled with a dielectric medium to insulate the high voltage elements of the tube from the operating area. The accelerator and ion source high voltages are provided by external power supplies. The control console allows the operator to adjust the operating parameters of the neutron tube. The power supplies and control equipment are normally located within 10–30 feet of the accelerator head in laboratory instruments, but may be several kilometers away in well logging instruments.

In comparison with their predecessors, sealed neutron tubes do not require vacuum pumps and gas sources for operation. They are therefore more mobile and compact, while also durable and reliable. For example, sealed neutron tubes have replaced radioactive neutron initiators, in supplying a pulse of neutrons to the imploding core of modern nuclear weapons.

Examples of neutron tube ideas date as far back as the 1930s, pre-nuclear weapons era, by German scientists filing a 1938 German patent (March 1938, patent # 261,156) and obtaining a United States Patent (July 1941, USP#2,251,190); examples of present state of the art are given by developments such as the Neutristor,[3] a mostly solid state device, resembling a computer chip, invented at Sandia National Laboratories[4] in Albuquerque NM. Typical sealed designs are used in a pulsed mode[5] and can be operated at different output levels, depending on the life from the ion source and loaded targets.[6]

20.3   Ion sources

Main article: Ion source

A good ion source should provide a strong ion beam without consuming much of the gas. For hydrogen isotopes, production of atomic ions is favored over molecular ions, as atomic ions have higher neutron yield on collision. The ions generated in the ion source are then extracted by an electric field into the accelerator region, and accelerated towards the target. The gas consumption is chiefly caused by the pressure difference between the ion generating and ion accelerating spaces that has to be maintained. Ion currents of 10 mA at gas consumptions of 40 cm^3/hour are achievable.[2]

For a sealed neutron tube, the ideal ion source should use low gas pressure, give high ion current with large proportion of atomic ions, have low gas clean-up, use low power, have high reliability and high lifetime, its construction has to be simple and robust and its maintenance requirements have to be low.[2]

Gas can be efficiently stored in a replenisher, an electrically heated coil of zirconium wire. Its temperature determines the rate of absorption/desorption of hydrogen by the metal, which regulates the pressure in the enclosure.

20.3.1   Cold cathode (Penning)

The Penning source is a low gas pressure, cold cathode ion source which utilizes crossed electric and magnetic fields. The ion source anode is at a positive potential, either dc or pulsed, with respect to the source cathode. The ion source voltage is normally between 2 and 7 kilovolts. A magnetic field, oriented parallel to the source axis, is produced by a permanent magnet. A plasma is formed along the axis of the anode which traps electrons which, in turn, ionize gas in the source. The ions are extracted through the exit cathode. Under normal operation, the ion species produced by the Penning source are over 90% molecular ions. This disadvantage is however compensated for by the other advantages of the system.

One of the cathodes is a cup made of soft iron, enclosing most of the discharge space. The bottom of the cup has a hole through which most of the generated ions are ejected by the magnetic field into the acceleration space. The soft iron shields the acceleration space from the magnetic field, to prevent a breakdown.[2]

Ions emerging from the exit cathode are accelerated through the potential difference between the exit cathode and the accelerator electrode. The schematic indicates that the exit cathode is at ground potential and the target is at high (negative) potential. This is the case in many sealed tube neutron generators. However, in cases when it is desired to deliver the maximum flux to a sample, it is desirable to operate the neutron tube with the target grounded and the source floating at high (positive) potential. The accelerator voltage is normally between 80 and 180 kilovolts.

The accelerating electrode has the shape of a long hollow cylinder. The ion beam has a slightly diverging angle (about 0.1 radian). The electrode shape and distance from target can be chosen so the entire target surface is bombarded with ions. Acceleration voltages of up to 200 kV are achievable.

The ions pass through the accelerating electrode and strike the target. When ions strike the target, 2–3 electrons per ion are produced by secondary emission. In order to prevent these secondary electrons from being accelerated back into the ion source, the accelerator electrode is biased negative with respect to the target. This voltage, called the suppressor voltage, must be at least 500 volts and may be as high as a few kilovolts. Loss of suppressor voltage will result in damage, possibly catastrophic, to the neutron tube.

Some neutron tubes incorporate an intermediate electrode, called the focus or extractor electrode, to control the size of the beam spot on the target. The gas pressure in the source is regulated by heating or cooling the gas reservoir element.

20.3.2   Radio frequency (RF)

Ions can be created by electrons formed in high-frequency electromagnetic field. The discharge is formed in a tube located between electrodes, or inside a coil. Over 90% proportion of atomic ions is achievable.[2]

20.4   Targets

The targets used in neutron generators are thin films of metal such as titanium, scandium, or zirconium which are deposited onto a silver, copper or molybdenum substrate. Titanium, scandium, and zirconium form stable chemical compounds called metal hydrides when combined with hydrogen or its isotopes. These metal hydrides are made up of two hydrogen (deuterium or tritium) atoms per metal atom and allow the target to have extremely high densities of hydrogen. This is important to maximize the neutron yield of the neutron tube. The gas reservoir element also uses metal hydrides, e.g. uranium hydride, as the active material.

Titanium is preferred to zirconium as it can withstand higher temperatures (200 °C), and gives higher neutron yield as it captures deuterons better than zirconium. The maximum temperature allowed for the target, above which hydrogen isotopes undergo desorption and escape the material, limits the ion current per surface unit of the target; slightly divergent beams are therefore used. A 1 microampere ion beam accelerated at 200 kV to a titanium-tritium target can generate up to 10^8 neutrons per second. The neutron yield is mostly determined by the accelerating voltage and the ion current level.[2]

An example of a tritium target in use is a 0.2 mm thick silver disc with a 1 micrometer layer of titanium deposited on its surface; the titanium is then saturated with tritium.[2]

Metals with sufficiently low hydrogen diffusion can be turned into deuterium targets by bombardment of deuterons until the metal is saturated. Gold targets under such condition show four times higher efficiency than titanium. Even better results can be achieved with targets made of a thin film of a high-absorption high-diffusivity metal (e.g. titanium) on a substrate with low hydrogen diffusivity (e.g. silver), as the hydrogen is then concentrated on the top layer and can not diffuse away into the bulk of the material. Using a deuterium-tritium gas mixture, self-replenishing D-T targets can be made. The neutron yield of such targets is lower than of tritium-saturated targets in deuteron beams, but their advantage is much longer lifetime and constant level of neutron production. Self-replenishing targets are also tolerant to high-temperature bake-out of the tubes, as their saturation with hydrogen isotopes is performed after the bakeout and tube sealing.[2]

20.5   High voltage power supplies

One particularly interesting approach for generating the high voltage fields needed to accelerate ions in a neutron tube is to use a pyroelectric crystal. In April 2005 researchers at UCLA demonstrated the use of a thermally cycled pyroelectric crystal to generate high electric fields in a neutron generator application. In February 2006 researchers at Rensselaer Polytechnic Institute demonstrated the use of two oppositely poled crystals for this application. Using these low-tech power supplies it is possible to generate a sufficiently high electric field gradient across an accelerating gap to accelerate deuterium ions into a deuterated target to produce the D + D fusion reaction. These devices are similar in their operating principle to conventional sealed-tube neutron generators which typically use Cockcroft–Walton type high voltage power supplies. The novelty of this approach is in the simplicity of the high voltage source. Unfortunately, the relatively low accelerating current that pyroelectric crystals can generate, together with the modest pulsing frequencies that can be achieved (a few cycles per minute) limits their near-term application in comparison with today's commercial products (see below). Also see pyroelectric fusion.

20.6   Other technologies

In addition to the conventional neutron generator design described above several other approaches exist to use electrical systems for producing neutrons.

20.6.1   Inertial electrostatic confinement/fusor

Main article: Fusor

Another type of innovative neutron generator is the inertial electrostatic confinement fusion device. This neutron generator differs from the conventional ion beam onto solid target types because it avoids using a solid target which will be sputter eroded causing metalization of insulating surfaces. Depletion of the reactant gas within the solid target is also avoided. Far greater operational lifetime is achieved. Originally called a fusor, it was invented by Philo Farnsworth, the inventor of electronic television. This type of neutron generator is manufactured by NSD-Gradel-Fusion..

20.7 Manufacturers

- Adelphi Technology, Adelphi Technology (USA)

- Baker Hughes , Baker Hughes (USA)

- Sodern, Sodern (France)

- Halliburton, Halliburton (USA)

- Hotwell, Hotwell GmbH (Austria) *[Using neutron tubes from VNIIA]*

- Lawrence Berkeley National Laboratory, Lawrence Berkeley National Laboratory (USA)

- NSD-Gradel-Fusion, NSD-Gradel-Fusion (Luxembourg)

- Phoenix Nuclear Labs, Phoenix Nuclear Labs (USA)

- Sandia National Laboratories , Sandia National Laboratories (USA)

- Schlumberger, Schlumberger (USA)

- Starfire Industries, Starfire Industries (USA)

- Thermo Fisher Scientific, Thermo Fisher Scientific (USA)

- VNIIA, VNIIA All Russia Research Institute of Automatics (Russia)

20.8 See also

- Fast neutron

- Nuclear fission

- Nuclear fusion

- Neutron source

- Neutron moderator

- Radioactive decay

- Radioactivity

- Slow neutron

20.9 References

[1] Reijonen, J. "Compact Neutron Generators for Medical, Homeland Security, and Planetary Exploration" (PDF). *Proceedings of 2005 Particle Accelerator Conference, Knoxville, Tennessee*: 49–53.

[2] van der Horst, H. L. (1964). "VIIIc Neutron Generators". *Gas-Discharge Tubes* (pdf). Philips Technical Library **16**. Eindhoven, Netherlands: Philips Technical Library. pp. 281–295. OCLC 10391645. UDC No. 621.387.

[3] Elizondo-Decanini, J. M.; Schmale, D.; Cich, M.; Martinez, M.; Youngman, K.; Senkow, M.; Kiff, S.; Steele, J.; Goeke, R.; Wroblewski, B.; Desko, J.; Dragt, A. J. (2012). "Novel Surface-Mounted Neutron Generator". *IEEE Transactions on Plasma Sciences* **40** (9): 2145–2150. Bibcode:2012ITPS...40.2145E. doi:10.1109/TPS.2012.2204278.

[4] "Sandia National Laboratories".

[5] Gow, J. D.; Pollock, H. C. (1960). "Development of a Compact Evacuated Pulsed Neutron Source". *Review of Scientific Instruments* **31** (3): 235–240. Bibcode:1960RScI...31..235G. doi:10.1063/1.1716948.

[6] Walko, R. J.; Rochau, G. E. (1981). "A High Output Neutron Tube Using an Occluded Gas Ion Source". *IEEE Transactions on Nuclear Science* **28** (2): 1531–1534. Bibcode:1981ITNS...28.1531W. doi:10.1109/TNS.1981.4331459.

20.10   External links

- Chichester, D. L.; Simpson, J. D. (2003). "Compact Accelerator Neutron Generators" (pdf). *The Industrial Physicist* **9** (6): 22–25.

- Elizondo-Decanini, J. M.; Schmale, D.; Cich, M.; Martinez, M.; Youngman, K.; Senkow, M.; Kiff, S.; Steele, J.; Goeke, R.; Wroblewski, B.; Desko, J.; Dragt, A. J. (2012). "Novel Surface-Mounted Neutron Generator". *IEEE Transactions on Plasma Sciences* **40**(9):2145–2150.Bibcode:2012ITPS...40.2145E.doi:10.1109/TPS.2012.22048.

- "Sandia National Laboratories".

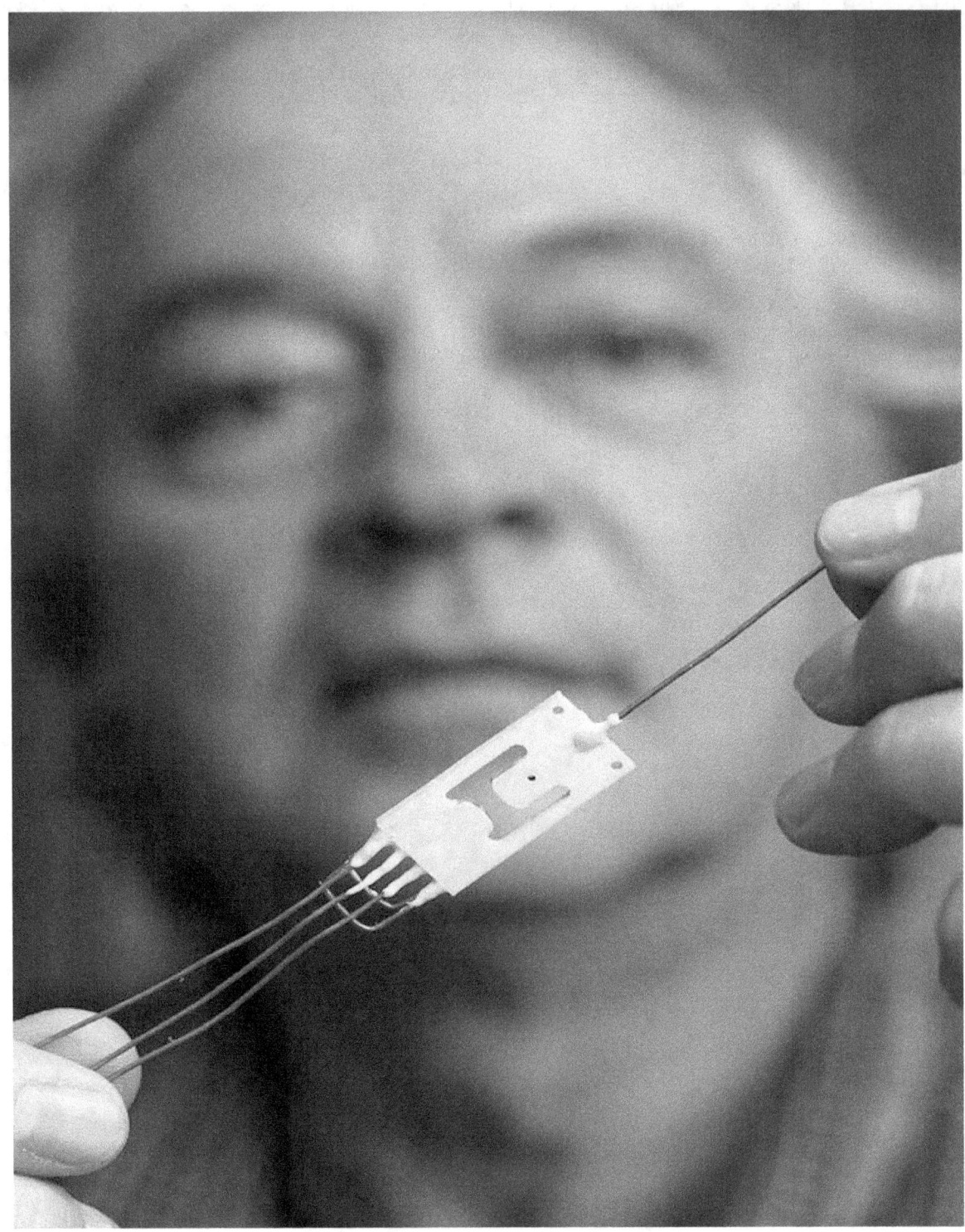

Neutristor in its simplest form as tested by the inventor at Sandia National Laboratories

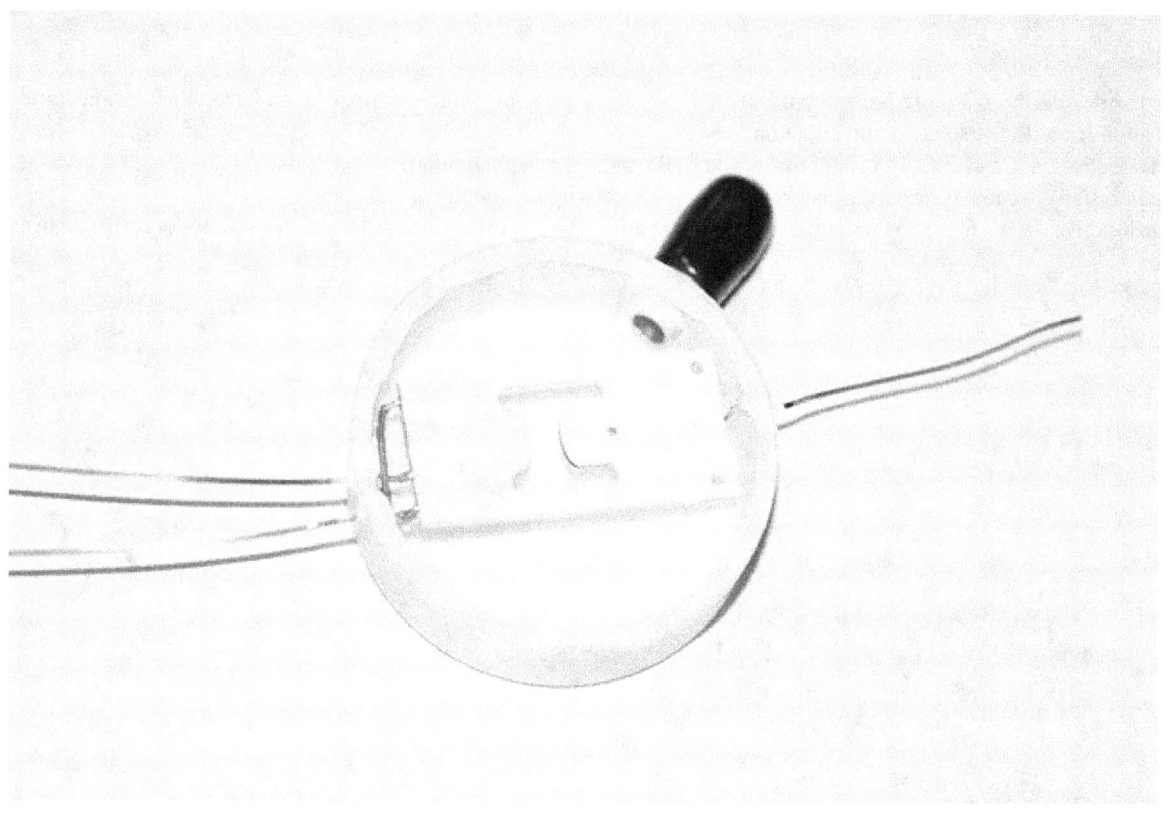

Neutristor in an inexpensive vacuum sealed package ready for testing

Chapter 21

Radioactive decay

For particle decay in a more general context, see Particle decay. For more information on hazards of various kinds of radiation from decay, see Ionizing radiation.

"Radioactive" redirects here. For other uses, see Radioactive (disambiguation).

"Radioactivity" redirects here. For other uses, see Radioactivity (disambiguation).

Radioactive decay, also known as **nuclear decay** or **radioactivity**, is the process by which a nucleus of an unstable atom

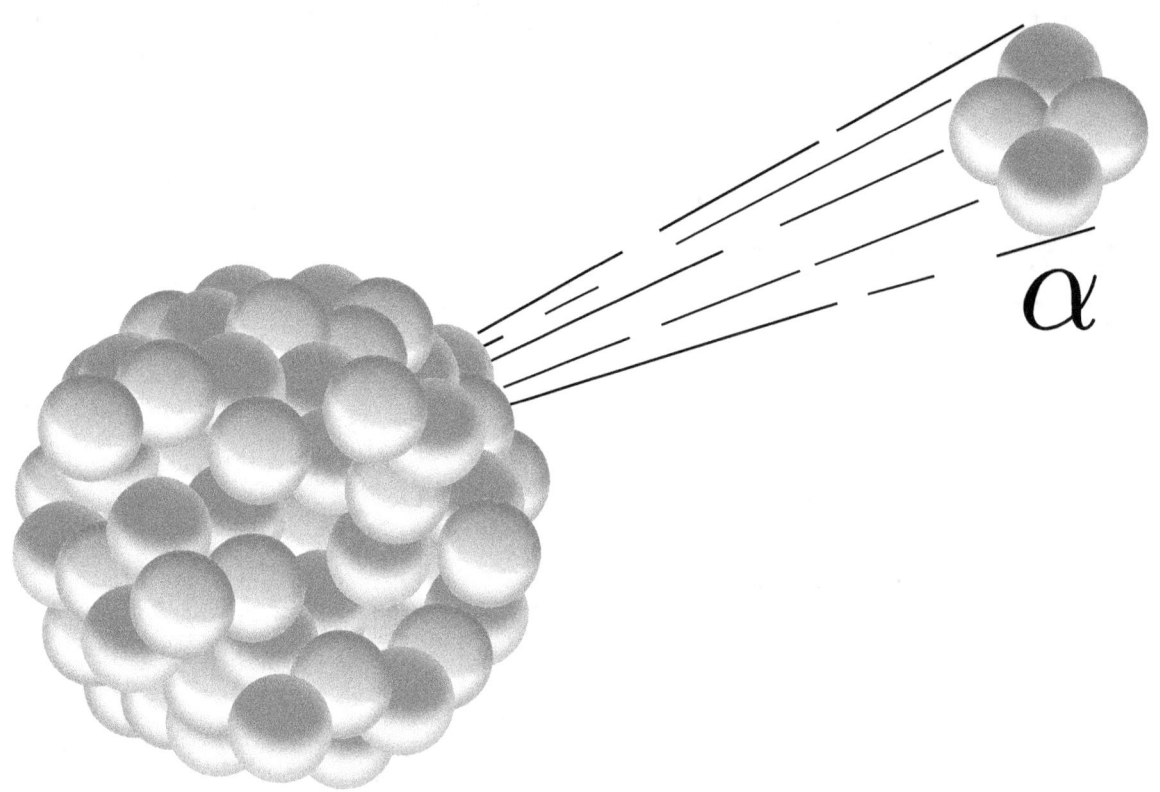

Alpha decay is one type of radioactive decay, in which an atomic nucleus emits an alpha particle, and thereby transforms (or "decays") into an atom with a mass number decreased by 4 and atomic number decreased by 2.

loses energy by emitting radiation. A material that spontaneously emits such radiation — which includes alpha particles, beta particles, gamma rays and conversion electrons — is considered **radioactive**.

Radioactive decay is a stochastic (i.e. random) process at the level of single atoms, in that, according to quantum theory, it is impossible to predict when a particular atom will decay.[1][2][3][4] The chance that a given atom will decay never changes, that is, it does not matter how long the atom has existed. For a large collection of atoms however, the decay rate for that collection can be calculated from their measured decay constants or half-lives. This is the basis of radiometric dating. The half-lives of radioactive atoms have no known limits for shortness or length of duration, and range over 55 orders of magnitude in time.

There are many types of radioactive decay (see table below). A decay, or loss of energy, results when an atom with one type of nucleus, called the *parent radionuclide* (or *parent radioisotope*[note 1]), transforms into an atom with a nucleus in a different state, or with a nucleus containing a different number of protons and neutrons. The product is called the *daughter nuclide*. In some decays, the parent and the daughter nuclides are different chemical elements, and thus the decay process results in the creation of an atom of a different element. This is known as a nuclear transmutation.

The first decay processes to be discovered were alpha decay, beta decay, and gamma decay. Alpha decay occurs when the nucleus ejects an alpha particle (helium nucleus). This is the most common process of emitting nucleons, but in rarer types of decays, nuclei can eject protons, or in the case of cluster decay specific nuclei of other elements. Beta decay occurs when the nucleus emits an electron or positron and a neutrino, in a process that changes a proton to a neutron or the other way about. The nucleus may capture an orbiting electron, causing a proton to convert into a neutron in a process called electron capture. All of these processes result in a nuclear transmutation.

By contrast, there are radioactive decay processes that do not result in a nuclear transmutation. The energy of an excited nucleus may be emitted as a gamma ray in a process called gamma decay, or be used to eject an orbital electron by its interaction with the excited nucleus, in a process called internal conversion. Highly excited neutron-rich nuclei, formed as the product of other types of decay, occasionally lose energy by way of neutron emission, resulting in a change of an element from one isotope to another. Another type of radioactive decay results in products that are not defined, but appear in a range of "pieces" of the original nucleus. This decay, called spontaneous fission, happens when a large unstable nucleus spontaneously splits into two (and occasionally three) smaller daughter nuclei, and generally leads to the emission of gamma rays, neutrons, or other particles from those products.

For a summary table showing the number of stable and radioactive nuclides in each category, see radionuclide. There exist twenty-nine chemical elements on Earth that are radioactive. They are those that contain thirty-four radionuclides that date before the time of formation of the solar system. Well-known examples are uranium and thorium, but also included are naturally occurring long-lived radioisotopes such as potassium-40. Another fifty or so shorter-lived radionuclides, such as radium and radon, found on Earth, are the products of decay chains that began with the primordial nuclides, and ongoing cosmogenic processes, such as the production of carbon-14 from nitrogen-14 by cosmic rays. Radionuclides may also be produced artificially in particle accelerators or nuclear reactors, resulting in 650 of these with half-lives of over an hour, and several thousand more with even shorter half-lives. See this list of nuclides for a list by half life.

21.1 History of discovery

Radioactivity was discovered in 1896 by the French scientist Henri Becquerel, while working with phosphorescent materials.[5] These materials glow in the dark after exposure to light, and he suspected that the glow produced in cathode ray tubes by X-rays might be associated with phosphorescence. He wrapped a photographic plate in black paper and placed various phosphorescent salts on it. All results were negative until he used uranium salts. The uranium salts caused a blackening of the plate in spite of the plate being wrapped in black paper. These radiations were given the name "Becquerel Rays".

It soon became clear that the blackening of the plate had nothing to do with phosphorescence, as the blackening was also produced by non-phosphorescent salts of uranium and metallic uranium. It became clear from these experiments that there was a form of invisible radiation that could pass through paper and was causing the plate to react as if exposed to light.

At first, it seemed as though the new radiation was similar to the then recently discovered X-rays. Further research by Becquerel, Ernest Rutherford, Paul Villard, Pierre Curie, Marie Curie, and others showed that this form of radioactivity was significantly more complicated. Rutherford was the first to realize that all such elements decay in accordance with the same mathematical exponential formula. Rutherford and his student Frederick Soddy were the first to realize that many decay processes resulted in the transmutation of one element to another. Subsequently, the radioactive displacement law

Pierre and Marie Curie in their Paris laboratory, before 1907

of Fajans and Soddy was formulated to describe the products of alpha and beta decay.[6][7]

The early researchers also discovered that many other chemical elements, besides uranium, have radioactive isotopes. A systematic search for the total radioactivity in uranium ores also guided Pierre and Marie Curie to isolate two new elements: polonium and radium. Except for the radioactivity of radium, the chemical similarity of radium to barium made these two elements difficult to distinguish.

21.2   Early health dangers

The dangers of ionizing radiation due to radioactivity and X-rays were not immediately recognized.

21.2.1   X-rays

The discovery of x-rays by Wilhelm Röntgen in 1895 led to widespread experimentation by scientists, physicians, and inventors. Many people began recounting stories of burns, hair loss and worse in technical journals as early as 1896. In February of that year, Professor Daniel and Dr. Dudley of Vanderbilt University performed an experiment involving X-raying Dudley's head that resulted in his hair loss. A report by Dr. H.D. Hawks, of his suffering severe hand and chest burns in an X-ray demonstration, was the first of many other reports in *Electrical Review*.[8]

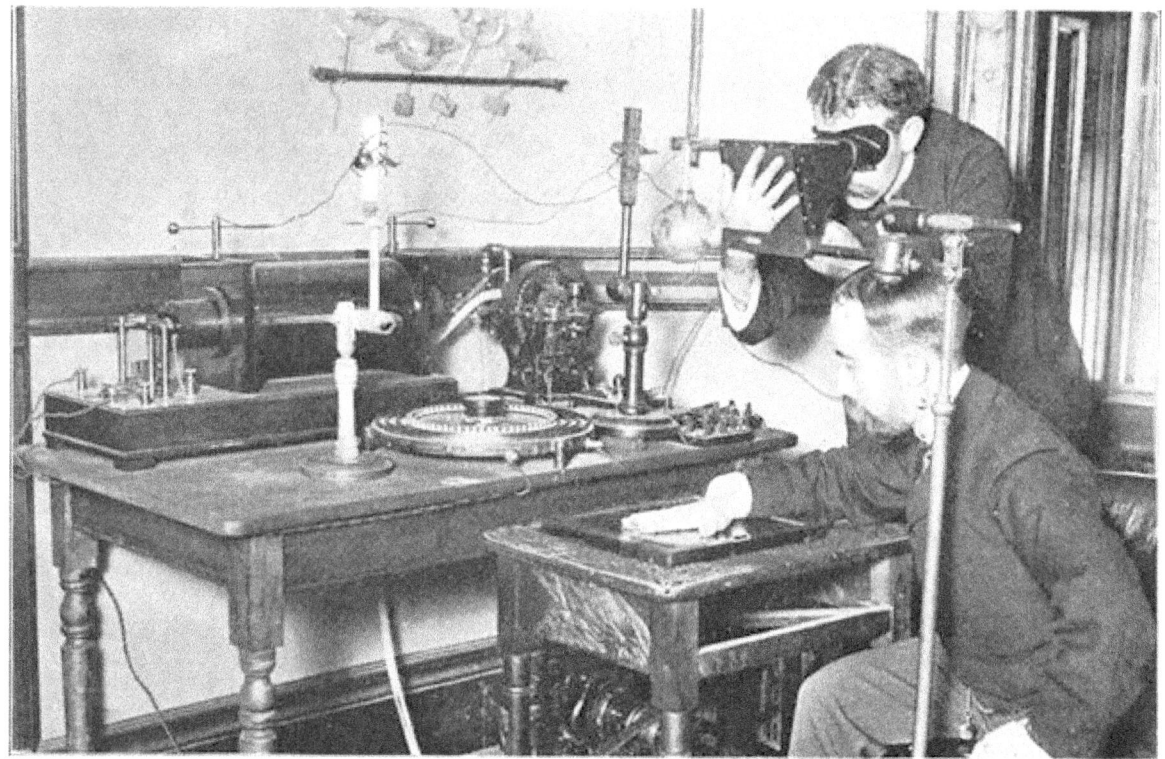

Taking an X-ray image with early Crookes tube apparatus in 1896. The Crookes tube is visible in the centre. The standing man is viewing his hand with a fluoroscope screen; this was a common way of setting up the tube. No precautions against radiation exposure are being taken; its hazards were not known at the time.

Other experimenters including Elihu Thomson, and Nikola Tesla also reported burns. Thomson deliberately exposed a finger to an X-ray tube over a period of time and suffered pain, swelling, and blistering.[9] Other effects, including ultraviolet rays and ozone were sometimes blamed for the damage,[10] and many physicians still claimed that there were no effects from X-ray exposure at all.[9]

Despite this, there were some early systematic hazard investigations, and as early as 1902 William Herbert Rollins wrote almost despairingly that his warnings about the dangers involved in careless use of X-rays was not being heeded, either by industry or by his colleagues. By this time Rollins had proved that X-rays could kill experimental animals, could cause a pregnant guinea pig to abort, and that they could kill a fetus.[11] He also stressed that "animals vary in susceptibility to the external action of X-light" and warned that these differences be considered when patients were treated by means of X-rays.

21.2.2   Radioactive substances

However, the biological effects of radiation due to radioactive substances were less easy to gauge. This gave the opportunity for many physicians and corporations to market radioactive substances as patent medicines. Examples were radium enema treatments, and radium-containing waters to be drunk as tonics. Marie Curie protested against this sort of treatment, warning that the effects of radiation on the human body were not well understood. Curie later died from aplastic anaemia, likely caused by exposure to ionizing radiation. By the 1930s, after a number of cases of bone necrosis and death of radium treatment enthusiasts, radium-containing medicinal products had been largely removed from the market (radioactive quackery).

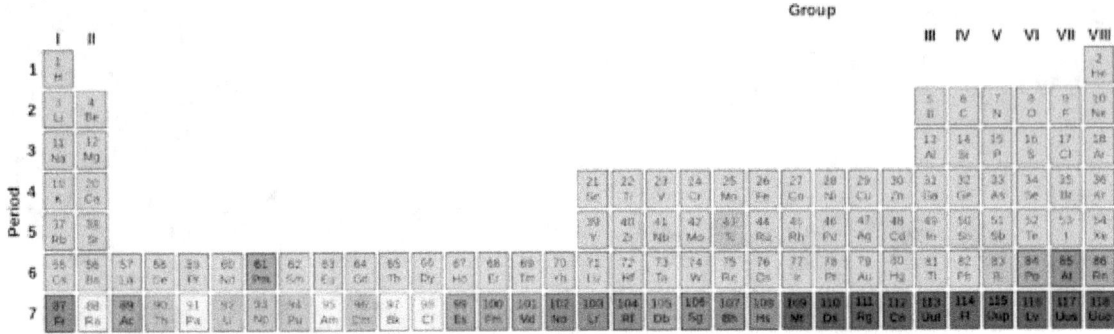

Radioactivity is characteristic of elements with large atomic number. Elements with at least one stable isotope are shown in light blue. Green shows elements whose most stable isotope has a half-life measured in millions of years. Yellow and orange are progressively more unstable, with half-lives in thousands or hundreds of years, down toward one day. Red and purple show highly and extremely radioactive elements where the most stable isotopes exhibit half-lives measured on the order of one day and much less.

21.2.3 Radiation protection

Main article: Radiation protection
See also: Sievert and Ionizing radiation

Only a year after Röntgen's discovery of X rays, the American engineer Wolfram Fuchs (1896) gave what is probably the first protection advice, but it was not until 1925 that the first International Congress of Radiology (ICR) was held and considered establishing international protection standards. The effects of radiation on genes, including the effect of cancer risk, were recognized much later. In 1927, Hermann Joseph Muller published research showing genetic effects and, in 1946, was awarded the Nobel prize for his findings.

The second ICR was held in Stockholm in 1928 and proposed the adoption of the rontgen unit, and the 'International X-ray and Radium Protection Committee' (IXRPC) was formed. Rolf Sievert was named Chairman, but a driving force was George Kaye of the British National Physical Laboratory. The committee met in 1931, 1934 and 1937.

After World War II the increased range and quantity of radioactive substances being handled as a result of military and civil nuclear programmes led to large groups of occupational workers and the public being potentially exposed to harmful levels of ionising radiation. This was considered at the first post-war ICR convened in London in 1950, when the present International Commission on Radiological Protection (ICRP) was born.[12] Since then the ICRP has developed the present international system of radiation protection, covering all aspects of radiation hazard.

21.3 Units of radioactivity

The International System of Units (SI) unit of radioactive activity is the becquerel (Bq), named in honour of the scientist Henri Becquerel. One Bq is defined as one transformation (or decay or disintegration) per second.

An older unit of radioactivity is the curie, Ci, which was originally defined as "the quantity or mass of radium emanation in equilibrium with one gram of radium (element)".[13] Today, the curie is defined as 3.7×10^{10} disintegrations per second, so that 1 curie (Ci) = 3.7×10^{10} Bq. For radiological protection purposes, although the United States Nuclear Regulatory Commission permits the use of the unit curie alongside SI units,[14] the European Union European units of measurement directives required that its use for "public health ... purposes" be phased out by 31 December 1985.[15]

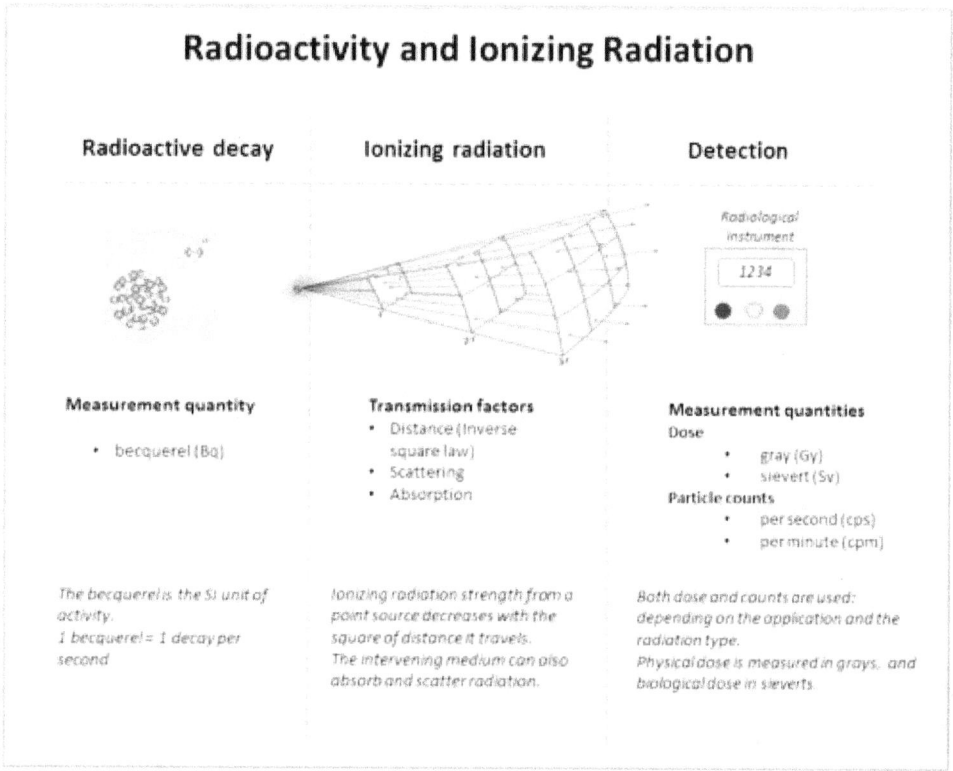

Graphic showing relationships between radioactivity and detected ionizing radiation

21.4   Types of decay

Early researchers found that an electric or magnetic field could split radioactive emissions into three types of beams. The rays were given the names alpha, beta, and gamma, in order of their ability to penetrate matter. While alpha decay was seen only in heavier elements of atomic number 52 (tellurium) and greater, the other two types of decay were produced by all of the elements. Lead, atomic number 82, is the heaviest element to have any isotopes stable (to the limit of measurement) to radioactive decay. Radioactive decay is seen in all isotopes of all elements of atomic number 83 (bismuth) or greater. Bismuth, however, is only very slightly radioactive.

In analysing the nature of the decay products, it was obvious from the direction of the electromagnetic forces applied to the radiations by external magnetic and electric fields that alpha particles carried a positive charge, beta particles carried a negative charge, and gamma rays were neutral. From the magnitude of deflection, it was clear that alpha particles were much more massive than beta particles. Passing alpha particles through a very thin glass window and trapping them in a discharge tube allowed researchers to study the emission spectrum of the captured particles, and ultimately proved that alpha particles are helium nuclei. Other experiments showed beta radiation, resulting from decay and cathode rays, were high-speed electrons. Likewise, gamma radiation and X-rays were found to be high-energy electromagnetic radiation.

The relationship between the types of decays also began to be examined: For example, gamma decay was almost always found to be associated with other types of decay, and occurred at about the same time, or afterwards. Gamma decay as a separate phenomenon, with its own half-life (now termed isomeric transition), was found in natural radioactivity to be a result of the gamma decay of excited metastable nuclear isomers, which were in turn created from other types of decay.

Although alpha, beta, and gamma radiations were most commonly found, other types of emission were eventually discovered. Shortly after the discovery of the positron in cosmic ray products, it was realized that the same process that

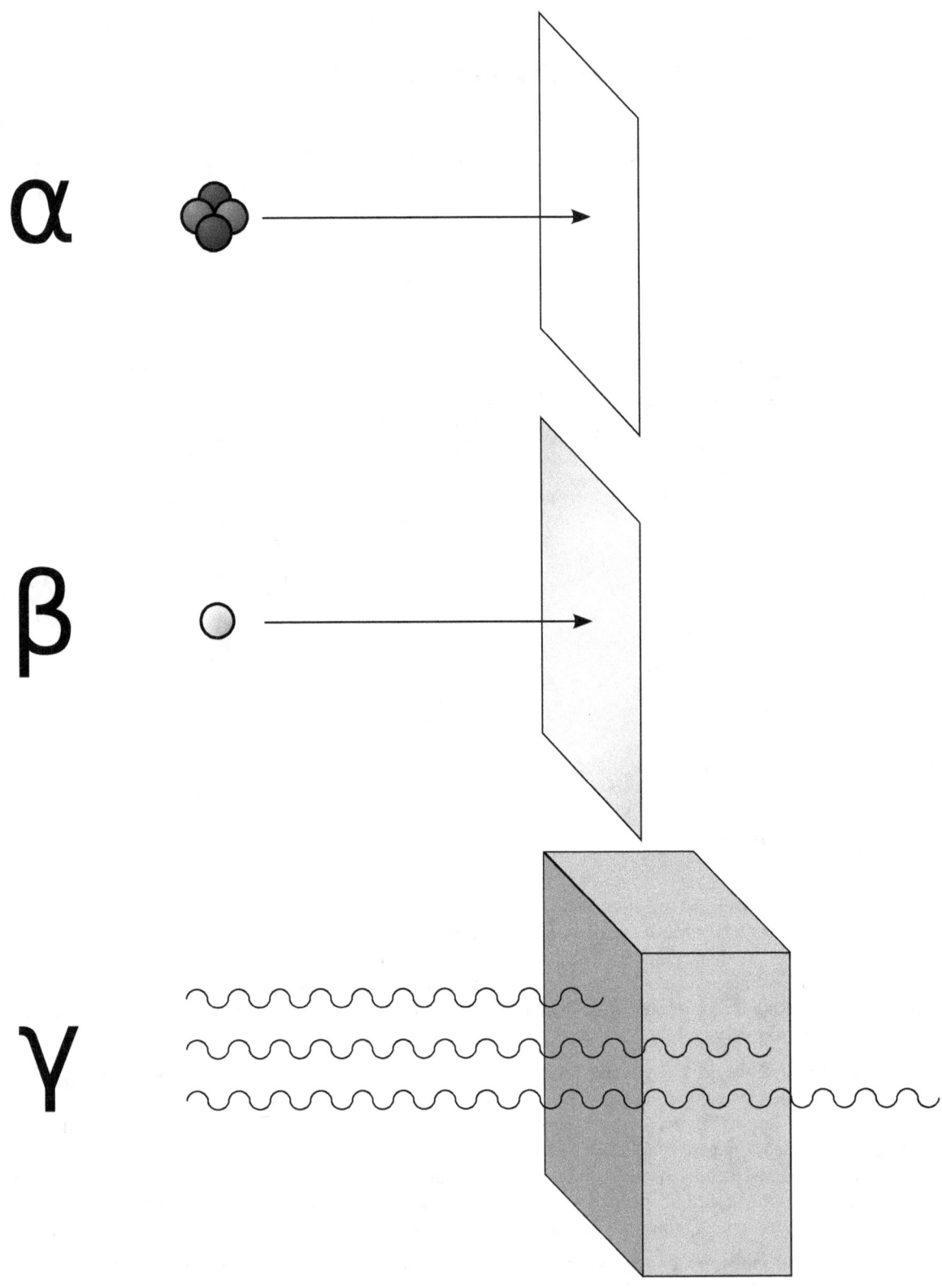

Alpha particles may be completely stopped by a sheet of paper, beta particles by aluminium shielding. Gamma rays can only be reduced by much more substantial mass, such as a very thick layer of lead.

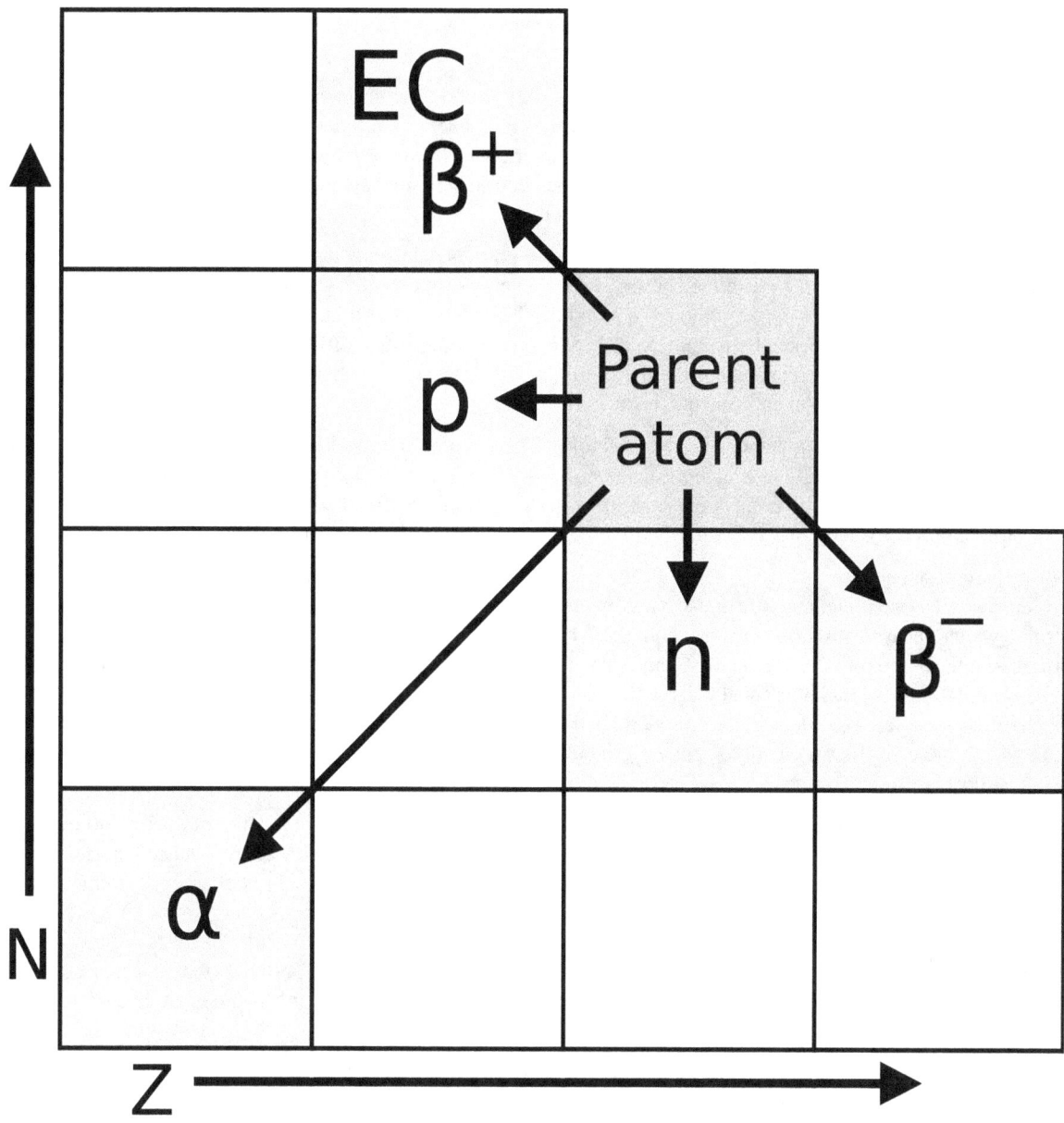

Transition diagram for decay modes of a radionuclide, with neutron number N *and atomic number* Z *(shown are* α, β^±, p^+, *and* n^0 *emissions, EC denotes electron capture).*

operates in classical beta decay can also produce positrons (positron emission). In an analogous process, instead of emitting positrons and neutrinos, some proton-rich nuclides were found to capture their own atomic electrons, a process called electron capture, and subsequently emit only a neutrino and usually also a gamma ray. Each of these types of decay involves the capture or emission of nuclear electrons or positrons, and acts to move a nucleus toward the ratio of neutrons to protons that has the least energy for a given total number of nucleons, consequently producing a more stable nucleus.

A theoretical process of positron capture, analogous to electron capture, is possible in antimatter atoms, but has not been observed as antimatter atoms are rarely available.[16] This would require antimatter atoms at least as complex as beryllium-7, which is the lightest known isotope of normal matter to undergo decay by electron capture.

Shortly after the discovery of the neutron in 1932, Enrico Fermi realized that certain rare beta-decay reactions immediately yield neutrons as a decay particle (neutron emission). Isolated proton emission was eventually observed in some elements. It was also found that some heavy elements may undergo spontaneous fission into products that vary in composition. In a phenomenon called cluster decay, specific combinations of neutrons and protons other than alpha particles (helium nuclei) were found to be spontaneously emitted from atoms.

Other types of radioactive decay of different mechanisms were found to emit previously-seen particles. An example is internal conversion, which results in electron and sometimes high-energy photon emission, although it involves neither beta nor gamma decay. A neutrino is not emitted, and neither the electron nor photon originate in the nucleus. Internal conversion decay, like isomeric transition gamma decay and neutron emission, involves the release of energy by an excited nuclide, without the transmutation of one element into another.

Rare events that involve a combination of two beta-decay type events happening simultaneously are known (see below). Any decay process that does not violate the conservation of energy or momentum laws (and perhaps other particle con-servation laws) is permitted to happen, although not all have been detected. An interesting example discussed in a final section, is bound state beta decay of rhenium-187. In this process, an inverse of electron capture, beta electron-decay of the parent nuclide is not accompanied by beta electron emission, because the beta particle has been captured into the K-shell of the emitting atom. An antineutrino, however, is emitted.

Radionuclides can undergo a number of different reactions. These are summarized in the following table. A nucleus with mass number A and atomic number Z is represented as (A, Z). The column "Daughter nucleus" indicates the difference between the new nucleus and the original nucleus. Thus, $(A - 1, Z)$ means that the mass number is one less than before, but the atomic number is the same as before.

If energy circumstances are favorable, a given radionuclide may undergo many competing types of decay, with some atoms decaying by one route, and others decaying by another. An example is copper-64, which has 29 protons, and 35 neutrons, which decays with a half-life of about 12.7 hours. This isotope has one unpaired proton and one unpaired neutron, so either the proton or the neutron can decay to the opposite particle. This particular nuclide (though not all nuclides in this situation) is almost equally likely to decay through proton decay, producing a positron emission (18%), or through electron capture (43%), as it does through neutron decay by electron emission (39%). The excited energy states resulting from these decays which fail to end in a ground energy state, also produce later internal conversion and gamma decay in almost 0.5% of the time.

Radioactive decay results in a reduction of summed rest mass, once the released energy (the *disintegration energy*) has escaped in some way. Although decay energy is sometimes defined as associated with the difference between the mass of the parent nuclide products and the mass of the decay products, this is true only of rest mass measurements, where some energy has been removed from the product system. This is true because the decay energy must always carry mass with it, wherever it appears (see mass in special relativity) according to the formula $E = mc^2$. The decay energy is initially released as the energy of emitted photons plus the kinetic energy of massive emitted particles (that is, particles that have rest mass). If these particles come to thermal equilibrium with their surroundings and photons are absorbed, then the decay energy is transformed to thermal energy, which retains its mass.

Decay energy therefore remains associated with a certain measure of mass of the decay system, called invariant mass, which does not change during the decay, even though the energy of decay is distributed among decay particles. The energy of photons, the kinetic energy of emitted particles, and, later, the thermal energy of the surrounding matter, all contribute to the invariant mass of the system. Thus, while the sum of the rest masses of the particles is not conserved in radioactive decay, the *system* mass and system invariant mass (and also the system total energy) is conserved throughout any decay process. This is a restatement of the equivalent laws of conservation of energy and conservation of mass.

21.5 Radioactive decay rates

The *decay rate*, or *activity*, of a radioactive substance is characterized by:

Constant quantities:

- The *half-life*—$t_1/2$, is the time taken for the activity of a given amount of a radioactive substance to decay to half of its initial value; see List of nuclides.

- The *decay constant*— λ, "lambda" the inverse of the mean lifetime, sometimes referred to as simply *decay rate*.

- The *mean lifetime*— τ, "tau" the average lifetime of a radioactive particle before decay.

Although these are constants, they are associated with the statistical behavior of populations of atoms. In consequence, predictions using these constants are less accurate for minuscule samples of atoms.

In principle a half-life, a third-life, or even a $(1/\sqrt{2})$-life, can be used in exactly the same way as half-life; but the mean life and half-life $t_{1/2}$ have been adopted as standard times associated with exponential decay.

Time-variable quantities:

- *Total activity*— A, is the number of decays per unit time of a radioactive sample.

- *Number of particles*—N, is the total number of particles in the sample.

- *Specific activity*—SA, number of decays per unit time per amount of substance of the sample at time set to zero ($t = 0$). "Amount of substance" can be the mass, volume or moles of the initial sample.

These are related as follows:

$$t_{1/2} = \frac{\ln(2)}{\lambda} = \tau \ln(2)$$

$$A = -\frac{\mathrm{d}N}{\mathrm{d}t} = \lambda N$$

$$S_A a_0 = -\frac{\mathrm{d}N}{\mathrm{d}t}\bigg|_{t=0} = \lambda N_0$$

where N_0 is the initial amount of active substance — substance that has the same percentage of unstable particles as when the substance was formed.

21.6   Mathematics of radioactive decay

For the mathematical details of exponential decay in general context, see exponential decay.
For related derivations with some further details, see half-life.
For the analogous mathematics in 1st order chemical reactions, see Consecutive reactions.

21.6.1   Universal law of radioactive decay

Radioactivity is one very frequently given example of exponential decay. The law describes the statistical behaviour of a large number of nuclides, rather than individual atoms. In the following formalism, the number of nuclides or the nuclide population N, is of course a discrete variable (a natural number)—but for any physical sample N is so large that it can be treated as a continuous variable. Differential calculus is needed to set up differential equations for the modelling the behaviour of the nuclear decay.

The mathematics of radioactive decay depend on a key assumption that a nucleus of a radionuclide has no "memory" or way of translating its history into its present behavior. A nucleus does not "age" with the passage of time. Thus, the probability of its breaking down does not increase with time, but stays constant no matter how long the nucleus has existed. This constant probability may vary greatly between different types of nuclei, leading to the many different observed decay rates. However, whatever the probability is, it does not change. This is in marked contrast to complex objects which do show aging, such as automobiles and humans. These systems do have a chance of breakdown per unit of time, that increases from the moment they begin their existence.

One-decay process

Consider the case of a nuclide A that decays into another B by some process $A \to B$ (emission of other particles, like electron neutrinos ν
e and electrons e^- as in beta decay, are irrelevant in what follows). The decay of an unstable nucleus is entirely random and it is impossible to predict when a particular atom will decay.[1] However, it is equally likely to decay at any instant in time. Therefore, given a sample of a particular radioisotope, the number of decay events $-dN$ expected to occur in a small interval of time dt is proportional to the number of atoms present N, that is[17]

$$-\frac{dN}{dt} \propto N.$$

Particular radionuclides decay at different rates, so each has its own decay constant λ. The expected decay $-dN/N$ is proportional to an increment of time, dt:

The negative sign indicates that N decreases as time increases, as the decay events follow one after another. The solution to this first-order differential equation is the function:

$$N(t) = N_0\, e^{-\lambda t} = N_0\, e^{-t/\tau},$$

where N_0 is the value of N at time $t = 0$.[17]

We have for all time t:

$$N_A + N_B = N_{\text{total}} = N_{A0},$$

where N_{total} is the constant number of particles throughout the decay process, which is equal to the initial number of A nuclides since this is the initial substance.

If the number of non-decayed A nuclei is:

$$N_A = N_{A0} e^{-\lambda t}$$

then the number of nuclei of B, i.e. the number of decayed A nuclei, is

$$N_B = N_{A0} - N_A = N_{A0} - N_{A0} e^{-\lambda t} = N_{A0} \left(1 - e^{-\lambda t}\right).$$

The number of decays observed over a given interval obeys Poisson statistics. If the average number of decays is $<N>$, the probability of a given number of decays N is[17]

$$P(N) = \frac{\langle N \rangle^N \exp(-\langle N \rangle)}{N!}.$$

Chain-decay processes

Chain of two decays

Now consider the case of a chain of two decays: one nuclide A decaying into another B by one process, then B decaying into another C by a second process, i.e. $A \rightarrow B \rightarrow C$. The previous equation cannot be applied to the decay chain, but can be generalized as follows. Since A decays into B, *then* B decays into C, the activity of A adds to the total number of B nuclides in the present sample, *before* those B nuclides decay and reduce the number of nuclides leading to the later sample. In other words, the number of second generation nuclei B increases as a result of the first generation nuclei decay of A, and decreases as a result of its own decay into the third generation nuclei C.[18] The sum of these two terms gives the law for a decay chain for two nuclides:

$$\frac{dN_B}{dt} = -\lambda_B N_B + \lambda_A N_A.$$

The rate of change of NB, that is dNB/dt, is related to the changes in the amounts of A and B, NB can increase as B is produced from A and decrease as B produces C.

Re-writing using the previous results:

The subscripts simply refer to the respective nuclides, i.e. NA is the number of nuclides of type A, NA_0 is the initial number of nuclides of type A, λA is the decay constant for A - and similarly for nuclide B. Solving this equation for NB gives:

$$N_B = \frac{N_{A0}\lambda_A}{\lambda_B - \lambda_A} \left(e^{-\lambda_A t} - e^{-\lambda_B t} \right).$$

In the case where B is a stable nuclide ($\lambda B = 0$), this equation reduces to the previous solution:

$$\lim_{\lambda_B \to 0} \left[\frac{N_{A0}\lambda_A}{\lambda_B - \lambda_A} \left(e^{-\lambda_A t} - e^{-\lambda_B t} \right) \right] = \frac{N_{A0}\lambda_A}{0 - \lambda_A} \left(e^{-\lambda_A t} - 1 \right) = N_{A0} \left(1 - e^{-\lambda_A t} \right),$$

as shown above for one decay. The solution can be found by the integration factor method, where the integrating factor is $e^{\lambda_B t}$. This case is perhaps the most useful, since it can derive both the one-decay equation (above) and the equation for multi-decay chains (below) more directly.

Chain of any number of decays

For the general case of any number of consecutive decays in a decay chain, i.e. $A_1 \rightarrow A_2 \cdots \rightarrow Ai \cdots \rightarrow AD$, where D is the number of decays and i is a dummy index ($i = 1, 2, 3, ...D$), each nuclide population can be found in terms of the previous population. In this case $N_2 = 0$, $N_3 = 0,..., ND = 0$. Using the above result in a recursive form:

$$\frac{dN_j}{dt} = -\lambda_j N_j + \lambda_{j-1} N_{(j-1)0} e^{-\lambda_{j-1} t}.$$

The general solution to the recursive problem is given by ***Bateman's equations***:[19]

Alternative decay modes

In all of the above examples, the initial nuclide decays into only one product.[20] Consider the case of one initial nuclide that can decay into either of two products, that is $A \rightarrow B$ and $A \rightarrow C$ in parallel. For example, in a sample of potassium-40, 89.3% of the nuclei decay to calcium-40 and 10.7% to argon-40. We have for all time t:

$$N = N_A + N_B + N_C$$

which is constant, since the total number of nuclides remains constant. Differentiating with respect to time:

$$\frac{dN_A}{dt} = -\left(\frac{dN_B}{dt} + \frac{dN_C}{dt}\right)$$
$$-\lambda N_A = -N_A\left(\lambda_B + \lambda_C\right)$$

defining the *total decay constant* λ in terms of the sum of *partial decay constants* λB and λC:

$$\lambda = \lambda_B + \lambda_C.$$

Notice that

$$\frac{dN_A}{dt} < 0, \frac{dN_B}{dt} > 0, \frac{dN_C}{dt} > 0.$$

Solving this equation for NA:

$$N_A = N_{A0}e^{-\lambda t}.$$

where NA_0 is the initial number of nuclide A. When measuring the production of one nuclide, one can only observe the total decay constant λ. The decay constants λB and λC determine the probability for the decay to result in products B or C as follows:

$$N_B = \frac{\lambda_B}{\lambda}N_{A0}\left(1 - e^{-\lambda t}\right),$$

$$N_C = \frac{\lambda_C}{\lambda}N_{A0}\left(1 - e^{-\lambda t}\right).$$

because the fraction $\lambda B/\lambda$ of nuclei decay into B while the fraction $\lambda C/\lambda$ of nuclei decay into C.

21.6.2 Corollaries of the decay laws

The above equations can also be written using quantities related to the number of nuclide particles N in a sample;

- The activity: $A = \lambda N$.

- The amount of substance: $n = N/L$.

- The mass: $M = Arn = ArN/L$.

where $L = 6.022 \times 10^{23}$ is Avogadro's constant, Ar is the relative atomic mass number, and the amount of the substance is in moles.

21.6.3   Decay timing: definitions and relations

Time constant and mean-life

For the one-decay solution $A \to B$:

$$N = N_0 \, e^{-\lambda t} = N_0 \, e^{-t/\tau},$$

the equation indicates that the decay constant λ has units of t^{-1}, and can thus also be represented as $1/\tau$, where τ is a characteristic time of the process called the *time constant*.

In a radioactive decay process, this time constant is also the mean lifetime for decaying atoms. Each atom "lives" for a finite amount of time before it decays, and it may be shown that this mean lifetime is the arithmetic mean of all the atoms' lifetimes, and that it is τ, which again is related to the decay constant as follows:

$$\tau = \frac{1}{\lambda}.$$

This form is also true for two-decay processes simultaneously $A \to B + C$, inserting the equivalent values of decay constants (as given above)

$$\lambda = \lambda_B + \lambda_C$$

into the decay solution leads to:

$$\frac{1}{\tau} = \lambda = \lambda_B + \lambda_C = \frac{1}{\tau_B} + \frac{1}{\tau_C}$$

Half-life

A more commonly used parameter is the half-life. Given a sample of a particular radionuclide, the half-life is the time taken for half the radionuclide's atoms to decay. For the case of one-decay nuclear reactions:

$$N = N_0 \, e^{-\lambda t} = N_0 \, e^{-t/\tau},$$

the half-life is related to the decay constant as follows: set $N = N_0/2$ and $t = T_{1/2}$ to obtain

$$t_{1/2} = \frac{\ln 2}{\lambda} = \tau \ln 2.$$

This relationship between the half-life and the decay constant shows that highly radioactive substances are quickly spent, while those that radiate weakly endure longer. Half-lives of known radionuclides vary widely, from more than 10^{19} years, such as for the very nearly stable nuclide ^{209}Bi, to 10^{-23} seconds for highly unstable ones.

The factor of ln(2) in the above relations results from the fact that concept of "half-life" is merely a way of selecting a different base other than the natural base e for the lifetime expression. The time constant τ is the $e - 1$ -life, the time until only $1/e$ remains, about 36.8%, rather than the 50% in the half-life of a radionuclide. Thus, τ is longer than $t_{1/2}$. The following equation can be shown to be valid:

$$N(t) = N_0 \, e^{-t/\tau} = N_0 \, 2^{-t/t_{1/2}}.$$

Since radioactive decay is exponential with a constant probability, each process could as easily be described with a different constant time period that (for example) gave its "(1/3)-life" (how long until only 1/3 is left) or "(1/10)-life" (a time period until only 10% is left), and so on. Thus, the choice of τ and $t1/2$ for marker-times, are only for convenience, and from convention. They reflect a fundamental principle only in so much as they show that the *same proportion* of a given radioactive substance will decay, during any time-period that one chooses.

Mathematically, the nth life for the above situation would be found in the same way as above—by setting $N = N_0/n$, {{{1}}} and substituting into the decay solution to obtain

$$t_{1/n} = \frac{\ln n}{\lambda} = \tau \ln n.$$

21.6.4 Example

A sample of ^{14}C has a half-life of 5,730 years and a decay rate of 14 disintegration per minute (dpm) per gram of natural carbon.

If an artifact is found to have radioactivity of 4 dpm per gram of its present C, we can find the approximate age of the object using the above equation:

$$N = N_0 \, e^{-t/\tau},$$

where: $\frac{N}{N_0} = 4/14 \approx 0.286,$

$$\tau = \frac{T_{1/2}}{\ln 2} \approx 8267$$

$$t = -\tau \ln \frac{N}{N_0} \approx 10360$$

21.7 Changing decay rates

The radioactive decay modes of electron capture and internal conversion are known to be slightly sensitive to chemical and environmental effects that change the electronic structure of the atom, which in turn affects the presence of **1s** and **2s** electrons that participate in the decay process. A small number of mostly light nuclides are affected. For example, chemical bonds can affect the rate of electron capture to a small degree (in general, less than 1%) depending on the proximity of electrons to the nucleus. In ^{7}Be, a difference of 0.9% has been observed between half-lives in metallic and insulating environments.[21] This relatively large effect is because beryllium is a small atom whose valence electrons are in **2s** atomic orbitals, which are subject to electron capture in ^{7}Be because (like all **s** atomic orbitals in all atoms) they naturally penetrate into the nucleus.

In 1992, Jung et al. of the Darmstadt Heavy-Ion Research group observed an accelerated β decay of ^{163}Dy^{66+}. Although neutral ^{163}Dy is a stable isotope, the fully ionized ^{163}Dy^{66+} undergoes β decay into the K and L shells with a half-life of 47 days.[22]

Rhenium-187 is another spectacular example. ^{187}Re normally beta decays to ^{187}Os with a half-life of 41.6×10^9 years,[23] but studies using fully ionised ^{187}Re atoms (bare nuclei) have found that this can decrease to only 33 years. This is attributed to "bound-state β$^-$ decay" of the fully ionised atom – the electron is emitted into the "K-shell" (**1s** atomic orbital), which cannot occur for neutral atoms in which all low-lying bound states are occupied.[24]

A number of experiments have found that decay rates of other modes of artificial and naturally occurring radioisotopes are, to a high degree of precision, unaffected by external conditions such as temperature, pressure, the chemical environment, and electric, magnetic, or gravitational fields.[25] Comparison of laboratory experiments over the last century, studies of the Oklo natural nuclear reactor (which exemplified the effects of thermal neutrons on nuclear decay), and astrophysical observations of the luminosity decays of distant supernovae (which occurred far away so the light has taken a great deal of time to reach us), for example, strongly indicate that unperturbed decay rates have been constant (at least to within the limitations of small experimental errors) as a function of time as well.

Recent results suggest the possibility that decay rates might have a weak dependence on environmental factors. It has been suggested that measurements of decay rates of silicon-32, manganese-54, and radium-226 exhibit small seasonal variations (of the order of 0.1%),[26][27][28] while the decay of Radon-222 exhibit large 4% peak-to-peak seasonal variations,[29] proposed to be related to either solar flare activity or distance from the Sun. However, such measurements are highly susceptible to systematic errors, and a subsequent paper[30] has found no evidence for such correlations in seven other isotopes (^{22}Na, ^{44}Ti, ^{108}Ag, ^{121}Sn, ^{133}Ba, ^{241}Am, ^{238}Pu), and sets upper limits on the size of any such effects.

21.8   Theoretical basis of decay phenomena

The neutrons and protons that constitute nuclei, as well as other particles that approach close enough to them, are governed by several interactions. The strong nuclear force, not observed at the familiar macroscopic scale, is the most powerful force over subatomic distances. The electrostatic force is almost always significant, and, in the case of beta decay, the weak nuclear force is also involved.

The interplay of these forces produces a number of different phenomena in which energy may be released by rearrangement of particles in the nucleus, or else the change of one type of particle into others. These rearrangements and transformations may be hindered energetically, so that they do not occur immediately. In certain cases, random quantum vacuum fluctuations are theorized to promote relaxation to a lower energy state (the "decay") in a phenomenon known as quantum tunneling. Radioactive decay half-life of nuclides has been measured over timescales of 55 orders of magnitude, from 2.3 x 10^{-23} seconds (for hydrogen-7) to 6.9 x 10^{31} seconds (for tellurium-128).[31] The limits of these timescales are set by the sensitivity of instrumentation only, and there are no known natural limits to how brief or long a decay half life for radioactive decay of a radionuclide may be.

The decay process, like all hindered energy transformations, may be analogized by a snowfield on a mountain. While friction between the ice crystals may be supporting the snow's weight, the system is inherently unstable with regard to a state of lower potential energy. A disturbance would thus facilitate the path to a state of greater entropy: The system will move towards the ground state, producing heat, and the total energy will be distributable over a larger number of quantum states. Thus, an avalanche results. The *total* energy does not change in this process, but, because of the second law of thermodynamics, avalanches have only been observed in one direction and that is toward the "ground state" — the state with the largest number of ways in which the available energy could be distributed.

Such a collapse (a *decay event*) requires a specific activation energy. For a snow avalanche, this energy comes as a disturbance from outside the system, although such disturbances can be arbitrarily small. In the case of an excited atomic nucleus, the arbitrarily small disturbance comes from quantum vacuum fluctuations. A radioactive nucleus (or any excited system in quantum mechanics) is unstable, and can, thus, *spontaneously* stabilize to a less-excited system. The resulting transformation alters the structure of the nucleus and results in the emission of either a photon or a high-velocity particle that has mass (such as an electron, alpha particle, or other type).

21.9   Occurrence and applications

According to the Big Bang theory, stable isotopes of the lightest five elements (H, He, and traces of Li, Be, and B) were produced very shortly after the emergence of the universe, in a process called Big Bang nucleosynthesis. These lightest stable nuclides (including deuterium) survive to today, but any radioactive isotopes of the light elements produced in the Big Bang (such as tritium) have long since decayed. Isotopes of elements heavier than boron were not produced at all in the Big Bang, and these first five elements do not have any long-lived radioisotopes. Thus, all radioactive nuclei are, therefore, relatively young with respect to the birth of the universe, having formed later in various other types of nucleosynthesis in stars (in particular, supernovae), and also during ongoing interactions between stable isotopes and energetic particles. For example, carbon-14, a radioactive nuclide with a half-life of only 5,730 years, is constantly produced in Earth's upper atmosphere due to interactions between cosmic rays and nitrogen.

Nuclides that are produced by radioactive decay are called radiogenic nuclides, whether they themselves are stable or not. There exist stable radiogenic nuclides that were formed from short-lived extinct radionuclides in the early solar system.[32][33] The extra presence of these stable radiogenic nuclides (such as Xe-129 from primordial I-129) against the background of primordial stable nuclides can be inferred by various means.

Radioactive decay has been put to use in the technique of radioisotopic labeling, which is used to track the passage of a chemical substance through a complex system (such as a living organism). A sample of the substance is synthesized with a high concentration of unstable atoms. The presence of the substance in one or another part of the system is determined by detecting the locations of decay events.

On the premise that radioactive decay is truly random (rather than merely chaotic), it has been used in hardware random-number generators. Because the process is not thought to vary significantly in mechanism over time, it is also a valuable tool in estimating the absolute ages of certain materials. For geological materials, the radioisotopes and some of their decay products become trapped when a rock solidifies, and can then later be used (subject to many well-known qualifications) to estimate the date of the solidification. These include checking the results of several simultaneous processes and their products against each other, within the same sample. In a similar fashion, and also subject to qualification, the rate of formation of carbon-14 in various eras, the date of formation of organic matter within a certain period related to the isotope's half-life may be estimated, because the carbon-14 becomes trapped when the organic matter grows and incorporates the new carbon-14 from the air. Thereafter, the amount of carbon-14 in organic matter decreases according to decay processes that may also be independently cross-checked by other means (such as checking the carbon-14 in individual tree rings, for example).

21.10   Origins of radioactive nuclides

Main article: nucleosynthesis

Radioactive primordial nuclides found in the Earth are residues from ancient supernova explosions which occurred before the formation of the solar system. They are the long-lived fraction of radionuclides surviving in the primordial solar nebula through planet accretion until the present. The naturally occurring short-lived radiogenic radionuclides found in rocks are the daughters of these radioactive primordial nuclides. Another minor source of naturally occurring radioactive nuclides are cosmogenic nuclides, formed by cosmic ray bombardment of material in the Earth's atmosphere or crust. The radioactive decay of these radionuclides in rocks within Earth's mantle and crust contribute significantly to Earth's internal heat budget.

21.11   Decay chains and multiple modes

The daughter nuclide of a decay event may also be unstable (radioactive). In this case, it will also decay, producing radiation. The resulting second daughter nuclide may also be radioactive. This can lead to a sequence of several decay events. Eventually, a stable nuclide is produced. This is called a *decay chain* (see this article for specific details of important natural decay chains).

An example is the natural decay chain of ^{238}U, which is as follows:

- decays, through alpha-emission, with a half-life of 4.5 billion years to thorium-234

- which decays, through beta-emission, with a half-life of 24 days to protactinium-234

- which decays, through beta-emission, with a half-life of 1.2 minutes to uranium-234

- which decays, through alpha-emission, with a half-life of 240 thousand years to thorium-230

- which decays, through alpha-emission, with a half-life of 77 thousand years to radium-226

- which decays, through alpha-emission, with a half-life of 1.6 thousand years to radon-222

- which decays, through alpha-emission, with a half-life of 3.8 days to polonium-218

- which decays, through alpha-emission, with a half-life of 3.1 minutes to lead-214

- which decays, through beta-emission, with a half-life of 27 minutes to bismuth-214

- which decays, through beta-emission, with a half-life of 20 minutes to polonium-214

- which decays, through alpha-emission, with a half-life of 160 microseconds to lead-210

- which decays, through beta-emission, with a half-life of 22 years to bismuth-210

- which decays, through beta-emission, with a half-life of 5 days to polonium-210

- which decays, through alpha-emission, with a half-life of 140 days to lead-206, which is a stable nuclide.

Some radionuclides may have several different paths of decay. For example, approximately 36% of bismuth-212 decays, through alpha-emission, to thallium-208 while approximately 64% of bismuth-212 decays, through beta-emission, to polonium-212. Both thallium-208 and polonium-212 are radioactive daughter products of bismuth-212, and both decay directly to stable lead-208.

21.12   Associated hazard warning signs

- The trefoil symbol used to indicate ionising radiation.

- 2007 ISO radioactivity danger symbol intended for IAEA Category 1, 2 and 3 sources defined as dangerous sources capable of death or serious injury.[1]

- The dangerous goods transport classification sign for radioactive materials

1. ^ IAEA news release Feb 2007

21.13   See also

- Actinides in the environment

- Background radiation

- Chernobyl disaster

- Crimes involving radioactive substances

- Decay chain

- Fallout shelter

- Half-life

- Lists of nuclear disasters and radioactive incidents

- National Council on Radiation Protection and Measurements

- Nuclear engineering
- Nuclear medicine
- Nuclear pharmacy
- Nuclear physics
- Nuclear power
- Particle decay
- Poisson process
- Radiation
- Radiation therapy
- Radioactive contamination
- Radioactivity in biology
- Radiometric dating
- Radionuclide a.k.a. "radio-isotope"
- Secular equilibrium
- Transient equilibrium

21.14 Notes

[1] Radionuclide is the more correct term, but radioisotope is also used. The difference between isotope and nuclide is explained at Isotope#Isotope vs. nuclide.

21.15 References

21.15.1 Inline

[1] "Decay and Half Life". Retrieved 2009-12-14.

[2] Stabin, Michael G. (2007). "3". *Radiation Protection and Dosimetry: An Introduction to Health Physics*. Springer. doi:10.1007/978-0-387-49983-3. ISBN 978-0387499826.

[3] Best, Lara; Rodrigues, George; Velker, Vikram (2013). "1.3". *Radiation Oncology Primer and Review*. Demos Medical Publishing. ISBN 978-1620700044.

[4] Loveland, W.; Morrissey, D.; Seaborg, G.T. (2006). *Modern Nuclear Chemistry*. Wiley-Interscience. p. 57. ISBN 0-471-11532-0.

[5] Mould, Richard F. (1995). *A century of X-rays and radioactivity in medicine : with emphasis on photographic records of the early years* (Reprint. with minor corr ed.). Bristol: Inst. of Physics Publ. p. 12. ISBN 9780750302241.

[6] Kasimir Fajans, "Radioactive transformations and the periodic system of the elements". Berichte der Deutschen Chemischen Gesellschaft, Nr. 46, 1913, p. 422–439

[7] Frederick Soddy, "The Radio Elements and the Periodic Law", Chem. News, Nr. 107, 1913, p.97–99

[8] Sansare, K.; Khanna, V.; Karjodkar, F. (2011). "Early victims of X-rays: a tribute and current perception". *Dentomaxillofacial Radiology* **40** (2): 123–125. doi:10.1259/dmfr/73488299. ISSN 0250-832X. PMC 3520298. PMID 21239576.

[9] Ronald L. Kathern and Paul L. Ziemer, he First Fifty Years of Radiation Protection, physics.isu.edu

[10] Hrabak, M.; Padovan, R. S.; Kralik, M.; Ozretic, D.; Potocki, K. (July 2008). "Nikola Tesla and the Discovery of X-rays". *RadioGraphics* **28** (4): 1189–92. doi:10.1148/rg.284075206. PMID 18635636.

[11] Geoff Meggitt (2008), *Taming the Rays - A history of Radiation and Protection.*, Lulu.com, ISBN 978-1-4092-4667-1

[12] Clarke, R.H.; J. Valentin (2009). "The History of ICRP and the Evolution of its Policies" (PDF). *Annals of the ICRP*. ICRP Publication 109 **39** (1): pp. 75–110. doi:10.1016/j.icrp.2009.07.009. Retrieved 12 May 2012.

[13] Rutherford, Ernest (6 October 1910). "Radium Standards and Nomenclature". *Nature* **84** (2136): 430–431.

[14] *10 CFR 20.1005*. US Nuclear Regulatory Commission. 2009.

[15] The Council of the European Communities (1979-12-21). "Council Directive 80/181/EEC of 20 December 1979 on the approximation of the laws of the Member States relating to Unit of measurement and on the repeal of Directive 71/354/EEC". Retrieved 19 May 2012.

[16] Radioactive Decay

[17] Patel, S.B. (2000). *Nuclear physics : an introduction*. New Delhi: New Age International. pp. 62–72. ISBN 9788122401257.

[18] Introductory Nuclear Physics, K.S. Krane, 1988, John Wiley & Sons Inc, ISBN 978-0-471-80553-3

[19] Cetnar, Jerzy (May 2006). "General solution of Bateman equations for nuclear transmutations". *Annals of Nuclear Energy* **33** (7): 640–645. doi:10.1016/j.anucene.2006.02.004.

[20] K.S. Krane (1988). *Introductory Nuclear Physics*. John Wiley & Sons Inc. p. 164. ISBN 978-0-471-80553-3.

[21] Wang, B.; Yan, S.; Limata, B. et al. (2006). "Change of the 7Be electron capture half-life in metallic environments". *The European Physical Journal A* **28** (3): 375–377. Bibcode:2006EPJA...28..375W. doi:10.1140/epja/i2006-10068-x. ISSN 1434-6001.

[22] Jung, M.; Bosch, F.; Beckert, K. et al. (1992). "First observation of bound-state β⁻ decay". *Physical Review Letters* **69** (15): 2164–2167. Bibcode:1992PhRvL..69.2164J. doi:10.1103/PhysRevLett.69.2164. ISSN 0031-9007. PMID 10046415.

[23] Smoliar, M.I.; Walker, R.J.; Morgan, J.W. (1996). "Re-Os ages of group IIA, IIIA, IVA, and IVB iron meteorites". *Science* **271** (5252): 1099–1102. Bibcode:1996Sci...271.1099S. doi:10.1126/science.271.5252.1099.

[24] Bosch, F.; Faestermann, T.; Friese, J.; Heine, F.; Kienle, P.; Wefers, E.; Zeitelhack, K.; Beckert, K.; Franzke, B.; Klepper, O.; Kozhuharov, C.; Menzel, G.; Moshammer, R.; Nolden, F.; Reich, H.; Schlitt, B.; Steck, M.; Stöhlker, T.; Winkler, T.; Takahashi, K. (1996). "Observation of bound-state β– decay of fully ionized ^{187}Re:^{187}Re-^{187}Os Cosmochronometry". *Physical Review Letters* **77** (26): 5190–5193. Bibcode:1996PhRvL..77.5190B. doi:10.1103/PhysRevLett.77.5190. PMID 10062738.

[25] Emery, G.T. (1972). "Perturbation of Nuclear Decay Rates" (PDF). *Annual Review of Nuclear Science* (ACS Publications) **22**: 165–202. Bibcode:1972ARNPS..22..165E. doi:10.1146/annurev.ns.22.120172.001121. Retrieved 6 August 2012.

[26] "The mystery of varying nuclear decay". *Physics World*. 2 October 2008.

[27] Jenkins, Jere H.; Fischbach, Ephraim (2009). "Perturbation of Nuclear Decay Rates During the Solar Flare of 13 December 2006".*Astroparticle Physics*31(6):407–411.arXiv:0808.3156.Bibcode:2009APh....31..407J.doi:10.1016/j.astropartphys.5.

[28] Jenkins, J. H.; Buncher, John B.; Gruenwald, John T.; Krause, Dennis E.; Mattes, Joshua J. et al. (2009). "Evidence of correlations between nuclear decay rates and Earth–Sun distance". *Astroparticle Physics* **32** (1): 42–46. arXiv:0808.3283. Bibcode:2009APh....32...42J. doi:10.1016/j.astropartphys.2009.05.004.

[29] Peter A. Sturrock, Gideon Steinitz, Ephraim Fischbach, Daniel Javorsek, II, Jere H. Jenkins, Analysis of Gamma Radiation from a Radon Source: Indications of a Solar Influence, Accessed on line September 2, 2012.

[30] Norman, E. B.; Shugart, Howard A.; Joshi, Tenzing H.; Firestone, Richard B. et al. (2009). "Evidence against correlations between nuclear decay rates and Earth–Sun distance" (PDF). *Astroparticle Physics* **31** (2): 135–137. arXiv:0810.3265. Bibcode:2009APh....31..135N. doi:10.1016/j.astropartphys.2008.12.004.

[31] NUBASE evaluation of nuclear and decay properties

[32] Clayton, Donald D. (1983). *Principles of Stellar Evolution and Nucleosynthesis* (2nd ed.). University of Chicago Press. p. 75. ISBN 0-226-10953-4.

[33] Bolt, B. A.; Packard, R. E.; Price, P. B. (2007). "John H. Reynolds, Physics: Berkeley". The University of California, Berkeley. Retrieved 2007-10-01.

21.15.2 General

- "Radioactivity", Encyclopædia Britannica. 2006. Encyclopædia Britannica Online. December 18, 2006

- Radio-activity by Ernest Rutherford Phd, Encyclopædia Britannica Eleventh Edition

21.16 External links

- The Lund/LBNL Nuclear Data Search – Contains tabulated information on radioactive decay types and energies.

- Nomenclature of nuclear chemistry

- Specific activity and related topics.

- The Live Chart of Nuclides – IAEA

- Health Physics Society Public Education Website

- Beach, Chandler B., ed. (1914). "Becquerel Rays". *The New Student's Reference Work*. Chicago: F. E. Compton and Co.

- Annotated bibliography for radioactivity from the Alsos Digital Library for Nuclear Issues

- Stochastic Java applet on the decay of radioactive atoms by Wolfgang Bauer

- Stochastic Flash simulation on the decay of radioactive atoms by David M. Harrison

- "Henri Becquerel: The Discovery of Radioactivity", Becquerel's 1896 articles online and analyzed on *BibNum* [click 'à télécharger' for English version].

- "Radioactive change", Rutherford & Soddy article (1903), online and analyzed on *Bibnum* [click 'à télécharger' for English version].

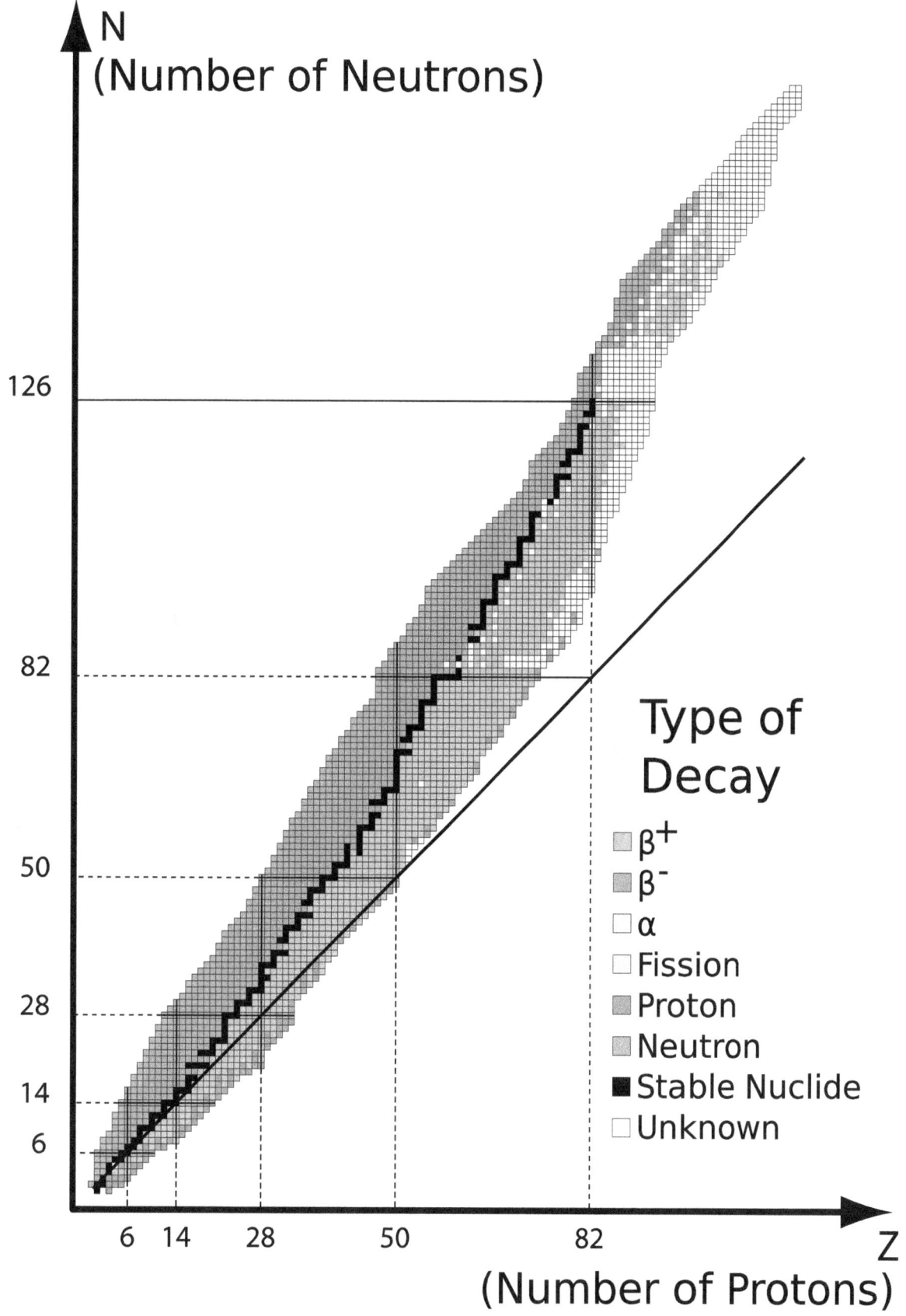

Types of radioactive decay related to N and Z numbers

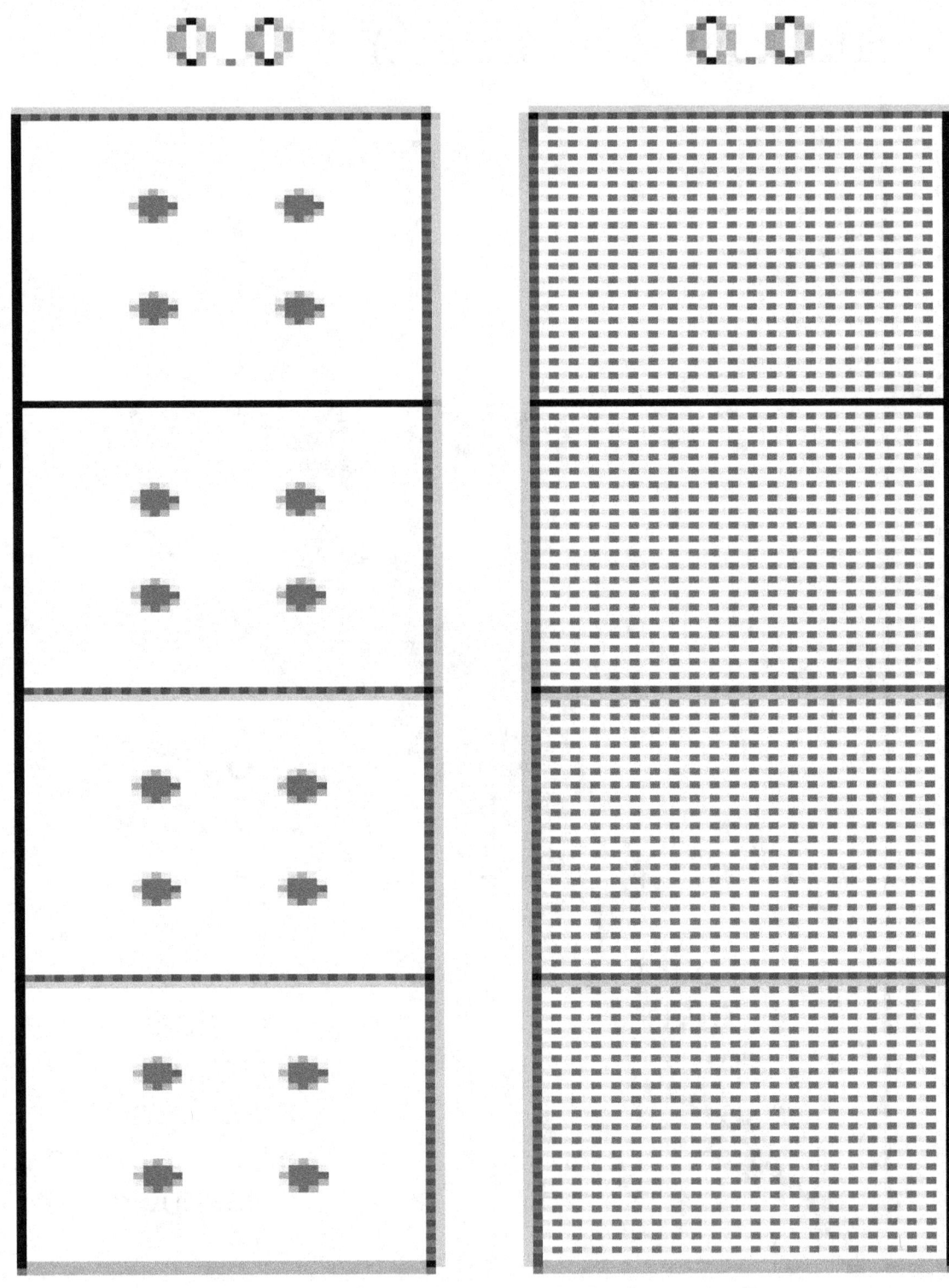

Alpha particles may be completely stopped by a sheet of paper, beta particles by aluminium shielding. Gamma rays can only be reduced by much more substantial mass, such as a very thick layer of lead.

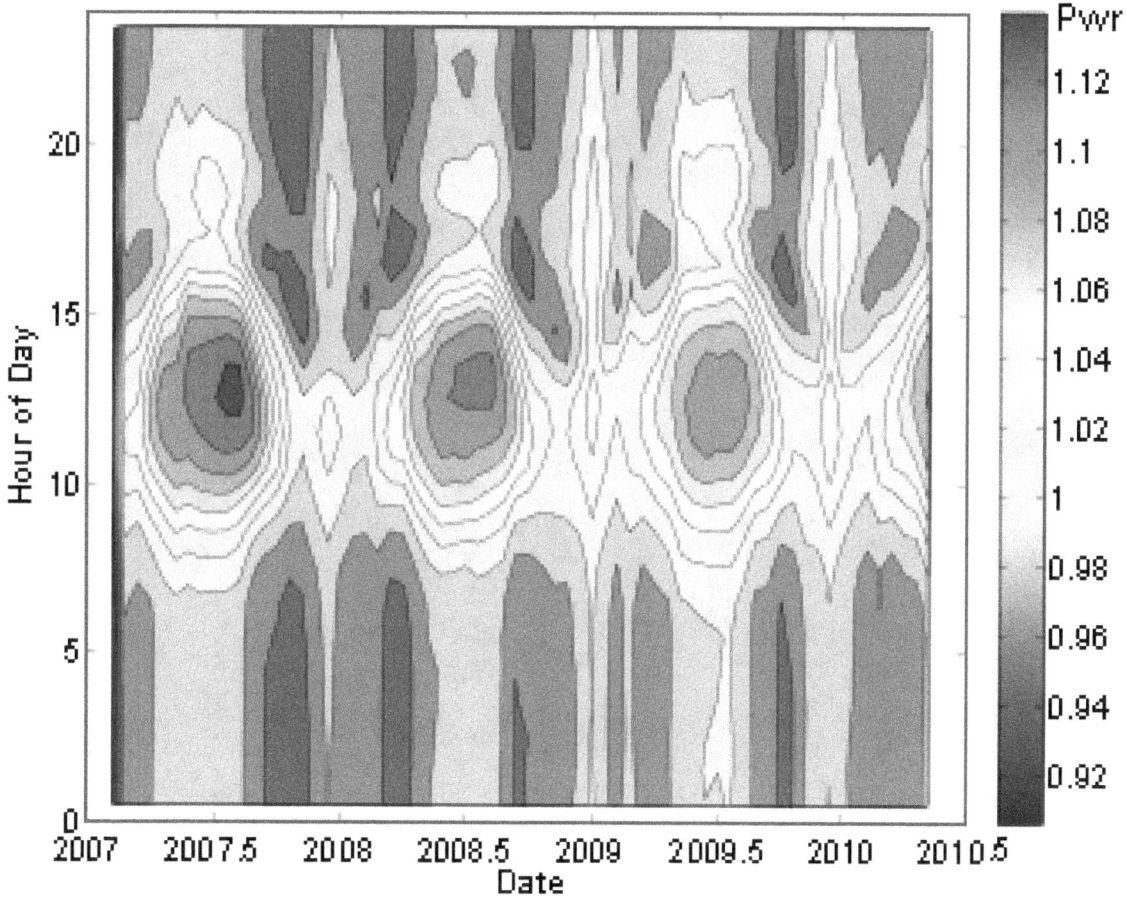

Decay Rate of Radon-222 as a function of date and time of day. The color-bar gives the power of the observed signal and represents ~4% seasonal decay rate variation.

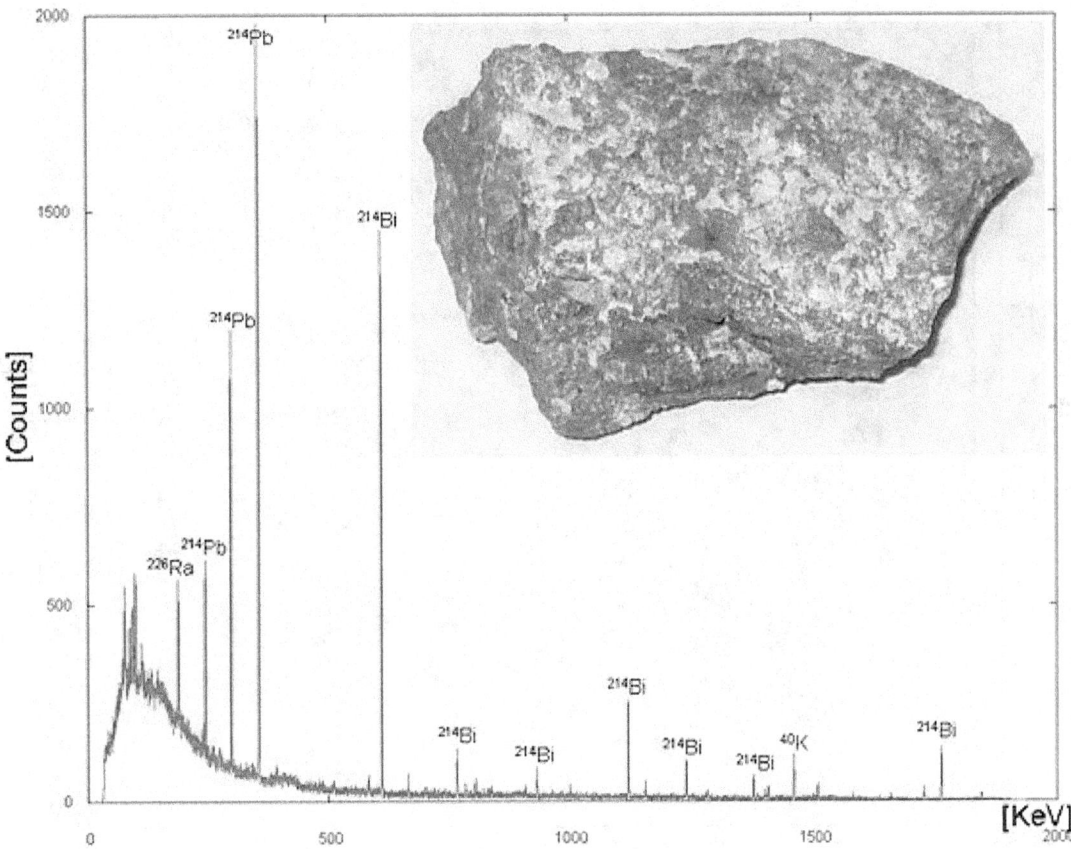

Gamma-ray energy spectrum of uranium ore (inset). Gamma-rays are emitted by decaying nuclides, and the gamma-ray energy can be used to characterize the decay (which nuclide is decaying to which). Here, using the gamma-ray spectrum, several nuclides that are typical of the decay chain of ^{238}U have been identified: ^{226}Ra, ^{214}Pb, ^{214}Bi.

Chapter 22

Neutron activation

Neutron activation is the process in which neutron radiation induces radioactivity in materials, and occurs when atomic nuclei capture free neutrons, becoming heavier and entering excited states. The excited nucleus often decays immediately by emitting gamma rays, or particles such as beta particles, alpha particles, fission products and neutrons (in nuclear fission). Thus, the process of neutron capture, even after any intermediate decay, often results in the formation of an unstable activation product. Such radioactive nuclei can exhibit half-lives ranging from small fractions of a second to many years.

Neutron activation is the only common way that a stable material can be induced into becoming intrinsically radioactive. All naturally-occurring materials, including air, water, and soil, can be induced (activated) by neutron capture into some amount of radioactivity in varying degrees, as a result of production of neutron-rich radioisotopes. Some atoms require more than one neutron to become unstable, which makes them harder to activate because the probability of a double or triple capture by a nucleus is below that of single capture. Water, for example, is made up of hydrogen and oxygen. Hydrogen requires a double capture to attain instability as hydrogen-3, tritium, while natural oxygen (oxygen-16) requires three captures to become unstable oxygen-19. Thus water is relatively difficult to activate, as compared to sea salt (NaCl), in which both the sodium and chlorine atoms become unstable with a single capture each. These facts were realized first-hand at the *Operation Crossroads* atomic test series in 1946.

22.1 Examples

Main article: neutron capture

An example of this kind of a nuclear reaction occurs in the production of cobalt-60 within a nuclear reactor:

$$^{59}_{27}\text{Co} + ^{1}_{0}\text{n} \rightarrow ^{60}_{27}\text{Co}$$

The cobalt-60 then decays by the emission of a beta particle plus gamma rays into nickel−60. This reaction has a half-life of about 5.27 years; and due to the availability of cobalt-59 (100% of its natural abundance) this neutron bombarded isotope of cobalt is a valuable source of nuclear radiation (namely gamma radiation) for radiotherapy.[1]

In other cases, and depending on the kinetic energy of the neutron, the capture of a neutron can cause nuclear fission— the splitting of the atomic nucleus into two smaller nuclei. If the fission requires an input of energy, that comes from the kinetic energy of the neutron. An example of this kind of fission in a light element can occur when the stable isotope of lithium, lithium-7, is bombarded with fast neutrons and undergoes the following nuclear reaction:

7
3Li + 1
0n → 4
2He + 3
1H + 1
0n + gamma rays + kinetic energy

In other words, the capture of a neutron by lithium-7 causes it to split into an energetic helium nucleus (alpha particle), a hydrogen-3 (tritium) nucleus and a free neutron. The Castle Bravo accident, in which the thermonuclear bomb test at Enewetak Atoll in 1954 exploded with 2.5 times the expected yield, was caused by the unexpectedly high probability of this reaction.

In the areas around a pressurized water reactors or boiling water reactors during normal operation, a significant amount of radiation is produced due to the fast neutron activation of coolant water oxygen via a (n,p) reaction. The activated oxygen-16 nucleus emits a proton (hydrogen nucleus), and transmutes to nitrogen-16, which has a very short life before decaying back to oxygen-16.[2]

16
8O + 1
0n → 1
1p + 16
7N (Decays rapidly)

16
7N → γ + e− → 16
8O

This activation of the coolant water requires extra biological shielding around the nuclear reactor plant. It is the high energy gamma ray in the second reaction that causes the major concern. This is why water that has recently been inside a nuclear reactor must be shielded until this radiation subsides. One to two minutes is generally sufficient.

22.2 Occurrence

Neutron activation is the only common way that a stable material can be induced into becoming intrinsically radioactive. Neutrons are only free in quantity in the microseconds of a nuclear weapon's explosion and in an active nuclear reactor.

In an atomic weapon neutrons are only generated for from 1 to 50 microseconds, but in huge numbers. Most are absorbed by the metallic bomb casing, which has not yet or only just started to be affected by the explosion within it. The neutron activation of the soon-to-be vaporized metal is responsible for a significant portion of the nuclear fallout in nuclear bursts high in the atmosphere. In other types of activation neutrons may irradiate soil that is dispersed in a mushroom cloud at or near the Earth's surface, resulting in fallout from activation of soil chemical elements.

22.3 Effects on materials over time

In any location with high neutron fluxes, such as within the cores of nuclear reactors, neutron activation contributes to material erosion; periodically the lining materials themselves must be disposed of, as low-level radioactive waste. Some materials are more subject to neutron activation than others, so a suitably chosen low-activation material can significantly reduce this problem (see International Fusion Materials Irradiation Facility). For example Chromium-51 will form by neutron activation in Chrome steel (which contains Cr-50) that is exposed to a typical reactor neutron flux.[3]

Carbon-14, most frequently but not solely, generated by the neutron activation of atmospheric nitrogen-14 with thermal neutron, is (together with its dominant natural production pathway from cosmic ray-air interactions and historical production from atmospheric nuclear testing) also generated in comparatively minute amounts inside many designs of nuclear reactors which contain nitrogen gas impurities in their fuel cladding, coolant water and by neutron activation of the oxygen contained in the water itself. Fast breeder reactors (FBR) produce about an order of magnitude less C-14 than the most common reactor type, the Pressurized water reactor, as FBRs do not use water as a primary coolant.[4]

22.4   Uses

22.4.1   Radiation safety

For physicians and radiation safety officers, activation of sodium in the human body to sodium-24, and phosphorus to phosphorus-32, can give a good immediate estimate of acute accidental neutron exposure.[5]

Neutron detection

One way to demonstrate that nuclear fusion has occurred inside a fusor device is to use a Geiger counter to measure the gamma ray radioactivity that is produced from a sheet of aluminum foil.

In the ICF fusion approach, the fusion yield of the experiment (directly proportional to neutron production) is usually determined by measuring the gamma-ray emissions of aluminum or copper neutron activation targets.[6] Aluminum can capture a neutron and generate radioactive sodium-24, which has a half life of 15 hours[7][8] and a beta decay energy of 5.514 MeV.[9]

The activation of a number of test target elements such as sulfur, copper, tantalum and gold have been used to determine the yield of both pure fission[10][11] and thermonuclear[12] weapons.

Materials analysis

Main article: neutron activation analysis

Neutron activation analysis is one of the most sensitive and accurate methods of trace element analysis. It requires no sample preparation or solubilization and can therefore be applied to objects that need to be kept intact such as a valuable piece of art. Although the activation induces radioactivity in the object, its level is typically low and its lifetime may be short, so that its effects soon disappear. In this sense, neutron activation is a non-destructive analysis method.

Neutron activation analysis can be done in situ. For example, aluminum (Al-27) can be activated by capturing relatively low-energy neutrons to produce the isotope Al-28, which decays with a half-life of 2.3 minutes with a decay energy of 4.642 MeV.[13] This activated isotope is used in oil drilling to determine the clay content (clay is generally an aluminosilicate) of the underground area under exploration.[14]

Historians can use accidental neutron activation to authenticate atomic artifacts and materials subjected to neutron fluxes from fission incidents. For example, one of the fairly unique isotopes found in trinitite, and therefore with its absence likely signifying a fake sample of the mineral, is a barium neutron activation product, the barium in the trinity device coming from the slow explosive lens employed in the device, known as Baratol.[15]

22.5   See also

- Induced radioactivity

- Neutron activation analysis

- Phosphorus-32 produced when sulfur captures a neutron.

- Table of nuclides

- Salted bomb

22.6 References

[1] Manual for reactor produced radioisotopes from the International Atomic Energy Agency

[2] Neeb, Karl Heinz (1997). *The Radiochemistry of Nuclear Power Plants with Light Water Reactors.* Berlin-New York: Walter de Gruyter. p. 227. ISBN 3-11-013242-7.

[3] http://ie.lbl.gov/toi/nuclide.asp?iZA=240051

[4] "IAEA Technical report series no.421, Management of Waste Containing Tritium and Carbon-14" (PDF).

[5] ORNL Report on determination of dose from criticality accidents

[6] Stephen Padalino, Heather Oliver and Joel Nyquist; LLE Collaborators: Vladimir Smalyukand, Nancy Rogers. "DT neutron yield measurements using neutron activation of aluminum".

[7] http://www.aanda.org/articles/aa/full/2001/10/aah2362/node4.html

[8] http://kubchemistry.weebly.com/uploads/6/9/8/7/6987088/chapter_22_nuclear_reactions.ppt

[9] http://www.site.uottawa.ca:4321/astronomy/index.html#sodium24

[10] Kerr, George D.; Young, Robert W.; Cullings, Harry M.; Christy, Robert F. (2005). "Bomb Parameters". In Robert W. Young, George D. Kerr. *Reassessment of the Atomic Bomb Radiation Dosimetry for Hiroshima and Nagasaki – Dosimetry System 2002* (PDF). The Radiation Effects Research Foundation. pp. 42–43.

[11] Malik, John (September 1985). "The Yields of the Hiroshima and Nagasaki Explosions" (PDF). Los Alamos National Laboratory. Retrieved March 9, 2014.

[12] US Army (1952). *Operation Ivy Final Report Joint Task Force 132* (PDF).

[13] http://www.site.uottawa.ca:4321/astronomy/index.html#aluminium28

[14] "Aluminum activation log".

[15] "Radioactivity in Trinitite six decades later. Journal of Environmental Radioactivity Volume 85, Issue 1, 2006, Pages 103–120".

22.7 External links

- Neutron Activation Analysis web

- Handbook on Nuclear Activation Cross-Sections, IAEA, 1974

- Decay Data in MIRD Format from the National Nuclear Data Center at Brookhaven National Laboratory

- neutron capture as it relates to nucleosynthesis

- neutron capture and the Chart of the nuclides

- the chart of the Nuclides

- Discovery of the Chromium isotopes, Chromium-55 by Cr-54 neutron capture

22.8 Further reading

- US Army (1952). *Operation Ivy Final Report Joint Task Force 132* (PDF).

Chapter 23

Neutron temperature

The **neutron detection temperature**, also called the **neutron energy**, indicates a free neutron's kinetic energy, usually given in electron volts. The term *temperature* is used, since hot, thermal and cold neutrons are moderated in a medium with a certain temperature. The neutron energy distribution is then adopted to the Maxwellian distribution known for thermal motion. Qualitatively, the higher the temperature, the higher the kinetic energy is of the free neutron. Kinetic energy, speed and wavelength of the neutron are related through the De Broglie relation.

23.1 Neutron energy distribution ranges

But different ranges with different names are observed in other sources. For example,

> Epithermal neutrons have energies between 1 eV and 10 keV and smaller nuclear cross sections than thermal neutrons.

H. Tomita, C. Shoda, J. Kawarabayashi, T. Matsumoto, J. Hori, S. Uno, M. Shoji, T. Uchida, N. Fukumotoa and T. Iguchia-"Development of epithermal neutron camera based on resonance-energy-filtered imaging with GEM" (2012)

The following is a detailed classification:

23.1.1 Thermal

- A **thermal neutron** is a free neutron with a kinetic energy of about 0.025 eV (about 4.0×10^{-21} J or 2.4 MJ/kg, hence a speed of 2.2 km/s), which is the energy corresponding to the most probable velocity at a temperature of 290 K (17 °C or 62 °F), the mode of the Maxwell–Boltzmann distribution for this temperature.

After a number of collisions with nuclei (scattering) in a medium (neutron moderator) at this temperature, neutrons arrive at about this energy level, provided that they are not absorbed.

Thermal neutrons have a different and often much larger effective neutron absorption cross-section for a given nuclide than fast neutrons, and can therefore often be absorbed more easily by an atomic nucleus, creating a heavier, often unstable isotope of the chemical element as a result (neutron activation).

23.1.2 Epithermal

- Neutrons of energy greater than thermal
- Greater than 0.2 eV

184

23.1.3 Cadmium

- Neutrons which are strongly absorbed by cadmium
- Less than 0.4 eV.

23.1.4 Epicadmium

- Neutrons which are not strongly absorbed by cadmium
- Greater than 0.6 eV.

23.1.5 Slow

- Neutrons of energy slightly greater than epicadmium neutrons.
- Less than 1 to 10 eV.

23.1.6 Resonance

- Refers to neutrons which are strongly susceptible non-fission capture by U-238.
- 1 eV to 300 eV

23.1.7 Intermediate

- Neutrons that are between slow and fast
- Few hundred eV to 0.5 MeV.

23.1.8 Fast

A **fast neutron** is a free neutron with a kinetic energy level close to 1 MeV (100 TJ/kg), hence a speed of 14,000 km/s, or higher. They are named *fast* neutrons to distinguish them from lower-energy thermal neutrons, and high-energy neutrons produced in cosmic showers or accelerators.

Fast neutrons are produced by nuclear processes:

- nuclear fission produces neutrons with a mean energy of 2 MeV (200 TJ/kg, i.e. 20,000 km/s), which qualifies as "fast". However the range of neutrons from fission follows a Maxwell–Boltzmann distribution from 0 to about 14 MeV in the center of momentum frame of the disintegration, and the mode of the energy is only 0.75 MeV, meaning that fewer than half of fission neutrons qualify as "fast" even by the 1 MeV criterion.[2]

- nuclear fusion: deuterium–tritium fusion produces neutrons of 14.1 MeV (1400 TJ/kg, i.e. 52,000 km/s, 17.3% of the speed of light) that can easily fission uranium-238 and other non-fissile actinides.

Fast neutrons can be made into thermal neutrons via a process called moderation. This is done with a neutron moderator. In reactors, typically heavy water, light water, or graphite are used to moderate neutrons.

23.1.9 Ultrafast

- Relativistic
- Greater than 20 MeV

23.1.10 Other classifications

Pile   • Neutrons of all energies present in nuclear reactors

• 0.001 eV to 15 MeV.

23.1.11 Ultracold neutrons (UCN)

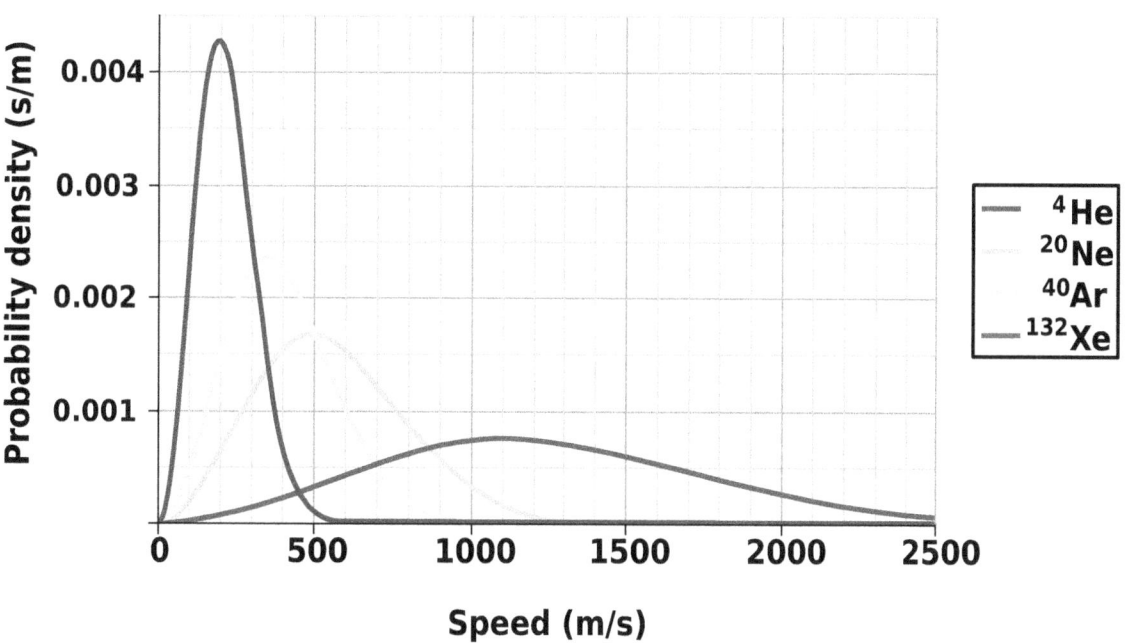

Maxwell-Boltzmann Molecular Speed Distribution for Noble Gases

A chart displaying the speed probability density functions of the speeds of a few noble gases at a temperature of 298.15 K (25 C). An explanation of the vertical axis label appears on the image page (click to see). Similar speed distributions are obtained for neutrons upon moderation.

Ultracold neutrons are free neutrons which can be stored in traps made from certain materials.

23.2   Fast reactor and thermal reactor compared

Most fission reactors are thermal reactors that use a neutron moderator to slow down (*"thermalize"*) the neutrons produced by nuclear fission. Moderation substantially increases the fission cross section for fissile nuclei such as uranium-235 or plutonium-239. In addition, uranium-238 has a much lower capture cross section for thermal neutrons, allowing more neutrons to cause fission of fissile nuclei and propagate the chain reaction, rather than being captured by ^{238}U. The combination of these effects allows light water reactors to use low-enriched uranium. Heavy water reactors and graphite-moderated reactors can even use natural uranium as these moderators have much lower neutron capture cross sections than light water.[3]

An increase in fuel temperature also raises U-238's thermal neutron absorption by Doppler broadening, providing negative feedback to help control the reactor. Also, when the moderator is also a circulating coolant (light water or heavy water), boiling of the coolant will reduce the moderator density and provide negative feedback (a negative void coefficient).

Intermediate-energy neutrons have poorer fission/capture ratios than either fast or thermal neutrons for most fuels. An exception is the uranium-233 of the thorium cycle, which has a good fission/capture ratio at all neutron energies.

Fast reactors use unmoderated fast neutrons to sustain the reaction and require the fuel to contain a higher concentration of fissile material relative to fertile material U-238. However, fast neutrons have a better fission/capture ratio for many nuclides, and each fast fission releases a larger number of neutrons, so a fast breeder reactor can potentially "breed" more fissile fuel than it consumes.

Fast reactor control cannot depend solely on Doppler broadening or on negative void coefficient from a moderator. However, thermal expansion of the fuel itself can provide quick negative feedback. Perennially expected to be the wave of the future, fast reactor development has been nearly dormant with only a handful of reactors built in the decades since the Chernobyl accident due to low prices in the uranium market, although there is now a revival with several Asian countries planning to complete larger prototype fast reactors in the next few years.

23.3 See also

- Absorption hardening

- Thermal-neutron reactor

- Fast-neutron reactor

- List of particles

- Neutron Detection

- Neutron source

- Nuclear reaction

- Thermal reactor

- Scintillator

23.4 References

[1] Carron, N.J. (2007). *An Introduction to the Passage of Energetic Particles Through Matter*. p. 308.

[2] Byrne, J. *Neutrons, Nuclei, and Matter*, Dover Publications, Mineola, New York, 2011, ISBN 978-0-486-48238-5 (pbk.) p. 259.

[3] Some Physics of Uranium. Accessed March 7, 2009

23.5 External links

- Language of the Nucleus

Chapter 24

Neutron diffraction

Neutron diffraction or **elastic neutron scattering** is the application of neutron scattering to the determination of the atomic and/or magnetic structure of a material. A sample to be examined is placed in a beam of thermal or cold neutrons to obtain a diffraction pattern that provides information of the structure of the material. The technique is similar to X-ray diffraction but due to their different scattering properties, neutrons and X-rays provide complementary information.

24.1 Instrumental and sample requirements

The technique requires a source of neutrons. Neutrons are usually produced in a nuclear reactor or spallation source. At a research reactor, other components are needed, including a crystal monochromators as well as filters to select the desired neutron wavelength. Some parts of the setup may also be movable. At a spallation source, the time of flight technique is used to sort the energies of the incident neutrons (higher energy neutrons are faster), so no monochromator is needed, but rather a series of aperture elements synchronized to filter neutron pulses with the desired wavelength.

The technique is most commonly performed as powder diffraction, which only requires a polycrystalline powder. For single crystal work, the crystals must be much larger than those used in X-ray crystallography. It is common to use crystals that are about 1 mm^3.[1]

Summarizing, the main disadvantage to neutron diffraction is the requirement for a nuclear reactor. For single crystal work, the technique requires relatively large crystals, which are usually challenging to grow. The main advantages to the technique are many - sensitivity to light atoms, ability to distinguish isotopes, absence of radiation damage.[1]

24.2 Nuclear scattering

Like all quantum particles, neutrons can exhibit wave phenomena typically associated with light or sound. Diffraction is one of these phenomena; it occurs when waves encounter obstacles whose size is comparable with the wavelength. If the wavelength of a quantum particle is short enough, atoms or their nuclei can serve as diffraction obstacles. When a beam of neutrons emanating from a reactor is slowed down and selected properly by their speed, their wavelength lies near one angstrom (0.1 nanometer), the typical separation between atoms in a solid material. Such a beam can then be used to perform a diffraction experiment. Impinging on a crystalline sample it will scatter under a limited number of well-defined angles according to the same Bragg's law that describes X-ray diffraction.

Neutrons and X-rays interact with matter differently. X-rays interact primarily with the electron cloud surrounding each atom. The contribution to the diffracted x-ray intensity is therefore larger for atoms with larger atomic number (Z). On the other hand, neutrons interact directly with the *nucleus* of the atom, and the contribution to the diffracted intensity depends on each isotope; for example, regular hydrogen and deuterium contribute differently. It is also often the case that light (low Z) atoms contribute strongly to the diffracted intensity even in the presence of large Z atoms. The scattering length varies from isotope to isotope rather than linearly with the atomic number. An element like vanadium is a strong scatterer of X-rays, but its nuclei hardly scatter neutrons, which is why it is often used as a container material. Non-magnetic neutron diffraction is directly sensitive to the positions of the nuclei of the atoms.

Unlike X-rays, neutrons scatter mostly from the nuclei of the atoms, which are tiny. Furthermore, there is no need for an atomic form factor to describe the shape of the electron cloud of the atom and the scattering power of an atom does not fall off with the scattering angle as it does for X-rays. Diffractograms therefore can show strong well defined diffraction peaks even at high angles, particularly if the experiment is done at low temperatures. Many neutron sources are equipped with liquid helium cooling systems that allow data collection at temperatures down to 4.2 K. The superb high angle (i.e. high *resolution*) information means that the atomic positions in the structure can be determined with high precision. On the other hand, Fourier maps (and to a lesser extent difference Fourier maps) derived from neutron data suffer from series termination errors, sometimes so much that the results are meaningless.

24.3   Magnetic scattering

Although neutrons are uncharged, they carry a spin, and therefore interact with magnetic moments, including those arising from the electron cloud around an atom. Neutron diffraction can therefore reveal the microscopic magnetic structure of a material.[2]

Magnetic scattering does require an atomic form factor as it is caused by the much larger electron cloud around the tiny nucleus. The intensity of the magnetic contribution to the diffraction peaks will therefore dwindle towards higher angles.

24.4   Uses

Neutron diffraction can be used to determine the static structure factor of gases, liquids or amorphous solids. Most experiments, however, aim at the structure of crystalline solids, making neutron diffraction an important tool of crystallography.

Neutron diffraction is closely related to X-ray powder diffraction.[3] In fact the single crystal version of the technique is less commonly used because currently available neutron sources require relatively large samples and large single crystals are hard or impossible to come by for most materials. Future developments, however, may well change this picture. Because the data is typically a 1D powder diffractogram they are usually processed using Rietveld refinement. In fact the latter found its origin in neutron diffraction (at Petten in the Netherlands) and was later extended for use in X-ray diffraction.

One practical application of elastic neutron scattering/diffraction is that the lattice constant of metals and other crystalline materials can be very accurately measured. Together with an accurately aligned micropositioner a map of the lattice constant through the metal can be derived. This can easily be converted to the stress field experienced by the material. This has been used to analyse stresses in aerospace and automotive components to give just two examples. This technique has led to the development of dedicated stress diffractometers, such as the ENGIN-X instrument at the ISIS neutron source.

Neutron Diffraction can also be employed to give insight into the 3D structure any material that diffracts.[4][5]

24.4.1   Hydrogen, null-scattering and contrast variation

Neutron diffraction can be used to establish the structure of low atomic number materials like proteins and surfactants much more easily with lower flux than at a synchrotron radiation source. This is because some low atomic number materials have a higher cross section for neutron interaction than higher atomic weight materials.

One major advantage of neutron diffraction over X-ray diffraction is that the latter is rather insensitive to the presence of hydrogen (H) in a structure, whereas the nuclei ^1H and ^2H (i.e. Deuterium, D) are strong scatterers for neutrons. The greater scattering power of protons and deuterons means that the position of hydrogen in a crystal and its thermal motions can be determined with greater precision by neutron diffraction. The structures of metal hydride complexes, e.g., Mg_2FeH_6 have been assessed by neutron diffraction.[6]

The neutron scattering lengths $bH = -3.7406(11)$ fm [7] and $bD = 6.671(4)$ fm,[7] for H and D respectively, have opposite sign, which allows the technique to distinguish them. In fact there is a particular isotope ratio for which the contribution of the element would cancel, this is called null-scattering.

It is undesirable to work with the relatively high concentration of H in a sample. The scattering intensity by H-nuclei has a large inelastic component, which creates a large continuous background that is more or less independent of scattering angle. The elastic pattern typically consists of sharp Bragg reflections if the sample is crystalline. They tend to drown in the inelastic background. This is even more serious when the technique is used for the study of liquid structure. Nevertheless, by preparing samples with different isotope ratios it is possible to vary the scattering contrast enough to highlight one element in an otherwise complicated structure. The variation of other elements is possible but usually rather expensive. Hydrogen is inexpensive and particularly interesting because it plays an exceptionally large role in biochemical structures and is difficult to study structurally in other ways.

24.5 History

The first neutron diffraction experiments were carried out in 1945 by Ernest O. Wollan using the Graphite Reactor at Oak Ridge. He was joined shortly thereafter (June 1946)[8] by Clifford Shull, and together they established the basic principles of the technique, and applied it successfully to many different materials, addressing problems like the structure of ice and the microscopic arrangements of magnetic moments in materials. For this achievement Shull was awarded one half of the 1994 Nobel Prize in Physics. Wollan had died in 1984. (The other half of the 1994 Nobel Prize for Physics went to Bert Brockhouse for development of the inelastic scattering technique at the Chalk River facility of AECL. This also involved the invention of the triple axis spectrometer). The delay between the achieved work (1946) and the Nobel Prize awarded to Brockhouse and Shull (1994) brings them close to the delay between the invention by Ernst Ruska of the electron microscope (1933) - also in the field of particle optics - and his own Nobel prize (1986). This in turn is near to the record of 55 years between the discoveries of Peyton Rous and his award of the Nobel Prize in 1966.

24.6 See also

- Crystallography
- Crystallographic database
- Electron diffraction

24.7 References

[1] Paula M. B. Piccoli, Thomas F. Koetzle, Arthur J. Schultz "Single Crystal Neutron Diffraction for the Inorganic Chemist—A Practical Guide" Comments on Inorganic Chemistry 2007, Volume 28, 3-38. doi:10.1080/02603590701394741

[2] Neutron diffraction of magnetic materials / Yu. A. Izyumov, V.E. Naish, and R.P. Ozerov ; translated from Russian by Joachim Büchner. New York : Consultants Bureau, c1991.ISBN 0-306-11030-X

[3] *Neutron powder diffraction* by Richard M. Ibberson and William I.F. David, Chapter 5 of Structure determination form powder diffraction data IUCr monographphs on crystallography, Oxford scientific publications 2002, ISBN 0-19-850091-2

[4] Ojeda-May, P.; Terrones, M.; Terrones, H.; Hoffman, D. et al. (2007), "Determination of chiralities of single-walled carbon nanotubes by neutron powder diffraction technique", *Diamond and Related Materials* 16: 473–476, Bibcode:2007DRM... O,doi:10.1016/j.diamond.2006.09.019

[5] Page, K.; Proffen, T.; Niederberger, M.; Seshadri, R. (2010), "Probing Local Dipoles and Ligand Structure in BaTiO3 Nanoparticles", *Chemistry of Materials* 22: 4386–4391, doi:10.1021/cm100440p

[6] Robert Bau, Mary H. Drabnis "Structures of transition metal hydrides determined by neutron diffraction" Inorganica Chimica Acta 1997, vol. 259, pp/ 27-50. doi:10.1016/S0020-1693(97)89125-6

[7]Sears,V.F. (1992), "Neutron scattering lengths and cross sections",*Neutron News***3**:26–37,doi:10.1080/104639208218770

[8] Clifford Shull: Early development of neutron scattering. Rev. Mod. Phys. 67 (1995) 753–757

24.8 Further reading

- Lovesey, S. W. (1984). *Theory of Neutron Scattering from Condensed Matter; Volume 1: Neutron Scattering*. Oxford: Clarendon Press. ISBN 0-19-852015-8.

- Lovesey, S. W. (1984). *Theory of Neutron Scattering from Condensed Matter; Volume 2: Condensed Matter*. Oxford: Clarendon Press. ISBN 0-19-852017-4.

- Squires, G.L. (1996). *Introduction to the Theory of Thermal Neutron Scattering* (2nd ed.). Mineola, New York: Dover Publications Inc. ISBN 0-486-69447-X.

24.8.1 Applied Computational Powder Diffraction Data Analysis

- Young, R.A., ed. (1993). *The Rietveld Method*. Oxford: Oxford University Press & International Union of Crystallography. ISBN 0-19-855577-6.

24.9 External links

- National Institute of Standards and Technology Center for Neutron Research

- Stress determination in crystalline materials

- From Bragg's law to neutron diffraction

- Neutronsources.org - Collaboration website of all neutron facilities

- Integrated Infrastructure Initiative for Neutron Scattering and Muon Spectroscopy (NMI3) - a European consortium of 18 partner organisations from 12 countries, including all major facilities in the fields of neutron scattering and muon spectroscopy

Chapter 25

Inelastic neutron scattering

Inelastic neutron scattering is an experimental technique commonly used in condensed matter research to study atomic and molecular motion as well as magnetic and crystal field excitations. It distinguishes itself from other neutron scattering techniques by resolving the change in kinetic energy that occurs when the collision between neutrons and the sample is an inelastic one. Results are generally communicated as the dynamic structure factor (also called inelastic scattering law) $S(\mathbf{Q}, \omega)$, sometimes also as the dynamic susceptibility $\chi''(\mathbf{Q}, \omega)$ where the scattering vector $\mathbf{Q}$ is the difference between incoming and outgoing wave vector, and $\hbar\omega$ is the energy change experienced by the sample (negative that of the scattered neutron). When results are plotted as function of ω, they can often be interpreted in the same way as spectra obtained by conventional spectroscopic techniques; insofar as inelastic neutron scattering can be seen as a special spectroscopy.

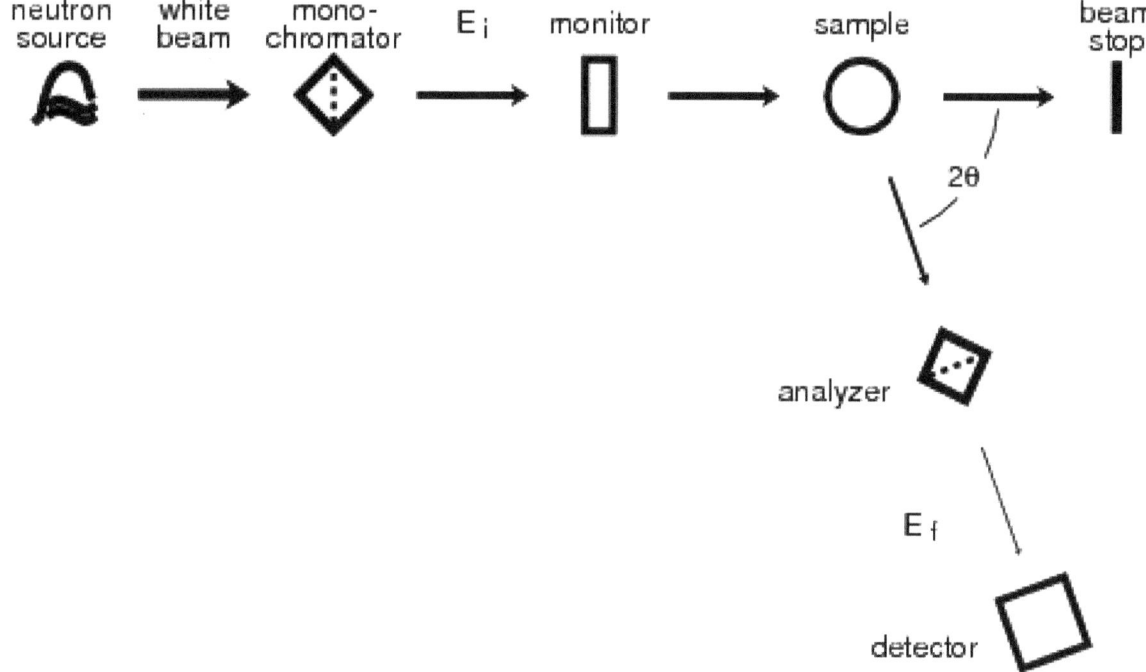

Generic layout of an inelastic neutron scattering experiment

Inelastic scattering experiments normally require a monochromatization of the incident or outgoing beam and an energy analysis of the scattered neutrons. This can be done either through time-of-flight techniques (neutron time-of-flight scattering) or through Bragg reflection from single crystals (neutron triple-axis spectroscopy, neutron backscattering). Monochromatization is not needed in echo techniques (neutron spin echo, neutron resonance spin echo), which use the quantum mechanical phase of the neutrons in addition to their amplitudes.

25.1 See also

- Inelastic scattering

25.2 Further Information

Literature:

- G L Squires *Introduction to the Theory of Thermal Neutron Scattering* Dover 1997 (reprint?)

25.3 External links

- Joachim Wuttke: Introduction to Inelastic Crystal Spectrometers

Chapter 26

Neutron tomography

Neutron tomography is a form of computed tomography involving the production of three-dimensional images by the detection of the absorbance of neutrons produced by a neutron source.[1] It created a three-dimensional image of an object by combining multiple planar images with a known separation.[2] It has a resolution of around 200–500 μm.[3] Whilst its resolution is lower than that of X-ray tomography, it can be useful for specimens containing low contrast between the matrix and object of interest; for instance, fossils with a high carbon content, such as plants or vertebrate remains.[4]

Neutron tomography can have the unfortunate side-effect of leaving imaged samples radioactive if they contain appreciable levels of certain elements.[4]

26.1   See also

- Winkler, B. (2006). "Applications of Neutron Radiography and Neutron Tomography". *Reviews in Mineralogy and Geochemistry* **63**: 459. doi:10.2138/rmg.2006.63.17.

- Schwarz, D.; Vontobel, P. L., Eberhard, H., Meyer, C. A. & Bongartz, G. (2005). "Neutron tomography of internal structures of vertebrate remains: a comparison with X-ray computed tomography" (PDF). *Paleontol. Electronica* **8** (30).

26.2   References

[1] Grünauer, F.; Schillinger, B.; Steichele, E. (2004). "Optimization of the beam geometry for the cold neutron tomography facility at the new neutron source in Munich". *Applied radiation and isotopes : including data, instrumentation and methods for use in agriculture, industry and medicine* **61** (4): 479–485. doi:10.1016/j.apradiso.2004.03.073. PMID 15246387.

[2] McClellan Nuclear Radiation Center

[3] "Neutron Tomography". *Paul Scherrer Institut*.

[4] Sutton, M. D. (2008). "Tomographic techniques for the study of exceptionally preserved fossils". *Proceedings of the Royal Society B: Biological Sciences* **275** (1643): 1587–1593. doi:10.1098/rspb.2008.0263. PMC 2394564. PMID 18426749.

Chapter 27

Fast neutron therapy

Fast neutron therapy utilizes high energy neutrons typically between 50 and 70 MeV to treat cancer. Most fast neutron therapy beams are produced by reactors, cyclotrons (d+Be) and linear accelerators. Neutron therapy is currently available in Germany, Russia, South Africa and the United States. In the US three treatment centers operate in Seattle, Washington, Detroit, Michigan and Batavia, Illinois. The Detroit and Seattle centers use a cyclotron which produces a proton beam impinging upon a beryllium target; the Batavia center at Fermilab uses a proton linear accelerator.

27.1 Advantages

Radiation therapy kills cancer cells in two ways depending on the effective energy of the radiative source. The amount of energy deposited as the particles traverse a section of tissue is referred to as the Linear Energy Transfer (LET). X-rays and protons produce low LET radiation, and neutrons produce high LET radiation. Low LET radiation damages cells predominantly through the generation of reactive oxygen species, see free radicals. The neutron is uncharged and damages cells by nuclear interactions. Malignant tumors tend to have low oxygen levels and thus can be resistant to low LET radiation. This gives an advantage to neutrons in certain situations. One advantage is a generally shorter treatment cycle. To kill the same number of cancerous cells, neutrons require one third the effective dose as protons.[1] Another advantage is the established ability of neutrons to better treat some cancers, such as salivary gland, adenoid cystic carcinomas and certain types of brain tumors, especially high-grade gliomas [2]

27.1.1 LET

When therapeutic energy X-rays (1 to 25 MeV) interact with cells in human tissue, they do so mainly by Compton interactions, and produce relatively high energy secondary electrons. These high energy electrons deposit their energy at about 1 keV/μm.[3] By comparison, the charged particles produced at a site of a neutron interaction may deliver their energy at a rate of 30-80 keV/μm. The amount of energy deposited as the particles traverse a section of tissue is referred to as the Linear Energy Transfer (LET). X-rays produce low LET radiation, and neutrons produce high LET radiation.

Because the electrons produced from X-rays have high energy and low LET, when they interact with a cell typically only a few ionizations will occur. It is likely then that the low LET radiation will cause only single strand breaks of the DNA helix. Single strand breaks of DNA molecules can be readily repaired, and so the effect on the target cell is not necessarily lethal. By contrast, the high LET charged particles produced from neutron irradiation cause many ionizations as they traverse a cell, and so double-strand breaks of the DNA molecule are possible. DNA repair of double-strand breaks are much more difficult for a cell to repair, and more likely to lead to cell death.

DNA repair mechanisms are quite efficient,[4] and during a cell's lifetime many thousands of single strand DNA breaks will be repaired. A sufficient dose of ionizing radiation, however, delivers so many DNA breaks that it overwhelms the capability of the cellular mechanisms to cope.

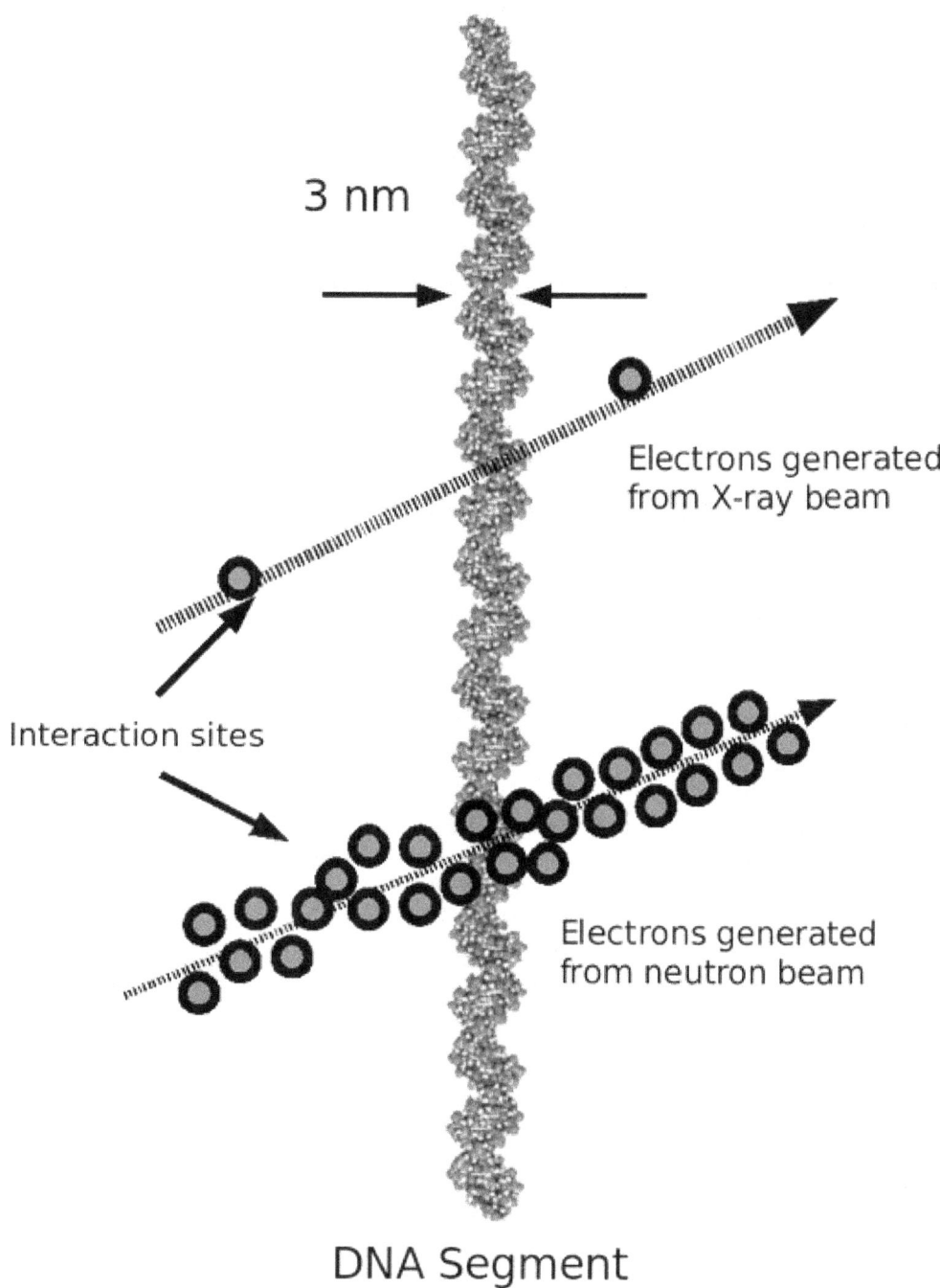

Comparison of Low LET electrons and High LET electrons

Heavy ion therapy (e.g. carbon ions) makes use of the similarly high LET of $^{12}C^{6+}$ ions.[5][6]

Because of the high LET, the relative radiation damage (relative biological effect or RBE) of fast neutrons is 4 times that of X-rays,[7] [8] meaning 1 rad of fast neutrons is equal to 4 rads of X-rays. The RBE of neutrons is also energy dependent, so neutron beams produced with different energy spectra at different facilities will have different RBE values.

27.1.2 Oxygen effect

The presence of oxygen in a cell acts as a radiosensitizer, making the effects of the radiation more damaging. Tumor cells typically have a lower oxygen content than normal tissue. This medical condition is known as tumor hypoxia and therefore the oxygen effect acts to decrease the sensitivity of tumor tissue.[9] The oxygen effect may be quantitatively described by the Oxygen Enhancement Ratio (OER). Generally it is believed that neutron irradiation overcomes the effect of tumor hypoxia,[10] although there are counterarguments[11]

27.2 Clinical uses

The efficacy of neutron beams for use on prostate cancer has been shown through randomized trials.[12][13][14]

Fast neutron therapy has been applied successfully against salivary gland tumors.[15][16][17][18][19][20][21] See also the NCI Salivary Cancer Page.

Adenoid cystic carcinomas have also been treated.[22][23]

Various other head and neck tumors have been examined.[24][25][26]

27.3 Side Effects

No cancer therapy is without the risk of side effects. Neutron therapy is a very powerful nuclear scalpel that has to be utilized with exquisite care. For instance, some of the most remarkable cures it has been able to achieve are with cancers of the head and neck. Many of these cancers cannot effectively be treated with other therapies. However, neutron damage to nearby vulnerable areas such as the brain and sensory neurons can produce irreversible brain atrophy, blindness, etc. The risk of these side effects can be greatly mitigated by several techniques, but they cannot be totally eliminated. Moreover, some patients are more susceptible to such side effects than others and this cannot be predicted in advance. The patient ultimately must decide whether the advantages of a possibly lasting cure outweigh the risks of this treatment when faced with an otherwise incurable cancer.[27]

27.4 Fast neutron centers

Several centers around the world have used fast neutrons for treating cancer. Due to lack of funding and support, at present only three are active in the USA. The University of Washington and the Gershenson Radiation Oncology Center operate fast neutron therapy beams and both are equipped with a Multi-Leaf Collimator (MLC) to shape the neutron beam.[28][29][30]

27.4.1 University of Washington

The Radiation Oncology Department[31] operates a proton cyclotron that produces fast neutrons from directing 50.5MeV protons onto a beryllium target.

The UW Cyclotron is equipped with a gantry mounted delivery system an MLC to produce shaped fields. The UW Neutron system is referred to as the Clinical Neutron Therapy System (CNTS).[32]

The CNTS is typical of most neutron therapy systems. A large, well shielded building is required to cut down on radiation exposure to the general public and to house the necessary equipment.

A beamline transports the proton beam from the cyclotron to a gantry system. The gantry system contains magnets for deflecting and focusing the proton beam onto the beryllium target. The end of the gantry system is referred to as the head, and contains dosimetry systems to measure the dose, along with the MLC and other beam shaping devices. The advantage of having a beam transport and gantry are that the cyclotron can remain stationary, and the radiation source can be rotated around the patient. Along with varying the orientation of the treatment couch which the patient is positioned on, variation of the gantry position allows radiation to be directed from virtually any angle, allowing sparing of normal tissue and maximum radiation dose to the tumor.

During treatment, only the patient remains inside the treatment room (called a vault) and the therapists will remotely control the treatment, viewing the patient via video cameras. Each delivery of a set neutron beam geometry is referred to as a treatment field or beam. The treatment delivery is planned to deliver the radiation as effectively as possible, and usually results in fields that conform to the shape of the gross target, with any extension to cover microscopic disease.

27.4.2   Karmanos Cancer Center / Wayne State University

The neutron therapy facility at the Gershenson Radiation Oncology Center at Karmanos Cancer Center/Wayne State University (KCC/WSU) in Detroit bears some similarities to the CNTS at the University of Washington, but also has many unique characteristics.

While the CNTS accelerates protons, the KCC facility produces its neutron beam by accelerating 48.5 MeV deuterons onto a beryllium target. This method produces a neutron beam with depth dose characteristics roughly similar to those of a 4MV photon beam. The deuterons are accelerated using a gantry mounted superconducting cyclotron (GMSCC), eliminating the need for extra beam steering magnets and allowing the neutron source to rotate a full 360° around the patient couch.

The KCC facility is also equipped with an MLC beam shaping device,[33] the only other neutron therapy center in the USA besides the CNTS. The MLC at the KCC facility has been supplemented with treatment planning software that allows for the implementation of Intensity Modulated Neutron Radiotherapy (IMNRT), a recent advance in neutron beam therapy which allows for more radiation dose to the targeted tumor site than 3-D neutron therapy.[34]

KCC/WSU has more experience than anyone in the world using neutron therapy for prostate cancer, having treated nearly 1,000 patients during the past 10 years.

27.4.3   Fermilab / Northern Illinois University

The Fermilab neutron therapy center first treated patients in 1976,[35] and since that time has treated over 3,000 patients. In 2004, the Northern Illinois University began managing the center. The neutrons produced by the linear accelerator at Fermilab have the highest energies available in the US and among the highest in the world [36][37][38]

27.5   See also

- Boron neutron capture therapy

27.6   References

[1] Keyhandokht Shahri, Laleh Motavalli, and Hashem Hakimabad."Neutron Applications in Cancer Treatment" Hellenic Journal of Nuclear Medicine 14:2(May–August 2011)

[2] Feng-Yi Yang,Wen-Yuan Chang, Jia-Je Li,Hsin-Ell Wang,Jyh-Cheng Chen,and Chi-Wei Chang."Pharmacokinetic Analysis and Uptake of 18F-FBPA-Fr After Ultrasound-Induced Blood–Brain Barrier Disruption for Potential Enhancement of Boron Delivery for Neutron Capture Therapy" Journal of Nuclear Medicine 55:616-621(2014)

[3] Johns HE and Cunningham JR. The Physics of Radiology. Charles C Thomas 3rd edition 1978

[4] Goodsell DS. Fundamentals of Cancer Medicine The Molecular Perspective: Double-Stranded DNA Breaks The Oncologist, Vol. 10, No. 5, 361-362, May 2005

[5] Kubota N, Suzuki M, Furusawa Y, Ando K, Koike S, Kanai T, Yatagai F, Ohmura M, Tatsuzaki H, Matsubara S, et al. A comparison of biological effects of modulated carbon-ions and fast neutrons in human osteosarcoma cells. International Journal of Radiation Oncology*Biology*Physics, Volume 33, Issue 1, 30 August 1995, Pages 135-141

[6] German Cancer Research Center

[7] Pignol JP, Slabbert J and Binns P. Monte Carlo simulation of fast neutron spectra: Mean lineal energy estimation with an effectiveness function and correlation to RBE. International Journal of Radiation Oncology*Biology*Physics, Volume 49, Issue 1, 1 January 2001, Pages 251-260

[8] Theron T, Slabbert J, Serafin A and Böhm L. The merits of cell kinetic parameters for the assessment of intrinsic cellular radiosensitivity to photon and high linear energy transfer neutron irradiation. International Journal of Radiation Oncology*Biology*Physics, Volume 37, Issue 2, 15 January 1997, Pages 423-428

[9] Vaupel P, Harrison L. Tumor Hypoxia: Causative Factors, Compensatory Mechanisms, and Cellular Response The Oncologist 2004;9(suppl 5):4-9

[10] Wambersie A, Richard F, Breteau N. Development of fast neutron therapy worldwide. Radiobiological, clinical and technical aspects. Acta Oncol. 1994;33(3):261-74.

[11] Warenius HM, White R, Peacock JH, Hanson J, Richard A. Britten, Murray D. The Influence of Hypoxia on the Relative Sensitivity of Human Tumor Cells to 62.5 MeV (p?Be) Fast Neutrons and 4 MeV Photons. Radiation Research 154, 54-63 (2000)

[12] Russell KJ, Caplan RJ, Laramore GE, et al. Photon versus fast neutron external beam radiotherapy in the treatment of locally advanced prostate cancer: results of a randomized prospective trial. International Journal of Radiation Oncology, Biology, Physics 28(1): 47-54, 1993.

[13] Haraf DJ, Rubin SJ, Sweeney P, Kuchnir FT, Sutton HG, Chodak GW and Weichselbaum RR. Photon neutron mixed-beam radiotherapy of locally advanced prostate cancer. International Journal of Radiation Oncology*Biology*Physics, Volume 33, Issue 1, 30 August 1995, Pages 3-14

[14] Forman J, Ben-Josef E, Bolton SE, Prokop S and Tekyi-Mensah S . A randomized prospective trial of sequential neutron-photon vs. photon-neutron irradiation in organ confined prostate cancer. International Journal of Radiation Oncology*Biology*Physics, Volume 54, Issue 2, Supplement 1, 1 October 2002, Pages 10-11

[15] Douglas JD, Koh WJ , Austin-Seymour, M, Laramore GE. Treatment of Salivary Gland Neoplasms with fast neutron Radiotherapy. Arch Otolaryngol Head Neck Surg Vol 129 944-948 Sep 2003

[16] Laramore GE, Krall JM, Griffin TW, Duncan W, Richter MP, Saroja KR, Maor MH, Davis LW. Neutron versus photon irradiation for unresectable salivary gland tumors: final report of an RTOG-MRC randomized clinical trial. Int J Radiat Oncol Biol Phys. 1993 Sep 30;27(2):235-40.

[17] Laramore GE. Fast neutron radiotherapy for inoperable salivary gland tumors: is it the treatment of choice?Int J Radiat Oncol Biol Phys. 1987 Sep;13(9):1421-3.

[18] Prott FJ, Micke O, Pötter R, Haverkamp U, Schüller P and Willich N. 2137 Results of fast neutron therapy of adenoid cystic carcinoma of the salivary glands. International Journal of Radiation Oncology*Biology*Physics, Volume 39, Issue 2, Supplement 1, 1997, Page 309

[19] Saroja KR, Mansell J, Hendrickson FR, et al.: An update on malignant salivary gland tumors treated with neutrons at Fermilab. Int J Radiat Oncol Biol Phys 13 (9): 1319-25, 1987.

[20] Buchholz TA, Laramore GE, Griffin BR, et al.: The role of fast neutron radiation therapy in the management of advanced salivary gland malignant neoplasms. Cancer 69 (11): 2779-88, 1992.

[21] Krüll A, Schwarz R, Engenhart R, et al.: European results in neutron therapy of malignant salivary gland tumors. Bull Cancer Radiother 83 (Suppl): 125-9s, 1996

[22] Adenoid Cystic Carcinoma Neutron Radiation Therapy

[23] Douglas JG, Laramore GE, Austin-Seymour M, Koh WJ, Lindsley KL, Cho P and Griffin TW. Neutron radiotherapy for adenoid cystic carcinoma of minor salivary glands. International Journal of Radiation Oncology*Biology*Physics, Volume 36, Issue 1, 1 August 1996, Pages 87-93

[24] MacDougall RH, Orr JA, Kerr GR, and Duncan W. Fast neutron treatment for squamous cell carcinoma of the head and neck: final report of Edinburgh randomised trial. BMJ. 1990 December 1; 301(6763): 1241-1242.

[25] Asgarali S, Errington RD, Jones AS. The treatment of recurrence following fast neutron therapy for head and neck malignancy. Clin Otolaryngol Allied Sci. 1996 Jun;21(3):274-7.

[26] K.J. Stelzer, K.L. Lindsley, P.S. Cho, G.E. Laramore and T.W. Griffin. Fast Neutron Radiotherapy: The University of Washington Experience and Potential Use of Concomitant Boost with Boron Neutron Capture. Radiation Protection Dosimetry 70:471-475 (1997)

[27] http://medicalphysicsweb.org/cws/article/opinion/32466

[28] Brahme A, Eenmaa J, Lindback S, Montelius A, Wootton P. Neutron beam characteristics from 50 MeV protons on beryllium using a continuously variable multi-leaf collimator. Radiother Oncol. 1983 Aug;1(1):65-76.

[29] Farr JB. A compact multileaf collimator for conventional and intensity modulated fast neutron therapy Medical Physics April 2004 Volume 31, Issue 4, p. 951

[30] Farr JB, Maughan RL, Yudelev M, Blosser E, Brandon J, Horste T Compact multileaf collimator for conformal and intensity modulated fast neutron therapy: Electromechanical design and validation Medical Physics -- September 2006 -- Volume 33, Issue 9, pp. 3313-3320

[31] University of Washington (UW) Radiation Oncology Department

[32] Clinical Neutron Therapy System (CNTS)

[33] Farr, J. B., R. L. Maughan, et al. (2007). "Radiologic validation of a fast neutron multileaf collimator." Med Phys 34(9): 3475-3484.

[34] Santanam, L., T. He, et al. (2007). "Intensity modulated neutron radiotherapy for the treatment of adenocarcinoma of the prostate." Int J Radiat Oncol Biol Phys 68(5): 1546-1556.

[35] Cohen L and Lennox A. Midwest Institute for Neutron Therapy at Fermilab. International Journal of Radiation Oncology*Biology*Physics, Volume 34, Issue 1, 1 January 1996, Page 269

[36] http://www.news-medical.net/news/2004/12/07/6746.aspx

[37] http://www.neutrontherapy.niu.edu/neutrontherapy/aboutus/index.shtml

[38] http://www.neutrontherapy.niu.edu/neutrontherapy/NIUINT_booklet.pdf

27.7 External links

- University of Washington Radiation Oncology - Neutron Therapy

- UWMC SCCA Site

- FermiLab Neutron Therapy

    - FermiLab Neutron Therapy overview

- Gershenson Radiation Oncology Center

- Michigan State Therapy NSCL

Chapter 28

Neutron capture therapy of cancer

Neutron capture therapy (NCT) is a noninvasive therapeutic modality for treating locally invasive malignant tumors such as primary brain tumors and recurrent head and neck cancer. It is a two step procedure: first, the patient is injected with a tumor localizing drug containing a non-radioactive isotope that has a high propensity or cross section (σ) to capture slow neutrons. The cross section of the capture agent is many times greater than that of the other elements present in tissues such as hydrogen, oxygen, and nitrogen. In the second step, the patient is radiated with epithermal neutrons, which after losing energy as they penetrate tissue, are absorbed by the capture agent which subsequently emits high-energy charged particles, thereby resulting in a biologically destructive nuclear reaction (Fig.1).

All of the clinical experience to date with NCT is with the non-radioactive isotope boron-10, and this is known as **boron neutron capture therapy** (**BNCT**).[2] At this time, the use of other non-radioactive isotopes, such as gadolinium, has been limited, and to date, it has not been used clinically. BNCT has been evaluated clinically as an alternative to conventional radiation therapy for the treatment of malignant brain tumors (gliomas), and more recently, recurrent, locally advanced head and neck cancer.[3]

28.1 Boron neutron capture therapy

28.1.1 History

After the initial discovery of the neutron in 1932 by Sir James Chadwick, H. J. Taylor in 1935 showed that boron-10 nuclei could capture thermal neutrons. Neutron capture resulted in fission of the resulting excited boron-11 nuclei into helium-4 (alpha particles) and lithium-7 ions. In 1936, G.L. Locher, a scientist at the Franklin Institute in Pennsylvania, realized the therapeutic potential of this discovery and suggested that neutron capture could be used to treat cancer. W. H. Sweet first suggested the technique for the most malignant brain tumors in 1951,[4] and a trial of the therapy against glioblastoma multiforme using borax as the delivery agent was reported first in a collaboration between Brookhaven National Laboratory and Massachusetts General Hospital in 1954.[5] In neutron capture therapy, a binary system uses two separate components for a therapeutic effect. Each component in itself is non-tumoricidal, but when combined together they produce highly lethal effect.

28.2 Basic principles

Boron neutron capture therapy (BNCT) is based on the nuclear capture and fission reactions that occur when non-radioactive boron-10, which makes up approx 20% of natural elemental boron, is irradiated with neutrons of the appropriate energy to yield excited boron-11 ($^{11}B^*$), which in turn decays into high energy alpha particles ("stripped" down ^4He nuclei) and high energy lithium-7 (^7Li) nuclei. The nuclear reaction is:

$$^{10}B + n_{th} \rightarrow [^{11}B]^* \rightarrow \alpha + {}^7Li + 2.31 \text{ MeV}$$

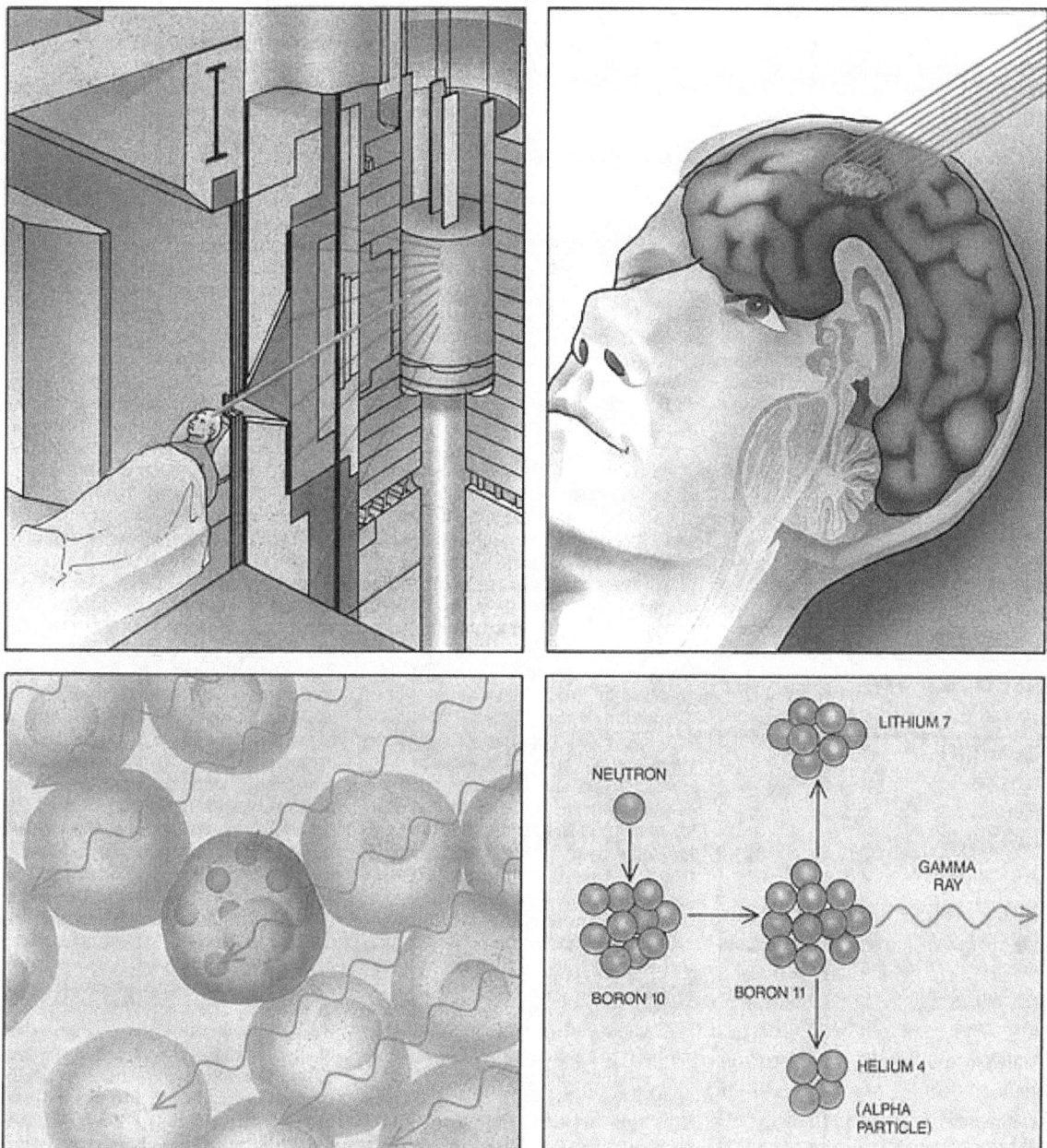

Fig.1 Boron neutron capture therapy (BNCT) can be performed at a facility with a nuclear reactor or at hospitals that have developed alternative neutron sources. A beam of epithermal neutrons penetrates the brain tissue, reaching the malignancy. Once there the epithermal neutrons slow down and these low-energy neutrons combine with boron-10 (delivered beforehand to the cancer cells by drugs or antibodies) to form boron-11, releasing lethal radiation (alpha particles and lithium ions) that can kill the tumor.[1]

Both the alpha particles and the lithium ions produce closely spaced ionizations in the immediate vicinity of the reaction, with a range of approximately 5–9 μm, or approximately the diameter of one cell. Their lethality is limited to boron containing cells. BNCT, therefore, can be regarded as both a biologically and a physically targeted type of radiation therapy. The success of BNCT is dependent upon the selective delivery of sufficient amounts of ^{10}B to the tumor with only small amounts localized in the surrounding normal tissues.[6] Thus, normal tissues, if they have not taken up boron-10, can be spared from the nuclear capture and fission reactions. Normal tissue tolerance is determined by the nuclear capture reactions that occur with normal tissue hydrogen and nitrogen.

Boron Neutron Capture Therapy (BNCT)

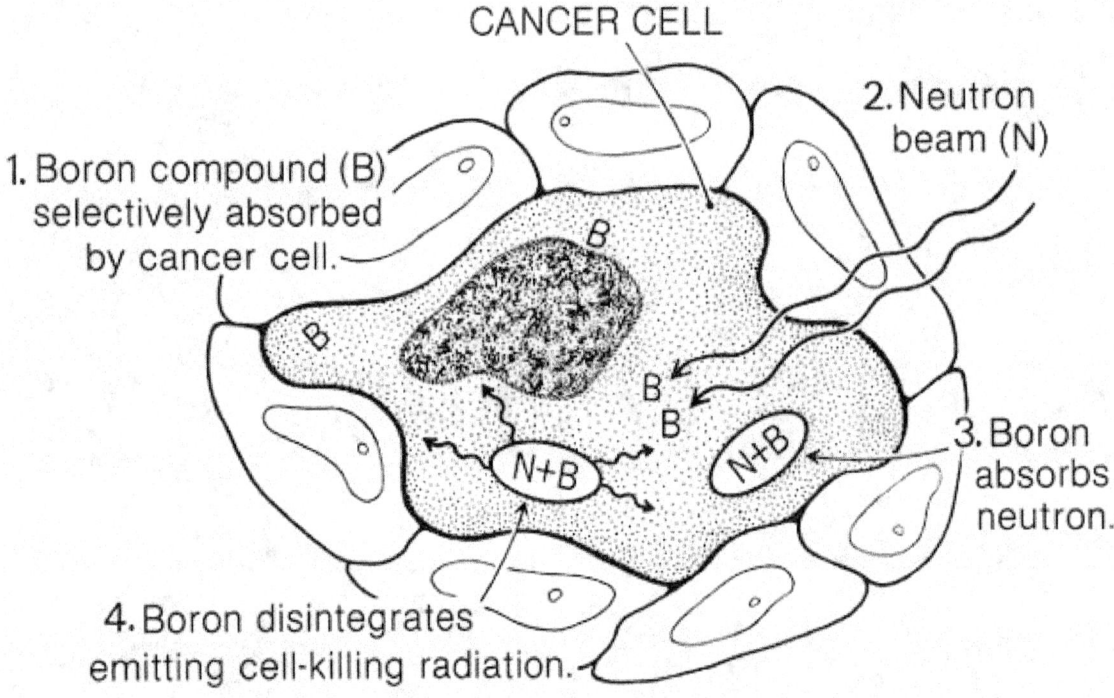

1) Boron compound (b) is selectively absorbed by cancer cell(s). 2) Neutron beam (n) is aimed at cancer site. 3) Boron absorbs neutron. 4) Boron disintegrates emitting cancer-killing radiation.

A wide variety of boron delivery agents have been synthesized,[7] but only two of these currently are being used in clinical trials. The first, which has been used primarily in Japan, is a polyhedral borane anion, sodium borocaptate or BSH ($Na_2B_{12}H_{11}SH$), and the second is a dihydroxyboryl derivative of phenylalanine, referred to as boronophenylalanine or BPA. The latter has been used in clinical trials in the United States, Europe, Japan and more recently, Argentina and Taiwan. Following administration of either BPA or BSH by intravenous infusion, the tumor site is irradiated with neutrons, the source of which has been specially modified nuclear reactors. Up to 1994, low-energy (< 0.5 eV) thermal neutron beams were used primarily in Japan,[8] but since they have a limited depth of penetration in tissues, higher energy (>.5eV<10 keV) epithermal neutron beams, which have a greater depth of penetration, have been used in clinical trials in the United States,[9][10] Europe,[11][12] and Japan.[13][14]

In theory BNCT is a highly selective type of radiation therapy that can selectively target the tumor at the cellular level without causing radiation damage to the adjacent normal cells and tissues. Doses up to 60–70 Gy can be delivered to the tumor cells in one or two applications compared to 6–7 weeks for conventional external beam photon irradiation. However, the effectiveness of BNCT is dependent upon a relatively homogeneous distribution of ^{10}B within the tumor, and this is still one of the key stumbling blocks that have limited its success.

28.3 Radiobiological considerations

The radiation doses delivered to tumor and normal tissues during BNCT are due to energy deposition from three types of directly ionizing radiation that differ in their linear energy transfer (LET), which is the rate of energy loss along the path of an ionizing particle:

1. low LET gamma rays, resulting primarily from the capture of thermal neutrons by normal tissue hydrogen atoms [^{1}H(n,γ)^{2}H];

2. high LET protons, produced by the scattering of fast neutrons and from the capture of thermal neutrons by nitrogen atoms [^{14}N(n,p)^{14}C]; and

3. high LET, heavier charged alpha particles (stripped down ^{4}He nuclei) and lithium-7 ions, released as products of the thermal neutron capture and fission reactions with ^{10}B [^{10}B(n,α)^{7}Li].

Since both tumor and surrounding normal tissues are present in the radiation field, even with an ideal epithermal neutron beam, there will be an unavoidable, nonspecific background dose, consisting of both high and low LET radiation. However, a higher concentration of ^{10}B in the tumor will result in it receiving a higher total dose than that of adjacent normal tissues, which is the basis for the therapeutic gain in BNCT.[15] The total radiation dose (Gy) delivered to any tissue can be expressed in photon-equivalent units as the sum of each of the high LET dose components multiplied by weighting factors (Gy$_w$), which depend on the increased radiobiological effectiveness of each of these components.

28.4   Clinical dosimetry

Biological weighting factors have been used in all of the recent clinical trials in patients with high grade gliomas, using boronophenylalanine (BPA) in combination with an epithermal neutron beam. The ^{10}B(n,α)^{7}Li component of the radiation dose to the scalp has been based on the measured boron concentration in the blood at the time of BNCT, assuming a blood: scalp boron concentration ratio of 1.5:1 and a compound biological effectiveness (CBE) factor for BPA in skin of 2.5. A relative biological effectiveness (RBE) factor of 3.2 has been used in all tissues for the high LET components of the beam, such as alpha particles. The RBE factor is used to compare the biologic effectiveness of different types of ionizing radiation. The high LET components include protons resulting from the capture reaction with nitrogen, and recoil protons resulting from the collision of fast neutrons with hydrogen.[15] It must be emphasized that the tissue distribution of the boron delivery agent in humans should be similar to that in the experimental animal model in order to use the experimentally derived values for estimation of the radiation "Gray" (Gy) doses for clinical radiations.[15] For more detailed information relating to computational dosimetry and treatment planning, interested readers are referred to a comprehensive review on this subject.[16]

28.5   Boron delivery agents

The development of boron delivery agents for BNCT began approximately 50 years ago and is an ongoing and difficult task of high priority. A number of boronated pharmaceuticals using boron-10, have been prepared for potential use in BNCT.[17] The most important requirements for a successful boron delivery agent are: 1. low systemic toxicity and normal tissue uptake with high tumor uptake and concomitantly high tumor: to brain (T:Br) and tumor: to blood (T:Bl) concentration ratios (> 3–4:1); 2. tumor concentrations in the range of ~20 μg ^{10}B/g tumor; 3. rapid clearance from blood and normal tissues and persistence in tumor during BNCT. However, it should be noted that at this time no single boron delivery agent fulfills all of these criteria. With the development of new chemical synthetic techniques and increased knowledge of the biological and biochemical requirements needed for an effective agent and their modes of delivery, a number of promising new boron agents has emerged (see examples in Table 1).

*These are the only two boron delivery agents that have been used clinically.

The major challenge in their development has been the requirement for selective tumor targeting in order to achieve boron concentrations sufficient to deliver therapeutic doses of radiation to the tumor with minimal normal tissue toxicity. The selective destruction of brain tumor (glioma) cells in the presence of normal cells represents an even greater challenge compared to malignancies at other sites in the body, since malignant gliomas are highly infiltrative of normal brain, histologically complex and heterogeneous in their cellular composition. In principle, NCT is a radiation therapy that could selectively deliver lethal doses of radiation to tumor cells while sparing adjacent normal cells.

28.6 Gadolinium neutron capture therapy (Gd NCT)

There also has been interest in the possible use of gadolinium-157 (^{157}Gd) as a capture agent for NCT for the following reasons:[18] *First*, and foremost, has been its very high neutron capture cross section of 254,000 barns. *Second*, gadolinium compounds, such as Gd-DTPA (gadopentetate dimeglumine Magnevist®), have been used routinely as contrast agents for magnetic resonance imaging (MRI) of brain tumors and have shown high uptake by brain tumor cells in tissue culture (*in vitro*).[19] *Third*, gamma rays and internal conversion and Auger electrons are products of the ^{157}Gd (n,γ)^{158}Gd capture reaction (^{157}Gd + n$_{th}$ (0.025eV) $\rightarrow$ [^{158}Gd] $\rightarrow$ ^{158}Gd + γ + 7.94 MeV).

Although the gamma rays have long pathlengths, orders of magnitude greater depths of penetration compared with the other radiations, the other radiation products (internal conversion and Auger electrons) have pathlengths of approximately one cell diameter and can directly cause DNA damage. Therefore, it would be highly advantageous for the production of DNA damage if the ^{157}Gd were localized within the cell nucleus. However, the possibility of incorporating gadolinium into biologically active molecules is very limited and only a small number of potential delivery agents for Gd NCT have been studied.[20][21]

The gadolinium studies in cells and animals may be compared to the relatively large number of boron containing compounds (Table 1) have been synthesized and evaluated *in vitro* and in experimental animals (*in vivo*). Although *in vitro* activity has been demonstrated using the Gd-containing MRI contrast agent Magnevist® as the Gd delivery agent,[22] there are very few studies demonstrating the efficacy of Gd NCT in experimental animal tumor models,[21][23] and Gd NCT has to date never been used clinically (i.e., in humans).

28.7 Neutron sources

28.7.1 Nuclear reactors

Neutron sources for NCT have been limited to nuclear reactors and in the present section we only will summarize information that is described in more detail in a recently published review.[24] Reactor derived neutrons are classified according to their energies as thermal (E$_n$ <0.5 eV), epithermal (0.5 eV <E$_n$ <10 keV) or fast (E$_n$ >10 keV). Thermal neutrons are the most important for BNCT since they usually initiate the ^{10}B(n,α)7Li capture reaction. However, because they have a limited depth of penetration, epithermal neutrons, which lose energy and fall into the thermal range as they penetrate tissues, are now preferred for clinical therapy.

A number of reactors with very good neutron beam quality have been developed and used clinically. These include:

1. Kyoto University Research Reactor (KURR) in Kumatori, Japan;

2. the Massachusetts Institute of Technology Research Reactor (MITR);

3. the RA-6 CNEA reactor in Bariloche, Argentina;

4. the High Flux Reactor (HFR) at Petten in the Netherlands; and

5. the FiR1 (Triga Mk II) research reactor at VTT Technical Research Centre, Espoo, Finland.

Although not currently being used for BNCT, the neutron irradiation facility at the MITR represented the state of the art in epithermal beams for NCT with the capability of completing a radiation field in 10–15 minutes with close to the theoretically maximum ratio of tumor to normal tissue dose. Unfortunately, however, no clinical studies currently are being carried out at the HFR and the MITR. The operation of the BNCT facility at the Finnish FiR1 research reactor (Triga Mk II), treating patients since 1999, was terminated in 2012 due to the BNCT operating company's bankruptcy. A new operator for the BNCT facility has not emerged, and consequently the owner of FiR1 (VTT Technical Research Centre of Finland) is planning to decommission the reactor in 2015. Finally, a low power "in-hospital" compact nuclear reactor has been designed and built in Beijing, China, and it has been undergoing performance evaluation over the past several years, but it is uncertain when and if it could ever be used for clinical BNCT.[25]

28.7.2 Accelerators

Main article: neutron generator

Accelerators also can be used to produce epithermal neutrons and accelerator-based neutron sources (ABNS) are being developed in a number of countries. Interested readers are referred to the recently published Proceedings of the 14th and 15th International Congresses on Neutron Capture Therapy and R. Moss's recent review[3] for information on this subject. For ABNSs, one of the more promising nuclear reactions involves bombarding a ^{7}Li target with high energy protons. An experimental BNCT facility, using a thick lithium solid target, has been in use since the early 1990s at the University of Birmingham in the UK. This facility makes use of a high current Dynamitron accelerator originally supplied by Radiation Dynamics.

Recently, a prototypic cyclotron-based neutron source (C-BENS) has been developed by Sumitomo Heavy Industries in Japan.[26] It has been installed at KURRI and now is being used in a Phase I clinical trial to evaluate its safety for treating patients with high grade gliomas. A second one has been constructed by Mitsubishi Heavy Industries for use at Tsukuba University in Japan, and should be ready for clinical use in 2015. A third one is being built by Hitachi for use in Tokyo. Finally, a fourth one is in the developmental stage and would utilize an accelerator fabricated by GT Advanced Technologies in Danvers, Massachusetts. This one will have a liquid lithium-7 target, designed by Osaka University, and it will be evaluated by a consortium of institutions, including Osaka University, as a demonstration project. Once clinical trials have been initiated, it will be important to determine how these ABNS compare to BNCT that has been carried out in the past using nuclear reactors as the neutron source.

28.8 Clinical studies of BNCT for brain tumors

28.8.1 Early studies in the US and Japan

It was not until the 1950s that the first clinical trials were initiated by Farr at the Brookhaven National Laboratory (BNL) in New York[5] and by Sweet and Brownell at the Massachusetts General Hospital (MGH) using the Massachusetts Institute of Technology (MIT) nuclear reactor (MITR)[27] and several different low molecular weight boron compounds (such as borate) as the boron containing drug. However, the results of these studies were disappointing, and no further clinical trials were carried out in the United States, until the 1990s.

Following a two-year fellowship in Sweet's laboratory, clinical studies were started by Hiroshi Hatanaka in Japan in 1967. He used a low energy thermal neutron beam, which has low tissue penetrating properties and sodium borocaptate (BSH). This had been developed as a boron delivery agent by Albert Soloway at the MGH. In Hatanaka's procedure,[28] as much of the tumor was surgically removed as possible ("debulking"), and at some time thereafter, sodium borocaptate (BSH) was administered by a slow infusion, usually intra-arterially, but later intravenously. Twelve to 14 hours later, BNCT was carried out at one or another of several different nuclear reactors using low energy thermal neutron beams. The less tissue-penetrating properties of the thermal neutron beams necessitated reflecting the skin and raising a bone flap in order to directly irradiate the exposed brain, a procedure first used by Sweet and his collaborators.

Approximately 200+ patients were treated by Hatanaka, and subsequently by his associate, Nakagawa.[8] Due to the heterogeneity of the patient population, in terms of the microscopic diagnosis of the tumor ("grade"), and its size, and the age and the ability of the patient to carry out normal daily activities ("performance status"), it was not possible to come up with definitive conclusions about therapeutic efficacy, as measured by a prolongation in the mean survival time (MST). However, the survival data were no worse than those obtained by standard therapy at the time, and there were several patients who were long-term survivors, and most probably they were cured of their brain tumors.[8]

28.8.2 More recent clinical studies in the US and Japan

BNCT of patients with brain tumors and a few with cutaneous melanoma was resumed in the United States in the mid-1990s at the Brookhaven National Laboratory Medical Research Reactor (BMRR) and at Harvard/Massachusetts Institute of Technology (MIT) using the MIT Research Reactor (MITR). For the first time, BPA was used as the boron delivery agent, and patients were irradiated with a collimated beam of higher energy epithermal neutrons, which had greater tissue-penetrating properties than thermal neutrons. This was well tolerated, but there were no significant differences in the MSTs compared to patients that had received conventional therapy.[9][10]

In Japan, Miyatake and Kawabata[13][14] have initiated several protocols employing the combination of BPA (500 mg/kg) and BSH (100 mg/kg), infused i.v. over 2 hrs, followed by neutron irradiation at Kyoto University Research Reactor Institute(KURRI). The MST of 10 patients was 15.6 months, with one long-term survivor (>5 years).[14] Based on experimental animal data,[29] which showed that BNCT in combination with X-irradiation produced enhanced survival compared to BNCT alone, Miyatake and Kawabata combined BNCT, as described above, with an X-ray boost.[13] A total dose of 20 to 30 Gy was administered, divided into 2 Gy daily fractions. The MST of this group of patients was 23.5 months and no significant toxicity was observed, other than hair loss (alopecia). These results suggest that the combination of BNCT with X-irradiation deserves further evaluation in a larger group of patients. In another Japanese trial, carried out by Yamamoto et al., BPA and BSH were infused over 1 hr, followed by BNCT at the Japan Research Reactor (JRR)−4 reactor.[30] Patients subsequently received an X-ray boost after completion of BNCT. The overall median survival time (MeST) was 27.1 months, and the 1 year and 2-year survival rates were 87.5 and 62.5%, respectively. Based on the reports of Miyatake, Kawabata, and Yamamoto, it appears that combining BNCT with an X-ray boost can produce a significant therapeutic gain. Further studies are needed to optimize this combined therapy and to evaluate it using a larger patient population.

28.8.3   Clinical studies in Finland

The team of clinicians and physicists at the Helsinki University Central Hospital and VTT Technical Research Center of Finland have treated a large number of patients with malignant gliomas (glioblastomas) and head and neck cancer who had undergone standard therapy, recurred, and subsequently received BNCT at the time of their recurrence using BPA as the boron delivery agent.[11][12] The median time to progression was 3 months, and the overall MeST was 7 months. It is difficult to compare these results with other reported results in patients with recurrent malignant gliomas, but they are a starting point for future studies using BNCT as salvage therapy in patients with recurrent tumors. Several hundred patients with recurrent head and neck cancers and brain tumors have been treated in Finland with BNCT using the Otaniemi nuclear reactor[31] which, as previously indicated, has now been closed down.

28.8.4   Clinical studies in Sweden

Finally, to conclude this section, the following is a brief summary of a clinical trial that was carried out in Sweden using BPA and an epithermal neutron beam, which had greater tissue penetration properties than the thermal beams originally used in Japan. This study differed significantly from all previous clinical trials in that the total amount of BPA administered was increased (900 mg/kg), and it was infused i.v. over 6 hours.[32][33] The longer infusion time of the drug was well tolerated by the 30 patients who were enrolled in this study. All were treated with 2 fields, and the average whole brain dose was 3.2–6.1 Gy (weighted), and the minimum dose to the tumor ranged from 15.4 to 54.3 Gy (w). There has been some disagreement among the Swedish investigators who carried out this study in terms of evaluation of the results. Based on incomplete survival data, the MeST was reported as 14.2 months and the time to tumor progression was 5.8 months.[32] Another group[33] had the complete survival data and concluded that the MeST was 17.7 months compared to 15.5 months that has been reported for patients who received standard therapy of surgery, followed by radiotherapy (RT) and the drug temozolomide (TMZ).[35] Furthermore, the frequency of adverse events were lower after BNCT (14%) than after RT alone (21%) and both of these were lower than those seen following RT in combination with TMZ. If this improved survival data, obtained using the higher dose of BPA and a 6-hour infusion time, can be confirmed by others, preferably in a randomized clinical trial, it could represent a significant step forward in BNCT of brain tumors, especially if combined with a photon boost.

28.9   Clinical Studies of BNCT for extracranial tumors

28.9.1 Head and neck cancers

The single most important clinical advance over the past 8 years in BNCT[36] has been the application of BNCT to treat patients with recurrent tumors of the head and neck region who had failed all other therapy. These studies were first initiated by Kato et al.[36] and subsequently followed by several other groups in Japan and by Kankaanranta and her co-workers in Finland.[12] All of these studies employed BPA as the boron delivery agent, either alone or in combination with BSH. A very heterogeneous group of patients with a variety of histopathologic types of tumors have been treated, the largest number of which had recurrent squamous cell carcinomas. Kato et al. have reported on a series of 26 patients with far-advanced cancer for whom there were no further treatment options.[36] Either BPA + BSH or BPA alone were administered by a 1 or 2 hr intravenous (i.v.) infusion, and this was followed by BNCT using an epithermal beam. In this series, there were complete regressions in 12 cases, 10 partial regressions, and progression in 3 cases. The MST was 13.6 months, and the 6-year survival was 24%. Significant treatment related complications ("adverse" events) included brain necrosis, osteomyelitis, transient mucositis, and alopecia.

Kankaanranta et al. have reported their results in a prospective Phase I/II study of 30 patients with inoperable, locally recurrent squamous cell carcinomas of the head and neck region.[12] Patients received either two or, in a few cases, one BNCT treatment using BPA (400 mg/kg), administered i.v. over 2 hours, followed by neutron irradiation. Of 29 evaluated patients, there were 13 complete and 9 partial remissions, with an overall response rate of 76%. The most common adverse event was oral mucositis, oral pain, and fatigue. Based on the clinical results, it was concluded that BNCT was effective for the treatment of inoperable, previously irradiated patients with head and neck cancer. Some responses were durable but progression was common, usually at the site of the previously recurrent tumor. As indicated in the section on neutron sources, all clinical studies have ended in Finland, based on economic difficulties of the two companies directly involved, VTT and Boneca.

Finally, a group in Taiwan has treated 12 patients with locally recurrent head and neck cancers at the Tsing Hua Open-pool Reactor (THOR) of the National Tsing Hua University.[37] Eleven of these patients received two fractions at 30-day intervals as part of a Phase I/II clinical trial with a total response rate of 58% with acceptable toxicity.

28.9.2 Other types of tumors

Melanoma

Other extracranial tumors that have been treated include malignant melanomas which originally was carried out in Japan by Yutaka Mishima at Kobe University and his clinical team[38] using BPA and a thermal neutron beam. Local control was achieved in almost all patients, and some were cured of their disease. More recently, Junichi Hiratsuka and his colleagues at Kawasaki Medical School Hospital have treated patients with melanoma of the head and neck region, vulva and vagina with impressive clinical results.[39] Finally, the first clinical trial of BNCT in Argentina was performed in October 2003[40] and several patients with cutaneous melanomas also have been treated.[40]

Colorectal cancer

Two patients with colon cancer, which had spread to the liver, have been treated by Zonta et al. in Italy.[41] The first was treated in 2001 and the second in mid-2003. The patients received an i.v. infusion of BPA, followed by removal of the liver (hepatectomy). This was treated out side of the body (extracorporeal) by BNCT and then re-transplanted into the patient. The first patient did remarkably well and survived for over 4 years after treatment, but the other died within a month of cardiac complications.[42] Clearly, this is a very challenging approach for the treatment of hepatic metastases, and it is unlikely that it will ever be widely used. Nevertheless, the good clinical results in the first patient established *proof of principle*. Finally Yanagie and his colleagues in Japan have treated several patients with recurrent rectal cancer using BNCT. Although no long-term results have been reported, there was evidence of short-term clinical responses.[43]

28.10 Conclusions

BNCT represents a joining together of nuclear technology, chemistry, biology, and medicine to treat malignant gliomas and recurrent head and neck cancers. Sadly, the lack of progress in developing more effective treatments for these tumors has been part of the driving force that continues to propel research in this field. BNCT may be best suited as an adjunctive treatment, used in combination with other modalities, including surgery, chemotherapy and external beam radiation therapy for those malignancies, whether primary or recurrent, for which there are no effective therapies. Clinical studies have demonstrated the safety of BNCT. The challenge facing clinicians and researchers is how to move forward. Advantages of BNCT include the potential ability to selectively deliver a radiation dose to the tumor with a much lower dose to surrounding normal tissues. This is an important feature that makes BNCT particularly attractive for salvage therapy of patients with a variety of malignancies who already have been heavily irradiated. Second, although it may be only palliative, it can produce striking clinical responses, as evidenced by the experiences of several groups treating patients with recurrent, therapeutically refractory head and neck cancers.

Problems with NCT and BNCT that need to be solved include:

1. The development of more tumor-selective boron delivery agents for BNCT. Similar problems are seen with Gd-NCT.

2. Accurate, real time dosimetry to better estimate the radiation doses delivered to the tumor and normal tissues.

3. Evaluation of recently constructed accelerator-based neutron sources as an alternative to nuclear reactors.

For a more detailed discussion of these problems and their solutions in BNCT, readers are referred to the published proceedings of the 13th and 14th International Congresses on Neutron Capture Therapy (2009 and 2011, 2014)[14][44][45] and a recently published review on the current status of BNCT of high grade gliomas and recurrent cancers of the head and neck region.[3] If the problems enumerated above can be solved BNCT could have an important role in twenty-first century cancer treatment of those malignancies that are loco-regional and that are presently incurable by other therapeutic modalities.[46]

28.11 See also

- Particle therapy, Neutron, proton or heavy ions (e.g. carbon)
  - Fast neutron therapy
  - Proton therapy

28.12 References

[1] Barth, R.F.; Soloway, A.H.; Fairchild, R.G. (1990). "Boron neutron capture therapy for cancer". *Scientific American* **263** (4): 100–3, 106–7. Bibcode:1990SciAm.263d.100B. doi:10.1038/scientificamerican1090-100. PMID 2173134.

[2] Barth, R.F.; Vicente, M.G.H.; Harling, O.K.; Kiger, W.S.; Riley, K.J.; Binns, P.J.; Wagner, F.M.; Suzuki, M.; Aihara, T.; Kato, I.; Kawabata, S. (2012). "Current status of boron neutron capture therapy of high grade gliomas and recurrent head and neck cancer". *Radiation Oncology* **7**: 146. doi:10.1186/1748-717X-7-146. PMC 3583064. PMID 22929110.

[3] Moss, R.L. (2014). "Critical review with an optimistic outlook on boron neutron capture therapy (BNCT)". *Applied Radiation and Isotopes* **88**: 2–11.

[4] Sweet, W.H. (1951). "The uses of nuclear disintegration in the diagnosis and treatment of brain tumor". *New England Journal of Medicine* **245** (23): 875–8. doi:10.1056/NEJM195112062452301. PMID 14882442.

[5] Farr, L.E.; Sweet, W.H.; Robertson, J.S.; Foster, C.G.; Locksley, H.B.; Sutherland, D.L.; Mendelsohn, M.L.; Stickley, E.E. (1954). "Neutron capture therapy with boron in the treatment of glioblastoma multiforme". *The American journal of roentgenology, radium therapy, and nuclear medicine* **71** (2): 279–93. PMID 13124616.

[6] Barth, R.F.; Coderre, J.A.; Vicente, M.G.H.; Blue, T.E. (2005). "Boron neutron capture therapy of cancer: Current status and future prospects". *Clinical Cancer Research* **11** (11): 3987–4002. doi:10.1158/1078-0432.CCR-05-0035. PMID 15930333.

[7] Vicente,M.G.H. (2006). "Boron in medicinal chemistry".*Anti-Cancer Agents in Medicinal Chemistry***6**(2):73.doi:10.2174/12.

[8] Nakagawa, Y.; Pooh, K.; Kobayashi, T.; Kageji, T.; Uyama, S.; Matsumura, A.; Kumada, H. (2003). "Clinical review of the Japanese experience with boron neutron capture therapy and a proposed strategy using epithermal neutron beams". *Journal of Neuro-Oncology* **62** (1–2): 87–99. doi:10.1023/A:1023234902479. PMID 12749705.

[9] Diaz, A.Z. (2003). "Assessment of the results from the phase I/II boron neutron capture therapy trials at the Brookhaven National Laboratory from a clinician's point of view". *Journal of Neuro-Oncology* **62** (1–2): 101–9. doi:10.1023/A:1023245123455. PMID 12749706.

[10] Busse, P.M.; Harling, O.K.; Palmer, M.R.; Kiger, W.S.; Kaplan, J.; Kaplan, I.; Chuang, C.F.; Goorley, J.T. et al. (2003). "A critical examination of the results from the Harvard-MIT NCT program phase I clinical trial of neutron capture therapy for intracranial disease". *Journal of Neuro-oncology* **62** (1–2): 111–21. doi:10.1007/BF02699938. PMID 12749707.

[11] Kankaanranta, L.; Seppälä, T.; Koivunoro, H.; Välimäki, P.; Beule, A.; Collan, J.; Kortesniemi, M.; Uusi-Simola, J. et al. (2011). "L-Boronophenylalanine-mediated boron neutron capture therapy for malignant glioma progressing after external beam radiation therapy: A Phase I study". *International Journal of Radiation Oncology • Biology • Physics* **80** (2): 369–76. doi:10.1016/j.ijrobp.2010.02.031. PMID 21236605.

[12] Kankaanranta, L.; Seppälä, T.; Koivunoro, H.; Saarilahti, K.; Atula, T.; Collan, J.; Salli, E.; Kortesniemi, M. et al. (2012). "Boron neutron capture therapy in the treatment of locally recurred head-and-neck cancer: Final analysis of a Phase I/II trial". *International Journal of Radiation Oncology • Biology • Physics* **82** (1): e67–75. doi:10.1016/j.ijrobp.2010.09.057. PMID 21300462.

[13] Kawabata, S.; Miyatake, S.-I.; Kuroiwa, T.; Yokoyama, K.; Doi, A.; Iida, K.; Miyata, S.; Nonoguchi, N. et al. (2009). "Boron neutron capture therapy for newly diagnosed glioblastoma". *Journal of Radiation Research* **50** (1): 51–60. doi:10.1269/jrr.08043. PMID 18957828.

[14] Miyatake, S.-I.; Kawabata, S.; Yokoyama, K.; Kuroiwa, T.; Michiue, H.; Sakurai, Y.; Kumada, H.; Suzuki, M. et al. (2008). "Survival benefit of boron neutron capture therapy for recurrent malignant gliomas". *Journal of Neuro-Oncology* **91** (2): 199–206. doi:10.1007/s11060-008-9699-x. PMID 18813875.

[15] Coderre, J.A.; Morris, G.M. (1999). "The radiation biology of boron neutron capture therapy". *Radiation research* **151** (1): 1–18. doi:10.2307/3579742. PMID 9973079.

[16] Nigg, D.W. (2003). "Computational dosimetry and treatment planning considerations for neutron capture therapy". *Journal of Neuro-Oncology* **62** (1–2): 75–86. doi:10.1023/A:1023241022546. PMID 12749704.

[17] Soloway, A.H., Tjarks, W., Barnum, B.A., Rong, F-G., Barth, R.F., Codogni, I.M., and Wilson, J.G.: The chemistry of neutron capture therapy. Chemical Rev 98: 1515-1562, 1998.

[18] Cerullo, N.; Bufalino, D.; Daquino, G. (2009). "Progress in the use of gadolinium for NCT". *Applied Radiation and Isotopes* **67** (7–8): S157–60. doi:10.1016/j.apradiso.2009.03.109. PMID 19410468.

[19] Yasui, L.S.; Andorf, C.; Schneider, L.; Kroc, T.; Lennox, A.; Saroja, K.R. (2008). "Gadolinium neutron capture in glioblastoma multiforme cells". *International Journal of Radiation Biology* **84** (12): 1130–9. doi:10.1080/09553000802538092. PMID 19061138.

[20] Nemoto, H.; Cai, J.; Nakamura, H.; Fujiwara, M.; Yamamoto, Y. (1999). "The synthesis of a carborane gadolinium–DTPA complex for boron neutron capture therapy".*Journal of Organometallic Chemistry***581**:170–5.doi:10.1016/S0022-328X(99-2.

[21] Tokumitsu, H.; Hiratsuka, J.; Sakurai, Y.; Kobayashi, T.; Ichikawa, H.; Fukumori, Y. (2000). "Gadolinium neutron-capture therapy using novel gadopentetic acid–chitosan complex nanoparticles: In vivo growth suppression of experimental melanoma solid tumor". *Cancer Letters* **150** (2): 177–82. doi:10.1016/S0304-3835(99)00388-2. PMID 10704740.

[22] De Stasio, G.; Rajesh, D.; Ford, J.M.; Daniels, M.J.; Erhardt, R.J.; Frazer, B.H.; Tyliszczak, T.; Gilles, M.K. et al. (2006). "Motexafin-gadolinium taken up in vitro by at least 90% of glioblastoma cell nuclei". *Clinical Cancer Research* **12** (1): 206–13. doi:10.1158/1078-0432.CCR-05-0743. PMID 16397044.

[23] Geninatti-Crich, S.; Alberti, D.; Szabo, I.; Deagostino, A.; Toppino, A.; Barge, A.; Ballarini, F.; Bortolussi, S. et al. (2011). "MRI-guided neutron capture therapy by use of a dual gadolinium/boron agent targeted at tumour cells through upregulated low-density lipoprotein transporters". *Chemistry* **17** (30): 8479–86. doi:10.1002/chem.201003741. PMID 21671294.

[24] Harling, O.K. (2009). "Fission reactor based epithermal neutron irradiation facilities for routine clinical application in BNCT—Hatanaka memorial lecture". *Applied Radiation and Isotopes* **67** (7–8): S7–11. doi:10.1016/j.apradiso.2009.03.095. PMID 19428265.

[25] Yiguo, L, Pu X, Xiao, W et al. *Start-up of the first in-hospital neutron irradiator (IHNI-1) & Presentation of the BNCT development status in China* (PDF). New Challenges in Neutron Capture Therapy 2010: Proceedings of the 14th International Congress on Neutron Capture Therapy. Buenos Aires. pp. 371–4.

[26] Mitsumoto,T, Yajima, S, Tsutsui, H et al. *Cyclotron-based neutron source for BNCT* (PDF). New Challenges in Neutron Capture Therapy 2010: Proceedings of the 14th International Congress on Neutron Capture Therapy. Buenos Aires. pp. 519–22.

[27] Sweet WH (1983). *Practical problems in the past in the use of boron-slow neutron capture therapy in the treatment of glioblastoma multiforme.* Proceedings of the First International Symposium on Neutron Capture Therapy. pp. 376–8.

[28] Hatanaka, H.; Nakagawa, Y. (1994). "Clinical results of long-surviving brain tumor patients who underwent boron neutron capture therapy". *International Journal of Radiation Oncology • Biology • Physics* **28** (5): 1061–6. doi:10.1016/0360-3016(94)90479-0. PMID 8175390.

[29] Barth, R.F.; Grecula, J.C.; Yang, W.; Rotaru, J.H.; Nawrocky, M.; Gupta, N.; Albertson, B.J.; Ferketich, A.K. et al. (2004). "Combination of boron neutron capture therapy and external beam radiotherapy for brain tumors". *International Journal of Radiation Oncology • Biology • Physics* **58** (1): 267–77. doi:10.1016/S0360-3016(03)01613-4. PMID 14697448.

[30] Yamamoto, T.; Nakai, K.; Nariai, T.; Kumada, H.; Okumura, T.; Mizumoto, M.; Tsuboi, K.; Zaboronok, A. et al. (2011). "The status of Tsukuba BNCT trial: BPA-based boron neutron capture therapy combined with X-ray irradiation". *Applied Radiation and Isotopes* **69** (12): 1817–8. doi:10.1016/j.apradiso.2011.02.013. PMID 21393005.

[31] http://www.boneca.fi/medical-info/referrals/17[][]

[32] Henriksson, R.; Capala, J.; Michanek, A.; Lindahl, S.-Å.; Salford, L.G.; Franzén, L.; Blomquist, E.; Westlin, J.-E. et al. (2008). "Boron neutron capture therapy (BNCT) for glioblastoma multiforme: A phase II study evaluating a prolonged high-dose of boronophenylalanine (BPA)". *Radiotherapy and Oncology* **88** (2): 183–91. doi:10.1016/j.radonc.2006.04.015. PMID 18336940.

[33] Sköld, K.; Gorlia, T.; Pellettieri, L.; Giusti, V.; H-Stenstam, B.; Hopewell, J.W. (2010). "Boron neutron capture therapy for newly diagnosed glioblastoma multiforme: An assessment of clinical potential". *British Journal of Radiology* **83** (991): 596–603. doi:10.1259/bjr/56953620. PMC 3473677. PMID 20603410.

[34] Wittig, A., Hideghety, K., Paquis, P. et al. (2002). Sauerwein, W.; Mass, R.; Wittig, A., eds. *Current clinical results of the EORTC – study 11961.* Research and Development in Neutron Capture Therapy Proc. 10th Intl. Congress on Neutron Capture Therapy. pp. 1117–22.

[35] Stupp, R.; Hegi, M.E.; Mason, W.P.; Van Den Bent, M.J.; Taphoorn, M.J.B.; Janzer, R.C.; Ludwin, S.K.; Allgeier, A. et al. (2009). "Effects of radiotherapy with concomitant and adjuvant temozolomide versus radiotherapy alone on survival in glioblastoma in a randomised phase III study: 5-year analysis of the EORTC-NCIC trial". *The Lancet Oncology* **10** (5): 459–66. doi:10.1016/S1470-2045(09)70025-7. PMID 19269895.

[36] Kato, I.; Fujita, Y.; Maruhashi, A.; Kumada, H.; Ohmae, M.; Kirihata, M.; Imahori, Y.; Suzuki, M. et al. (2009). "Effectiveness of boron neutron capture therapy for recurrent head and neck malignancies". *Applied Radiation and Isotopes* **67** (7–8): S37–42. doi:10.1016/j.apradiso.2009.03.103. PMID 19409799.

[37] Wang, L.W.; Wang, S.J.; Chu, P.Y.; Ho, C.Y.; Jiang, S.H.; Liu, Y.W.H.; Liu, Y.H.; Liu, H.M. et al. (2011). "BNCT for locally recurrent head and neck cancer: Preliminary clinical experience from a phase I/II trial at Tsing Hua Open-Pool Reactor". *Applied Radiation and Isotopes* **69** (12): 1803–6. doi:10.1016/j.apradiso.2011.03.008. PMID 21478023.

[38] Mishima, Y. (1996). "Selective thermal neutron capture therapy of cancer cells using their specific metabolic activities—melanoma as prototype". In Mishima, Y. *Cancer neutron capture therapy.* pp. 1–26. doi:10.1007/978-1-4757-9567-7_1. ISBN 978-1-4757-9569-1.

[39] Hiratsuka, J. Clinical results of BNCT for head and neck melanoma. 16th Intl' Congress on Neutron Capture Therapy, Helsinki, Finland, June 14–19, 2014

[40] "The BNCT Project at the National Atomic Energy Commission (CNEA)". Comision Nacional de Energia Atomica.

[41] Zonta, A.; Pinelli, T.; Prati, U.; Roveda, L.; Ferrari, C.; Clerici, A.M.; Zonta, C.; Mazzini, G. et al. (2009). "Extra-corporeal liver BNCT for the treatment of diffuse metastases: What was learned and what is still to be learned". *Applied Radiation and Isotopes* **67** (7–8): S67–75. doi:10.1016/j.apradiso.2009.03.087. PMID 19394837.

[42] Zonta, A.; Prati, U.; Roveda, L.; Ferrari, C.; Zonta, S.; Clerici, A.M.; Zonta, C.; Pinelli, T. et al. (2006). "Clinical lessons from the first applications of BNCT on unresectable liver metastases". *Journal of Physics: Conference Series* **41** (1): 484–95. Bibcode:2006JPhCS..41..484Z. doi:10.1088/1742-6596/41/1/054.

[43] Yanagie, H., Oyama, K., Hatae, R. et al. Clinical experinces of boron neutron capture therapy to recurrenced rectal cancers. Abstracts 16th Intl' Congress on Neutron Capture Therapy. Helsinki, Finland, June 14–19, 2014

[44] Altieri, S.; Bortolussi, S.; Barth, R.F.; Roveda, L.; Zonta, A. (2009). "Thirteenth International Congress on Neutron Capture Therapy". *Applied Radiation and Isotopes* **67** (7–8): S1–2. doi:10.1016/j.apradiso.2009.03.009. PMID 19395267.

[45] Yamamoto, T.; Nakai, K.; Matsumura, A. (2011). "15th International Congress on Neutron Capture Therapy: Impact of a new radiotherapy against cancer". *Applied Radiation and Isotopes* **88**: 1–246.

[46] Barth, R.F. (2009). "Boron neutron capture therapy at the crossroads: Challenges and opportunities". *Applied Radiation and Isotopes* **67** (7–8): S3–6. doi:10.1016/j.apradiso.2009.03.102. PMID 19467879.

28.13   External links

- Helsinki University Central Hospital and Technical Research Centre of Finland BNCT Project

- Boron and Gadolinium Neutron Capture Therapy for Cancer Treatment

- MIT Nuclear Reactor Lab overview of BNCT

- Washington State University Nuclear Radiation Center BNCT Overview

Chapter 29

Nuclear fission

"Splitting the atom" redirects here. For the EP, see Splitting the Atom.

  In nuclear physics and nuclear chemistry, **nuclear fission** is either a nuclear reaction or a radioactive decay process in which the nucleus of an atom splits into smaller parts (lighter nuclei). The fission process often produces free neutrons and photons (in the form of gamma rays), and releases a very large amount of energy even by the energetic standards of radioactive decay.

Nuclear fission of heavy elements was discovered on December 17, 1938 by German Otto Hahn and his assistant Fritz Strassmann, and explained theoretically in January 1939 by Lise Meitner and her nephew Otto Robert Frisch. Frisch named the process by analogy with biological fission of living cells. It is an exothermic reaction which can release large amounts of energy both as electromagnetic radiation and as kinetic energy of the fragments (heating the bulk material where fission takes place). In order for fission to produce energy, the total binding energy of the resulting elements must be less negative (higher energy) than that of the starting element.

Fission is a form of nuclear transmutation because the resulting fragments are not the same element as the original atom. The two nuclei produced are most often of comparable but slightly different sizes, typically with a mass ratio of products of about 3 to 2, for common fissile isotopes.[1][2] Most fissions are binary fissions (producing two charged fragments), but occasionally (2 to 4 times per 1000 events), *three* positively charged fragments are produced, in a ternary fission. The smallest of these fragments in ternary processes ranges in size from a proton to an argon nucleus.

Apart from fission induced by a neutron, harnessed and exploited by humans, a natural form of spontaneous radioactive decay (not requiring a neutron) is also referred to as fission, and occurs especially in very high-mass-number isotopes. Spontaneous fission was discovered in 1940 by Flyorov, Petrzhak and Kurchatov[3] in Moscow, when they decided to confirm that, without bombardment by neutrons, the fission rate of uranium was indeed negligible, as predicted by Niels Bohr; it wasn't.[3]

The unpredictable composition of the products (which vary in a broad probabilistic and somewhat chaotic manner) distinguishes fission from purely quantum-tunnelling processes such as proton emission, alpha decay and cluster decay, which give the same products each time. Nuclear fission produces energy for nuclear power and drives the explosion of nuclear weapons. Both uses are possible because certain substances called nuclear fuels undergo fission when struck by fission neutrons, and in turn emit neutrons when they break apart. This makes possible a self-sustaining nuclear chain reaction that releases energy at a controlled rate in a nuclear reactor or at a very rapid uncontrolled rate in a nuclear weapon.

The amount of free energy contained in nuclear fuel is millions of times the amount of free energy contained in a similar mass of chemical fuel such as gasoline, making nuclear fission a very dense source of energy. The products of nuclear fission, however, are on average far more radioactive than the heavy elements which are normally fissioned as fuel, and remain so for significant amounts of time, giving rise to a nuclear waste problem. Concerns over nuclear waste accumulation and over the destructive potential of nuclear weapons may counterbalance the desirable qualities of fission as an energy source, and give rise to ongoing political debate over nuclear power.

29.1 Physical overview

29.1.1 Mechanism

Nuclear fission can occur without neutron bombardment as a type of radioactive decay. This type of fission (called spontaneous fission) is rare except in a few heavy isotopes. In engineered nuclear devices, essentially all nuclear fission occurs as a "nuclear reaction" — a bombardment-driven process that results from the collision of two subatomic particles. In nuclear reactions, a subatomic particle collides with an atomic nucleus and causes changes to it. Nuclear reactions are thus driven by the mechanics of bombardment, not by the relatively constant exponential decay and half-life characteristic of spontaneous radioactive processes.

Many types of nuclear reactions are currently known. Nuclear fission differs importantly from other types of nuclear reactions, in that it can be amplified and sometimes controlled via a nuclear chain reaction (one type of general chain reaction). In such a reaction, free neutrons released by each fission event can trigger yet more events, which in turn release more neutrons and cause more fissions.

The chemical element isotopes that can sustain a fission chain reaction are called nuclear fuels, and are said to be *fissile*. The most common nuclear fuels are ^{235}U (the isotope of uranium with an atomic mass of 235 and of use in nuclear reactors) and ^{239}Pu (the isotope of plutonium with an atomic mass of 239). These fuels break apart into a bimodal range of chemical elements with atomic masses centering near 95 and 135 **u** (fission products). Most nuclear fuels undergo spontaneous fission only very slowly, decaying instead mainly via an alpha/beta decay chain over periods of millennia to eons. In a nuclear reactor or nuclear weapon, the overwhelming majority of fission events are induced by bombardment with another particle, a neutron, which is itself produced by prior fission events.

Nuclear fissions in fissile fuels are the result of the nuclear excitation energy produced when a fissile nucleus captures a neutron. This energy, resulting from the neutron capture, is a result of the attractive nuclear force acting between the neutron and nucleus. It is enough to deform the nucleus into a double-lobed "drop," to the point that nuclear fragments exceed the distances at which the nuclear force can hold two groups of charged nucleons together, and when this happens, the two fragments complete their separation and then are driven further apart by their mutually repulsive charges, in a process which becomes irreversible with greater and greater distance. A similar process occurs in fissionable isotopes (such as uranium-238), but in order to fission, these isotopes require additional energy provided by fast neutrons (such as those produced by nuclear fusion in thermonuclear weapons).

The liquid drop model of the atomic nucleus predicts equal-sized fission products as an outcome of nuclear deformation. The more sophisticated nuclear shell model is needed to mechanistically explain the route to the more energetically favorable outcome, in which one fission product is slightly smaller than the other. A theory of the fission based on shell model has been formulated by Maria Goeppert Mayer.

The most common fission process is binary fission, and it produces the fission products noted above, at 95±15 and 135±15 **u**. However, the binary process happens merely because it is the most probable. In anywhere from 2 to 4 fissions per 1000 in a nuclear reactor, a process called ternary fission produces three positively charged fragments (plus neutrons) and the smallest of these may range from so small a charge and mass as a proton (Z=1), to as large a fragment as argon (Z=18). The most common small fragments, however, are composed of 90% helium-4 nuclei with more energy than alpha particles from alpha decay (so-called "long range alphas" at ~ 16 MeV), plus helium-6 nuclei, and tritons (the nuclei of tritium). The ternary process is less common, but still ends up producing significant helium-4 and tritium gas buildup in the fuel rods of modern nuclear reactors.[4]

29.1.2 Energetics

Input

The fission of a heavy nucleus requires a total input energy of about 7 to 8 million electron volts (MeV) to initially overcome the nuclear force which holds the nucleus into a spherical or nearly spherical shape, and from there, deform it into a two-lobed ("peanut") shape in which the lobes are able to continue to separate from each other, pushed by their mutual positive charge, in the most common process of binary fission (two positively charged fission products + neutrons). Once the nuclear lobes have been pushed to a critical distance, beyond which the short range strong force can no longer hold them together, the process of their separation proceeds from the energy of the (longer range) electromagnetic repulsion between the fragments. The result is two fission fragments moving away from each other, at high energy.

About 6 MeV of the fission-input energy is supplied by the simple binding of an extra neutron to the heavy nucleus via the strong force; however, in many fissionable isotopes, this amount of energy is not enough for fission. Uranium-238, for example, has a near-zero fission cross section for neutrons of less than one MeV energy. If no additional energy is supplied by any other mechanism, the nucleus will not fission, but will merely absorb the neutron, as happens when U-238 absorbs slow and even some fraction of fast neutrons, to become U-239. The remaining energy to initiate fission can be supplied by two other mechanisms: one of these is more kinetic energy of the incoming neutron, which is increasingly able to fission a fissionable heavy nucleus as it exceeds a kinetic energy of one MeV or more (so-called fast neutrons). Such high energy neutrons are able to fission U-238 directly (see thermonuclear weapon for application, where the fast neutrons are supplied by nuclear fusion). However, this process cannot happen to a great extent in a nuclear reactor, as too small a fraction of the fission neutrons produced by any type of fission have enough energy to efficiently fission U-238 (fission neutrons have a mode energy of 2 MeV, but a median of only 0.75 MeV, meaning half of them have less than this insufficient energy).[5]

Among the heavy actinide elements, however, those isotopes that have an odd number of neutrons (such as U-235 with 143 neutrons) bind an extra neutron with an additional 1 to 2 MeV of energy over an isotope of the same element with an even number of neutrons (such as U-238 with 146 neutrons). This extra binding energy is made available as a result of the mechanism of neutron pairing effects. This extra energy results from the Pauli exclusion principle allowing an extra neutron to occupy the same nuclear orbital as the last neutron in the nucleus, so that the two form a pair. In such isotopes, therefore, no neutron kinetic energy is needed, for all the necessary energy is supplied by absorption of any neutron, either of the slow or fast variety (the former are used in moderated nuclear reactors, and the latter are used in fast neutron reactors, and in weapons). As noted above, the subgroup of fissionable elements that may be fissioned efficiently with their own fission neutrons (thus potentially causing a nuclear chain reaction in relatively small amounts of the pure material) are termed "fissile." Examples of fissile isotopes are U-235 and plutonium-239.

Output

Typical fission events release about two hundred million eV (200 MeV) of energy for each fission event. The exact isotope which is fissioned, and whether or not it is fissionable or fissile, has only a small impact on the amount of energy released. This can be easily seen by examining the curve of binding energy (image below), and noting that the average binding energy of the actinide nuclides beginning with uranium is around 7.6 MeV per nucleon. Looking further left on the curve of binding energy, where the fission products cluster, it is easily observed that the binding energy of the fission products tends to center around 8.5 MeV per nucleon. Thus, in any fission event of an isotope in the actinide's range of mass, roughly 0.9 MeV is released per nucleon of the starting element. The fission of U235 by a slow neutron yields nearly identical energy to the fission of U238 by a fast neutron. This energy release profile holds true for thorium and the various minor actinides as well.[6]

By contrast, most chemical oxidation reactions (such as burning coal or TNT) release at most a few eV per event. So, nuclear fuel contains at least ten million times more usable energy per unit mass than does chemical fuel. The energy of nuclear fission is released as kinetic energy of the fission products and fragments, and as electromagnetic radiation in the form of gamma rays; in a nuclear reactor, the energy is converted to heat as the particles and gamma rays collide with the atoms that make up the reactor and its working fluid, usually water or occasionally heavy water or molten salts.

When a uranium nucleus fissions into two daughter nuclei fragments, about 0.1 percent of the mass of the uranium nucleus[7] appears as the fission energy of ~200 MeV. For uranium-235 (total mean fission energy 202.5 MeV), typically ~169 MeV appears as the kinetic energy of the daughter nuclei, which fly apart at about 3% of the speed of light, due to Coulomb repulsion. Also, an average of 2.5 neutrons are emitted, with a mean kinetic energy per neutron of ~2 MeV (total of 4.8 MeV).[8] The fission reaction also releases ~7 MeV in prompt gamma ray photons. The latter figure means that a nuclear fission explosion or criticality accident emits about 3.5% of its energy as gamma rays, less than 2.5% of its energy as fast neutrons (total of both types of radiation ~ 6%), and the rest as kinetic energy of fission fragments (this appears almost immediately when the fragments impact surrounding matter, as simple heat). In an atomic bomb, this heat may serve to raise the temperature of the bomb core to 100 million kelvin and cause secondary emission of soft X-rays, which convert some of this energy to ionizing radiation. However, in nuclear reactors, the fission fragment kinetic energy remains as low-temperature heat, which itself causes little or no ionization.

So-called neutron bombs (enhanced radiation weapons) have been constructed which release a larger fraction of their energy as ionizing radiation (specifically, neutrons), but these are all thermonuclear devices which rely on the nuclear fusion stage to produce the extra radiation. The energy dynamics of pure fission bombs always remain at about 6% yield of the total in radiation, as a prompt result of fission.

The total *prompt fission* energy amounts to about 181 MeV, or ~ 89% of the total energy which is eventually released by fission over time. The remaining ~ 11% is released in beta decays which have various half-lives, but begin as a process in the fission products immediately; and in delayed gamma emissions associated with these beta decays. For example, in uranium-235 this delayed energy is divided into about 6.5 MeV in betas, 8.8 MeV in antineutrinos (released at the same time as the betas), and finally, an additional 6.3 MeV in delayed gamma emission from the excited beta-decay products (for a mean total of ~10 gamma ray emissions per fission, in all). Thus, about 6.5% of the total energy of fission is released some time after the event, as non-prompt or delayed ionizing radiation, and the delayed ionizing energy is about evenly divided between gamma and beta ray energy.

In a reactor that has been operating for some time, the radioactive fission products will have built up to steady state concentrations such that their rate of decay is equal to their rate of formation, so that their fractional total contribution to reactor heat (via beta decay) is the same as these radioisotopic fractional contributions to the energy of fission. Under these conditions, the 6.5% of fission which appears as delayed ionizing radiation (delayed gammas and betas from radioactive fission products) contributes to the steady-state reactor heat production under power. It is this output fraction which remains when the reactor is suddenly shut down (undergoes scram). For this reason, the reactor decay heat output begins at 6.5% of the full reactor steady state fission power, once the reactor is shut down. However, within hours, due to decay of these isotopes, the decay power output is far less. See decay heat for detail.

The remainder of the delayed energy (8.8 MeV/202.5 MeV = 4.3% of total fission energy) is emitted as antineutrinos, which as a practical matter, are not considered "ionizing radiation." The reason is that energy released as antineutrinos is not captured by the reactor material as heat, and escapes directly through all materials (including the Earth) at nearly the speed of light, and into interplanetary space (the amount absorbed is minuscule). Neutrino radiation is ordinarily not classed as ionizing radiation, because it is almost entirely not absorbed and therefore does not produce effects (although the very rare neutrino event is ionizing). Almost all of the rest of the radiation (6.5% delayed beta and gamma radiation) is eventually converted to heat in a reactor core or its shielding.

Some processes involving neutrons are notable for absorbing or finally yielding energy — for example neutron kinetic energy does not yield heat immediately if the neutron is captured by a uranium-238 atom to breed plutonium-239, but this energy is emitted if the plutonium-239 is later fissioned. On the other hand, so-called delayed neutrons emitted as radioactive decay products with half-lives up to several minutes, from fission-daughters, are very important to reactor control, because they give a characteristic "reaction" time for the total nuclear reaction to double in size, if the reaction is run in a "delayed-critical" zone which deliberately relies on these neutrons for a supercritical chain-reaction (one in which each fission cycle yields more neutrons than it absorbs). Without their existence, the nuclear chain-reaction would be prompt critical and increase in size faster than it could be controlled by human intervention. In this case, the first experimental atomic reactors would have run away to a dangerous and messy "prompt critical reaction" before their operators could have manually shut them down (for this reason, designer Enrico Fermi included radiation-counter-triggered control rods, suspended by electromagnets, which could automatically drop into the center of Chicago Pile-1). If these delayed neutrons are captured without producing fissions, they produce heat as well.[9]

29.1.3   Product nuclei and binding energy

Main articles: fission product and fission product yield

In fission there is a preference to yield fragments with even proton numbers, which is called the odd-even effect on the fragments charge distribution. However, no odd-even effect is observed on fragment **mass number** distribution. This result is attributed to nucleon pair breaking.

In nuclear fission events the nuclei may break into any combination of lighter nuclei, but the most common event is not fission to equal mass nuclei of about mass 120; the most common event (depending on isotope and process) is a slightly unequal fission in which one daughter nucleus has a mass of about 90 to 100 **u** and the other the remaining 130 to 140 **u**.[10] Unequal fissions are energetically more favorable because this allows one product to be closer to the energetic minimum near mass 60 **u** (only a quarter of the average fissionable mass), while the other nucleus with mass 135 **u** is still not far out of the range of the most tightly bound nuclei (another statement of this, is that the atomic binding energy curve is slightly steeper to the left of mass 120 **u** than to the right of it).

29.1.4 Origin of the active energy and the curve of binding energy

Nuclear fission of heavy elements produces energy because the specific binding energy (binding energy per mass) of intermediate-mass nuclei with atomic numbers and atomic masses close to ^{62}Ni and ^{56}Fe is greater than the nucleon-specific binding energy of very heavy nuclei, so that energy is released when heavy nuclei are broken apart. The total rest masses of the fission products (**Mp**) from a single reaction is less than the mass of the original fuel nucleus (**M**). The excess mass $\Delta m = M - Mp$ is the invariant mass of the energy that is released as photons (gamma rays) and kinetic energy of the fission fragments, according to the mass-energy equivalence formula $E = mc^2$.

The variation in specific binding energy with atomic number is due to the interplay of the two fundamental forces acting on the component nucleons (protons and neutrons) that make up the nucleus. Nuclei are bound by an attractive nuclear force between nucleons, which overcomes the electrostatic repulsion between protons. However, the nuclear force acts only over relatively short ranges (a few nucleon diameters), since it follows an exponentially decaying Yukawa potential which makes it insignificant at longer distances. The electrostatic repulsion is of longer range, since it decays by an inverse-square rule, so that nuclei larger than about 12 nucleons in diameter reach a point that the total electrostatic repulsion overcomes the nuclear force and causes them to be spontaneously unstable. For the same reason, larger nuclei (more than about eight nucleons in diameter) are less tightly bound per unit mass than are smaller nuclei; breaking a large nucleus into two or more intermediate-sized nuclei releases energy. The origin of this energy is the nuclear force, which intermediate-sized nuclei allows to act more efficiently, because each nucleon has more neighbors which are within the short range attraction of this force. Thus less energy is needed in the smaller nuclei and the difference to the state before is set free.

Also because of the short range of the strong binding force, large stable nuclei must contain proportionally more neutrons than do the lightest elements, which are most stable with a **1 to 1 ratio** of protons and neutrons. Nuclei which have more than 20 protons cannot be stable unless they have more than an equal number of neutrons. Extra neutrons stabilize heavy elements because they add to strong-force binding (which acts between all nucleons) without adding to proton–proton repulsion. Fission products have, on average, about the same ratio of neutrons and protons as their parent nucleus, and are therefore usually unstable to beta decay (which changes neutrons to protons) because they have proportionally too many neutrons compared to stable isotopes of similar mass.

This tendency for fission product nuclei to beta-decay is the fundamental cause of the problem of radioactive high level waste from nuclear reactors. Fission products tend to be beta emitters, emitting fast-moving electrons to conserve electric charge, as excess neutrons convert to protons in the fission-product atoms. See Fission products (by element) for a description of fission products sorted by element.

29.1.5 Chain reactions

Main article: Nuclear chain reaction

Several heavy elements, such as uranium, thorium, and plutonium, undergo both spontaneous fission, a form of radioactive decay and *induced fission*, a form of nuclear reaction. Elemental isotopes that undergo induced fission when struck by a free neutron are called fissionable; isotopes that undergo fission when struck by a thermal, slow moving neutron are also called fissile. A few particularly fissile and readily obtainable isotopes (notably ^{233}U, ^{235}U and ^{239}Pu) are called nuclear fuels because they can sustain a chain reaction and can be obtained in large enough quantities to be useful.

All fissionable and fissile isotopes undergo a small amount of spontaneous fission which releases a few free neutrons into any sample of nuclear fuel. Such neutrons would escape rapidly from the fuel and become a free neutron, with a mean lifetime of about 15 minutes before decaying to protons and beta particles. However, neutrons almost invariably impact and are absorbed by other nuclei in the vicinity long before this happens (newly created fission neutrons move at about 7% of the speed of light, and even moderated neutrons move at about 8 times the speed of sound). Some neutrons will impact fuel nuclei and induce further fissions, releasing yet more neutrons. If enough nuclear fuel is assembled in one place, or if the escaping neutrons are sufficiently contained, then these freshly emitted neutrons outnumber the neutrons that escape from the assembly, and a *sustained nuclear chain reaction* will take place.

An assembly that supports a sustained nuclear chain reaction is called a critical assembly or, if the assembly is almost entirely made of a nuclear fuel, a critical mass. The word "critical" refers to a cusp in the behavior of the differential equation that governs the number of free neutrons present in the fuel: if less than a critical mass is present, then the amount of neutrons is determined by radioactive decay, but if a critical mass or more is present, then the amount of neutrons is controlled instead by the physics of the chain reaction. The actual mass of a *critical mass* of nuclear fuel depends strongly on the geometry and surrounding materials.

Not all fissionable isotopes can sustain a chain reaction. For example, ^{238}U, the most abundant form of uranium, is fissionable but not fissile: it undergoes induced fission when impacted by an energetic neutron with over 1 MeV of kinetic energy. However, too few of the neutrons produced by ^{238}U fission are energetic enough to induce further fissions in ^{238}U, so no chain reaction is possible with this isotope. Instead, bombarding ^{238}U with slow neutrons causes it to absorb them (becoming ^{239}U) and decay by beta emission to ^{239}Np which then decays again by the same process to ^{239}Pu; that process is used to manufacture ^{239}Pu in breeder reactors. In-situ plutonium production also contributes to the neutron chain reaction in other types of reactors after sufficient plutonium-239 has been produced, since plutonium-239 is also a fissile element which serves as fuel. It is estimated that up to half of the power produced by a standard "non-breeder" reactor is produced by the fission of plutonium-239 produced in place, over the total life-cycle of a fuel load.

Fissionable, non-fissile isotopes can be used as fission energy source even without a chain reaction. Bombarding ^{238}U with fast neutrons induces fissions, releasing energy as long as the external neutron source is present. This is an important effect in all reactors where fast neutrons from the fissile isotope can cause the fission of nearby ^{238}U nuclei, which means that some small part of the ^{238}U is "burned-up" in all nuclear fuels, especially in fast breeder reactors that operate with higher-energy neutrons. That same fast-fission effect is used to augment the energy released by modern thermonuclear weapons, by jacketing the weapon with ^{238}U to react with neutrons released by nuclear fusion at the center of the device. But the explosive effects of nuclear fission chain reactions can be reduced by using substances like moderators which slow down the speed of secondary neutrons.[11]

29.1.6   Fission reactors

Critical fission reactors are the most common type of nuclear reactor. In a critical fission reactor, neutrons produced by fission of fuel atoms are used to induce yet more fissions, to sustain a controllable amount of energy release. Devices that produce engineered but non-self-sustaining fission reactions are subcritical fission reactors. Such devices use radioactive decay or particle accelerators to trigger fissions.

Critical fission reactors are built for three primary purposes, which typically involve different engineering trade-offs to take advantage of either the heat or the neutrons produced by the fission chain reaction:

- *power reactors* are intended to produce heat for nuclear power, either as part of a generating station or a local power system such as a nuclear submarine.

- *research reactors* are intended to produce neutrons and/or activate radioactive sources for scientific, medical, engineering, or other research purposes.

- *breeder reactors* are intended to produce nuclear fuels in bulk from more abundant isotopes. The better known fast breeder reactor makes ^{239}Pu (a nuclear fuel) from the naturally very abundant ^{238}U (not a nuclear fuel). Thermal breeder reactors previously tested using ^{232}Th to breed the fissile isotope ^{233}U (thorium fuel cycle) continue to be studied and developed.

While, in principle, all fission reactors can act in all three capacities, in practice the tasks lead to conflicting engineering goals and most reactors have been built with only one of the above tasks in mind. (There are several early counter-examples, such as the Hanford N reactor, now decommissioned). Power reactors generally convert the kinetic energy of fission products into heat, which is used to heat a working fluid and drive a heat engine that generates mechanical or electrical power. The working fluid is usually water with a steam turbine, but some designs use other materials such as gaseous helium. Research reactors produce neutrons that are used in various ways, with the heat of fission being treated as an unavoidable waste product. Breeder reactors are a specialized form of research reactor, with the caveat that the sample being irradiated is usually the fuel itself, a mixture of ^{238}U and ^{235}U. For a more detailed description of the physics and operating principles of critical fission reactors, see nuclear reactor physics. For a description of their social, political, and environmental aspects, see nuclear power.

29.1.7   Fission bombs

One class of nuclear weapon, a *fission bomb* (not to be confused with the *fusion bomb*), otherwise known as an *atomic bomb* or *atom bomb*, is a fission reactor designed to liberate as much energy as possible as rapidly as possible, before the released energy causes the reactor to explode (and the chain reaction to stop). Development of nuclear weapons was the motivation behind early research into nuclear fission: the Manhattan Project of the U.S. military during World War II carried out most of the early scientific work on fission chain reactions, culminating in the Trinity test bomb and the Little Boy and Fat Man bombs that were exploded over the cities Hiroshima, and Nagasaki, Japan in August 1945.

Even the first fission bombs were thousands of times more explosive than a comparable mass of chemical explosive. For example, Little Boy weighed a total of about four tons (of which 60 kg was nuclear fuel) and was 11 feet (3.4 m) long; it also yielded an explosion equivalent to about 15 kilotons of TNT, destroying a large part of the city of Hiroshima. Modern nuclear weapons (which include a thermonuclear *fusion* as well as one or more fission stages) are hundreds of times more energetic for their weight than the first pure fission atomic bombs (see nuclear weapon yield), so that a modern single missile warhead bomb weighing less than 1/8 as much as Little Boy (see for example W88) has a yield of 475,000 tons of TNT, and could bring destruction to about 10 times the city area.

While the fundamental physics of the fission chain reaction in a nuclear weapon is similar to the physics of a controlled nuclear reactor, the two types of device must be engineered quite differently (see nuclear reactor physics). A nuclear bomb is designed to release all its energy at once, while a reactor is designed to generate a steady supply of useful power. While overheating of a reactor can lead to, and has led to, meltdown and steam explosions, the much lower uranium enrichment makes it impossible for a nuclear reactor to explode with the same destructive power as a nuclear weapon. It is also difficult to extract useful power from a nuclear bomb, although at least one rocket propulsion system, Project Orion, was intended to work by exploding fission bombs behind a massively padded and shielded spacecraft.

The strategic importance of nuclear weapons is a major reason why the technology of nuclear fission is politically sensitive. Viable fission bomb designs are, arguably, within the capabilities of many being relatively simple from an engineering viewpoint. However, the difficulty of obtaining fissile nuclear material to realize the designs, is the key to the relative unavailability of nuclear weapons to all but modern industrialized governments with special programs to produce fissile materials (see uranium enrichment and nuclear fuel cycle).

29.2   History

29.2.1   Discovery of nuclear fission

The discovery of nuclear fission occurred in 1938 in the buildings of Kaiser Wilhelm Society for Chemistry, today part of the Free University of Berlin, following nearly five decades of work on the science of radioactivity and the elaboration of new nuclear physics that described the components of atoms. In 1911, Ernest Rutherford proposed a model of the atom in which a very small, dense and positively charged nucleus of protons (the neutron had not yet been discovered) was surrounded by orbiting, negatively charged electrons (the Rutherford model).[13] Niels Bohr improved upon this in 1913 by reconciling the quantum behavior of electrons (the Bohr model). Work by Henri Becquerel, Marie Curie, Pierre Curie, and Rutherford further elaborated that the nucleus, though tightly bound, could undergo different forms of radioactive decay, and thereby transmute into other elements. (For example, by alpha decay: the emission of an alpha particle—two protons and two neutrons bound together into a particle identical to a helium nucleus.)

Some work in nuclear transmutation had been done. In 1917, Rutherford was able to accomplish transmutation of nitrogen into oxygen, using alpha particles directed at nitrogen ^{14}N + α → ^{17}O + p. This was the first observation of a nuclear reaction, that is, a reaction in which particles from one decay are used to transform another atomic nucleus. Eventually, in 1932, a fully artificial nuclear reaction and nuclear transmutation was achieved by Rutherford's colleagues Ernest

Walton and John Cockcroft, who used artificially accelerated protons against lithium-7, to split this nucleus into two alpha particles. The feat was popularly known as "splitting the atom", although it was not the modern nuclear fission reaction later discovered in heavy elements, which is discussed below.[14] Meanwhile, the possibility of *combining* nuclei—nuclear fusion—had been studied in connection with understanding the processes which power stars. The first artificial fusion reaction had been achieved by Mark Oliphant in 1932, using two accelerated deuterium nuclei (each consisting of a single proton bound to a single neutron) to create a helium nucleus.[15]

After English physicist James Chadwick discovered the neutron in 1932,[16] Enrico Fermi and his colleagues in Rome studied the results of bombarding uranium with neutrons in 1934.[17] Fermi concluded that his experiments had created new elements with 93 and 94 protons, which the group dubbed ausonium and hesperium. However, not all were convinced by Fermi's analysis of his results. The German chemist Ida Noddack notably suggested in print in 1934 that instead of creating a new, heavier element 93, that "it is conceivable that the nucleus breaks up into several large fragments."[18][19] However, Noddack's conclusion was not pursued at the time.

After the Fermi publication, Otto Hahn, Lise Meitner, and Fritz Strassmann began performing similar experiments in Berlin. Meitner, an Austrian Jew, lost her citizenship with the "Anschluss", the occupation and annexation of Austria into Nazi Germany in March 1938, but she fled in July 1938 to Sweden and started a correspondence by mail with Hahn in Berlin. By coincidence, her nephew Otto Robert Frisch, also a refugee, was also in Sweden when Meitner received a letter from Hahn dated 19 December describing his chemical proof that some of the product of the bombardment of uranium with neutrons was barium. Hahn suggested a *bursting* of the nucleus, but he was unsure of what the physical basis for the results were. Barium had an atomic mass 40% less than uranium, and no previously known methods of radioactive decay could account for such a large difference in the mass of the nucleus. Frisch was skeptical, but Meitner trusted Hahn's ability as a chemist. Marie Curie had been separating barium from radium for many years, and the techniques were well-known. According to Frisch:

> Was it a mistake? No, said Lise Meitner; Hahn was too good a chemist for that. But how could barium be formed from uranium? No larger fragments than protons or helium nuclei (alpha particles) had ever been chipped away from nuclei, and to chip off a large number not nearly enough energy was available. Nor was it possible that the uranium nucleus could have been cleaved right across. A nucleus was not like a brittle solid that can be cleaved or broken; George Gamow had suggested early on, and Bohr had given good arguments that a nucleus was much more like a liquid drop. Perhaps a drop could divide itself into two smaller drops in a more gradual manner, by first becoming elongated, then constricted, and finally being torn rather than broken in two? We knew that there were strong forces that would resist such a process, just as the surface tension of an ordinary liquid drop tends to resist its division into two smaller ones. But nuclei differed from ordinary drops in one important way: they were electrically charged, and that was known to counteract the surface tension.

> The charge of a uranium nucleus, we found, was indeed large enough to overcome the effect of the surface tension almost completely; so the uranium nucleus might indeed resemble a very wobbly unstable drop, ready to divide itself at the slightest provocation, such as the impact of a single neutron. But there was another problem. After separation, the two drops would be driven apart by their mutual electric repulsion and would acquire high speed and hence a very large energy, about 200 MeV in all; where could that energy come from? ...Lise Meitner... worked out that the two nuclei formed by the division of a uranium nucleus together would be lighter than the original uranium nucleus by about one-fifth the mass of a proton. Now whenever mass disappears energy is created, according to Einstein's formula $E = mc^2$, and one-fifth of a proton mass was just equivalent to 200 MeV. So here was the source for that energy; it all fitted![20]

In short, Meitner and Frisch had correctly interpreted Hahn's results to mean that the nucleus of uranium had split roughly in half. Frisch suggested the process be named "nuclear fission," by analogy to the process of living cell division into two cells, which was then called binary fission. Just as the term nuclear "chain reaction" would later be borrowed from chemistry, so the term "fission" was borrowed from biology.

On 22 December 1938, Hahn and Strassmann sent a manuscript to *Naturwissenschaften* reporting that they had discovered the element barium after bombarding uranium with neutrons.[21] Simultaneously, they communicated these results to Meitner in Sweden. She and Frisch correctly interpreted the results as evidence of nuclear fission.[22] Frisch confirmed this experimentally on 13 January 1939.[23][24] For proving that the barium resulting from his bombardment of uranium with neutrons was the product of nuclear fission, Hahn was awarded the Nobel Prize for Chemistry in 1944 (the sole recipient) "for his discovery of the fission of heavy nuclei". (The award was actually given to Hahn in 1945, as "the Nobel Committee for Chemistry decided that none of the year's nominations met the criteria as outlined in the will of Alfred Nobel." In such cases, the Nobel Foundation's statutes permit that year's prize be reserved until the following year.)[25]

News spread quickly of the new discovery, which was correctly seen as an entirely novel physical effect with great scientific—and potentially practical—possibilities. Meitner's and Frisch's interpretation of the discovery of Hahn and Strassmann crossed the Atlantic Ocean with Niels Bohr, who was to lecture at Princeton University. I.I. Rabi and Willis Lamb, two Columbia University physicists working at Princeton, heard the news and carried it back to Columbia. Rabi said he told Enrico Fermi; Fermi gave credit to Lamb. Bohr soon thereafter went from Princeton to Columbia to see Fermi. Not finding Fermi in his office, Bohr went down to the cyclotron area and found Herbert L. Anderson. Bohr grabbed him by the shoulder and said: "Young man, let me explain to you about something new and exciting in physics."[26] It was clear to a number of scientists at Columbia that they should try to detect the energy released in the nuclear fission of uranium from neutron bombardment. On 25 January 1939, a Columbia University team conducted the first nuclear fission experiment in the United States,[27] which was done in the basement of Pupin Hall; the members of the team were Herbert L. Anderson, Eugene T. Booth, John R. Dunning, Enrico Fermi, G. Norris Glasoe, and Francis G. Slack. The experiment involved placing uranium oxide inside of an ionization chamber and irradiating it with neutrons, and measuring the energy thus released. The results confirmed that fission was occurring and hinted strongly that it was the isotope uranium 235 in particular that was fissioning. The next day, the Fifth Washington Conference on Theoretical Physics began in Washington, D.C. under the joint auspices of the George Washington University and the Carnegie Institution of Washington. There, the news on nuclear fission was spread even further, which fostered many more experimental demonstrations.[28]

During this period the Hungarian physicist Leó Szilárd, who was residing in the United States at the time, realized that the neutron-driven fission of heavy atoms could be used to create a nuclear chain reaction. Such a reaction using neutrons was an idea he had first formulated in 1933, upon reading Rutherford's disparaging remarks about generating power from his team's 1932 experiment using protons to split lithium. However, Szilárd had not been able to achieve a neutron-driven chain reaction with neutron-rich light atoms. In theory, if in a neutron-driven chain reaction the number of secondary neutrons produced was greater than one, then each such reaction could trigger multiple additional reactions, producing an exponentially increasing number of reactions. It was thus a possibility that the fission of uranium could yield vast amounts of energy for civilian or military purposes (i.e., electric power generation or atomic bombs).

Szilard now urged Fermi (in New York) and Frédéric Joliot-Curie (in Paris) to refrain from publishing on the possibility of a chain reaction, lest the Nazi government become aware of the possibilities on the eve of what would later be known as World War II. With some hesitation Fermi agreed to self-censor. But Joliot-Curie did not, and in April 1939 his team in Paris, including Hans von Halban and Lew Kowarski, reported in the journal *Nature* that the number of neutrons emitted with nuclear fission of ^{235}U was then reported at 3.5 per fission.[29] (They later corrected this to 2.6 per fission.) Simultaneous work by Szilard and Walter Zinn confirmed these results. The results suggested the possibility of building nuclear reactors (first called "neutronic reactors" by Szilard and Fermi) and even nuclear bombs. However, much was still unknown about fission and chain reaction systems.

29.2.2 Fission chain reaction realized

"Chain reactions" at that time were a known phenomenon in *chemistry*, but the analogous process in nuclear physics, using neutrons, had been foreseen as early as 1933 by Szilárd, although Szilárd at that time had no idea with what materials the process might be initiated. Szilárd considered that neutrons would be ideal for such a situation, since they lacked an electrostatic charge.

With the news of fission neutrons from uranium fission, Szilárd immediately understood the possibility of a nuclear chain reaction using uranium. In the summer, Fermi and Szilard proposed the idea of a nuclear reactor (pile) to mediate this process. The pile would use natural uranium as fuel. Fermi had shown much earlier that neutrons were far more effectively captured by atoms if they were of low energy (so-called "slow" or "thermal" neutrons), because for quantum reasons it made the atoms look like much larger targets to the neutrons. Thus to slow down the secondary neutrons released by the fissioning uranium nuclei, Fermi and Szilard proposed a graphite "moderator," against which the fast, high-energy secondary neutrons would collide, effectively slowing them down. With enough uranium, and with pure-enough graphite, their "pile" could theoretically sustain a slow-neutron chain reaction. This would result in the production of heat, as well as the creation of radioactive fission products.

In August 1939, Szilard and fellow Hungarian refugees physicists Teller and Wigner thought that the Germans might make use of the fission chain reaction and were spurred to attempt to attract the attention of the United States government to the issue. Towards this, they persuaded German-Jewish refugee Albert Einstein to lend his name to a letter directed to President Franklin Roosevelt. The Einstein–Szilárd letter suggested the possibility of a uranium bomb deliverable by ship, which would destroy "an entire harbor and much of the surrounding countryside." The President received the letter on 11 October 1939 — shortly after World War II began in Europe, but two years before U.S. entry into it. Roosevelt ordered that a scientific committee be authorized for overseeing uranium work and allocated a small sum of money for pile research.

In England, James Chadwick proposed an atomic bomb utilizing natural uranium, based on a paper by Rudolf Peierls with the mass needed for critical state being 30–40 tons. In America, J. Robert Oppenheimer thought that a cube of uranium deuteride 10 cm on a side (about 11 kg of uranium) might "blow itself to hell." In this design it was still thought that a moderator would need to be used for nuclear bomb fission (this turned out not to be the case if the fissile isotope was separated). In December, Werner Heisenberg delivered a report to the German Ministry of War on the possibility of a uranium bomb. Most of these models were still under the assumption that the bombs would be powered by slow neutron reactions—and thus be similar to a reactor undergoing a meltdown.

In Birmingham, England, Frisch teamed up with Peierls, a fellow German-Jewish refugee. They had the idea of using a purified mass of the uranium isotope ^{235}U, which had a cross section just determined, and which was much larger than that of ^{238}U or natural uranium (which is 99.3% the latter isotope). Assuming that the cross section for fast-neutron fission of ^{235}U was the same as for slow neutron fission, they determined that a pure ^{235}U bomb could have a critical mass of only 6 kg instead of tons, and that the resulting explosion would be tremendous. (The amount actually turned out to be 15 kg, although several times this amount was used in the actual uranium (Little Boy) bomb). In February 1940 they delivered the Frisch–Peierls memorandum. Ironically, they were still officially considered "enemy aliens" at the time. Glenn Seaborg, Joseph W. Kennedy, Arthur Wahl and Italian-Jewish refugee Emilio Segrè shortly thereafter discovered ^{239}Pu in the decay products of ^{239}U produced by bombarding ^{238}U with neutrons, and determined it to be a fissile material, like ^{235}U.

The possibility of isolating uranium-235 was technically daunting, because uranium-235 and uranium-238 are chemically identical, and vary in their mass by only the weight of three neutrons. However, if a sufficient quantity of uranium-235 could be isolated, it would allow for a fast neutron fission chain reaction. This would be extremely explosive, a true "atomic bomb." The discovery that plutonium-239 could be produced in a nuclear reactor pointed towards another approach to a fast neutron fission bomb. Both approaches were extremely novel and not yet well understood, and there was considerable scientific skepticism at the idea that they could be developed in a short amount of time.

On June 28, 1941, the Office of Scientific Research and Development was formed in the U.S. to mobilize scientific resources and apply the results of research to national defense. In September, Fermi assembled his first nuclear "pile" or reactor, in an attempt to create a slow neutron-induced chain reaction in uranium, but the experiment failed to achieve criticality, due to lack of proper materials, or not enough of the proper materials which were available.

Producing a fission chain reaction in natural uranium fuel was found to be far from trivial. Early nuclear reactors did not use isotopically enriched uranium, and in consequence they were required to use large quantities of highly purified graphite as neutron moderation materials. Use of ordinary water (as opposed to heavy water) in nuclear reactors requires enriched fuel — the partial separation and relative enrichment of the rare ^{235}U isotope from the far more common ^{238}U isotope. Typically, reactors also require inclusion of extremely chemically pure neutron moderator materials such as deuterium (in heavy water), helium, beryllium, or carbon, the latter usually as graphite. (The high purity for carbon is required because many chemical impurities such as the boron-10 component of natural boron, are very strong neutron absorbers and thus poison the chain reaction and end it prematurely.)

Production of such materials at industrial scale had to be solved for nuclear power generation and weapons production to be accomplished. Up to 1940, the total amount of uranium metal produced in the USA was not more than a few grams, and even this was of doubtful purity; of metallic beryllium not more than a few kilograms; and concentrated deuterium oxide (heavy water) not more than a few kilograms. Finally, carbon had never been produced in quantity with anything like the purity required of a moderator.

The problem of producing large amounts of high purity uranium was solved by Frank Spedding using the thermite or "Ames" process. Ames Laboratory was established in 1942 to produce the large amounts of natural (unenriched) uranium metal that would be necessary for the research to come. The critical nuclear chain-reaction success of the Chicago Pile-1 (December 2, 1942) which used unenriched (natural) uranium, like all of the atomic "piles" which produced the plutonium for the atomic bomb, was also due specifically to Szilard's realization that very pure graphite could be used for the moderator of even natural uranium "piles". In wartime Germany, failure to appreciate the qualities of very pure graphite led to reactor designs dependent on heavy water, which in turn was denied the Germans by Allied attacks in Norway, where heavy water was produced. These difficulties—among many others— prevented the Nazis from building a nuclear reactor capable of criticality during the war, although they never put as much effort as the United States into nuclear research, focusing on other technologies (see German nuclear energy project for more details).

29.2.3 Manhattan Project and beyond

See also: Manhattan Project

In the United States, an all-out effort for making atomic weapons was begun in late 1942. This work was taken over by the U.S. Army Corps of Engineers in 1943, and known as the Manhattan Engineer District. The top-secret Manhattan Project, as it was colloquially known, was led by General Leslie R. Groves. Among the project's dozens of sites were: Hanford Site in Washington state, which had the first industrial-scale nuclear reactors; Oak Ridge, Tennessee, which was primarily concerned with uranium enrichment; and Los Alamos, in New Mexico, which was the scientific hub for research on bomb development and design. Other sites, notably the Berkeley Radiation Laboratory and the Metallurgical Laboratory at the University of Chicago, played important contributing roles. Overall scientific direction of the project was managed by the physicist J. Robert Oppenheimer.

In July 1945, the first atomic bomb, dubbed "Trinity", was detonated in the New Mexico desert. It was fueled by plutonium created at Hanford. In August 1945, two more atomic bombs—"Little Boy", a uranium-235 bomb, and "Fat Man", a plutonium bomb—were used against the Japanese cities of Hiroshima and Nagasaki.

In the years after World War II, many countries were involved in the further development of nuclear fission for the purposes of nuclear reactors and nuclear weapons. The UK opened the first commercial nuclear power plant in 1956. In 2013, there are 437 reactors in 31 countries.

29.2.4 Natural fission chain-reactors on Earth

Criticality in nature is uncommon. At three ore deposits at Oklo in Gabon, sixteen sites (the so-called Oklo Fossil Reactors) have been discovered at which self-sustaining nuclear fission took place approximately 2 billion years ago. Unknown until 1972 (but postulated by Paul Kuroda in 1956[30]), when French physicist Francis Perrin discovered the Oklo Fossil Reactors, it was realized that nature had beaten humans to the punch. Large-scale natural uranium fission chain reactions, moderated by normal water, had occurred far in the past and would not be possible now. This ancient process was able to use normal water as a moderator only because 2 billion years before the present, natural uranium was richer in the shorter-lived fissile isotope ^{235}U (about 3%), than natural uranium available today (which is only 0.7%, and must be enriched to 3% to be usable in light-water reactors).

29.3 See also

- Hybrid fusion/fission

- Cold fission

- Nuclear propulsion

- Photofission

29.4   Notes

[1] M. G. Arora and M. Singh (1994). *Nuclear Chemistry*. Anmol Publications. p. 202. ISBN 81-261-1763-X.

[2] Gopal B. Saha (1 November 2010). *Fundamentals of Nuclear Pharmacy*. Springer. pp. 11–. ISBN 978-1-4419-5860-0.

[3] Петржак, Константин (1989). "Как было открыто спонтанное деление" [How spontaneous fission was discovered]. In Черникова, Вера. *Краткий Миг Торжества — О том, как делаются научные открытия* [*Brief Moment of Triumph — About making scientific discoveries*] (in Russian). Наука. pp. 108–112. ISBN 5-02-007779-8.

[4] S. Vermote, et al. (2008) "Comparative study of the ternary particle emission in 243-Cm (nth,f) and 244-Cm(SF)" in *Dynamical aspects of nuclear fission: proceedings of the 6th International Conference*. J. Kliman, M. G. Itkis, S. Gmuca (eds.). World Scientific Publishing Co. Pte. Ltd. Singapore.

[5] J. Byrne (2011) *Neutrons, Nuclei, and Matter*, Dover Publications, Mineola, NY, p. 259, ISBN 978-0-486-48238-5.

[6] Marion Brünglinghaus. "Nuclear fission". European Nuclear Society. Retrieved 2013-01-04.

[7] Hans A. Bethe (April 1950), "The Hydrogen Bomb", *Bulletin of the Atomic Scientists*, p. 99.

[8] These fission neutrons have a wide energy spectrum, with range from 0 to 14 MeV, with mean of 2 MeV and mode (statistics) of 0.75 Mev. See Byrne, op. cite.

[9] "Nuclear Fission and Fusion, and Nuclear Interactions". National Physical Laboratory. Retrieved 2013-01-04.

[10] L. Bonneau; P. Quentin. "Microscopic calculations of potential energy surfaces: fission and fusion properties" (PDF). Retrieved 2008-07-28.

[11] By R.D. Madan and Satya Prakash - *Modern Inorganic Chemistry*

[12] "Frequently Asked Questions #1". Radiation Effects Research Foundation. Retrieved September 18, 2007.

[13] E. Rutherford (1911). "The scattering of α and β particles by matter and the structure of the atom" (PDF). *Philosophical Magazine* **21** (4): 669–688. Bibcode:2012PMag...92..379R. doi:10.1080/14786435.2011.617037.

[14] "Cockcroft and Walton split lithium with high energy protons April 1932". Outreach.phy.cam.ac.uk. 1932-04-14. Retrieved 2013-01-04.

[15] "Sir Mark Oliphant (1901–2000)" (PDF). University of Adelaide. Retrieved 5 October 2013.

[16] Chadwick announced his initial findings in: J. Chadwick (1932). "Possible Existence of a Neutron" (PDF). *Nature* **129** (3252): 312. Bibcode:1932Natur.129Q.312C. doi:10.1038/129312a0. Subsequently he communicated his findings in more detail in: Chadwick, J. (1932). "The existence of a neutron". *Proceedings of the Royal Society A* **136** (830): 692–708. Bibcode:1932RSPSA.136..692C. doi:10.1098/rspa.1932.0112.; and Chadwick, J. (1933). "The Bakerian Lecture: The neutron". *Proceedings of the Royal Society A* **142** (846): 1–25. Bibcode:1933RSPSA.142....1C. doi:10.1098/rspa.1933.0152.

[17] E. Fermi, E. Amaldi, O. D'Agostino, F. Rasetti, and E. Segrè (1934) "Radioattività provocata da bombardamento di neutroni III," *La Ricerca Scientifica*, vol. 5, no. 1, pages 452–453.

[18] Ida Noddack (1934). "Über das Element 93". *Zeitschrift für Angewandte Chemie* **47** (37): 653. doi:10.1002/ange.19340473707.

[19] Tacke, Ida Eva. Astr.ua.edu. Retrieved on 2010-12-24.

[20] Bob Weintraub. *Lise Meitner (1878–1968): Protactinium, Fission, and Meitnerium*. Retrieved on June 8, 2009.

[21] O. Hahn and F. Strassmann (1939). "Über den Nachweis und das Verhalten der bei der Bestrahlung des Urans mittels Neutronen entstehenden Erdalkalimetalle ("On the detection and characteristics of the alkaline earth metals formed by irradiation of uranium with neutrons")". *Naturwissenschaften* **27** (1): 11–15. Bibcode:1939NW.....27...11H. doi:10.1007/BF01488241.. The authors were identified as being at the Kaiser-Wilhelm-Institut für Chemie, Berlin-Dahlem. Received 22 December 1938.

[22] L. Meitner and O. R. Frisch (1939). "Disintegration of Uranium by Neutrons: a New Type of Nuclear Reaction". *Nature* **143** (3615): 239. Bibcode:1939Natur.143..239M. doi:10.1038/143239a0.. The paper is dated 16 January 1939. Meitner is identified as being at the Physical Institute, Academy of Sciences, Stockholm. Frisch is identified as being at the Institute of Theoretical Physics, University of Copenhagen.

[23] O. R. Frisch (1939). "Physical Evidence for the Division of Heavy Nuclei under Neutron Bombardment". *Nature* **143** (3616): 276. Bibcode:1939Natur.143..276F. doi:10.1038/143276a0.

[24] "Physical Evidence for the Division of Heavy Nuclei under Neutron Bombardment". 17 January 1939. Archived from the original on 2008-01-08. The experiment for this letter to the editor was conducted on 13 January 1939; see Richard Rhodes (1986) *The Making of the Atomic Bomb*, Simon and Schuster. pp. 263 and 268, ISBN 0-671-44133-7.

[25] "The Nobel Prize in Chemistry 1944". Nobelprize.org. Retrieved 2008-10-06.

[26] Richard Rhodes. (1986) *The Making of the Atomic Bomb*, Simon and Schuster, p. 268, ISBN 0-671-44133-7.

[27] H. L. Anderson, E. T. Booth, J. R. Dunning, E. Fermi, G. N. Glasoe, and F. G. Slack (1939). "The Fission of Uranium". *Physical Review* **55** (5): 511. Bibcode:1939PhRv...55..511A. doi:10.1103/PhysRev.55.511.2.

[28] Richard Rhodes (1986). *The Making of the Atomic Bomb*, Simon and Schuster, pp. 267–270, ISBN 0-671-44133-7.

[29] H. Von Halban; F. Joliot and L. Kowarski (1939). "Number of Neutrons Liberated in the Nuclear Fission of Uranium". *Nature* **143** (3625): 680. Bibcode:1939Natur.143..680V. doi:10.1038/143680a0.

[30] P. K. Kuroda (1956). "On the Nuclear Physical Stability of the Uranium Minerals" (PDF). *The Journal of Chemical Physics* **25** (4): 781. Bibcode:1956JChPh..25..781K. doi:10.1063/1.1743058.

29.5 References

- *DOE Fundamentals Handbook: Nuclear Physics and Reactor Theory Volume 1* (PDF). U.S. Department of Energy. January 1993. Retrieved 2012-01-03.

- *DOE Fundamentals Handbook: Nuclear Physics and Reactor Theory Volume 2* (PDF). U.S. Department of Energy. January 1993. Retrieved 2012-01-03.

29.6 External links

- The Effects of Nuclear Weapons

- Annotated bibliography for nuclear fission from the Alsos Digital Library

- The Discovery of Nuclear Fission Historical account complete with audio and teacher's guides from the American Institute of Physics History Center

- atomicarchive.com Nuclear Fission Explained

- Nuclear Files.org What is Nuclear Fission?

- Nuclear Fission Animation

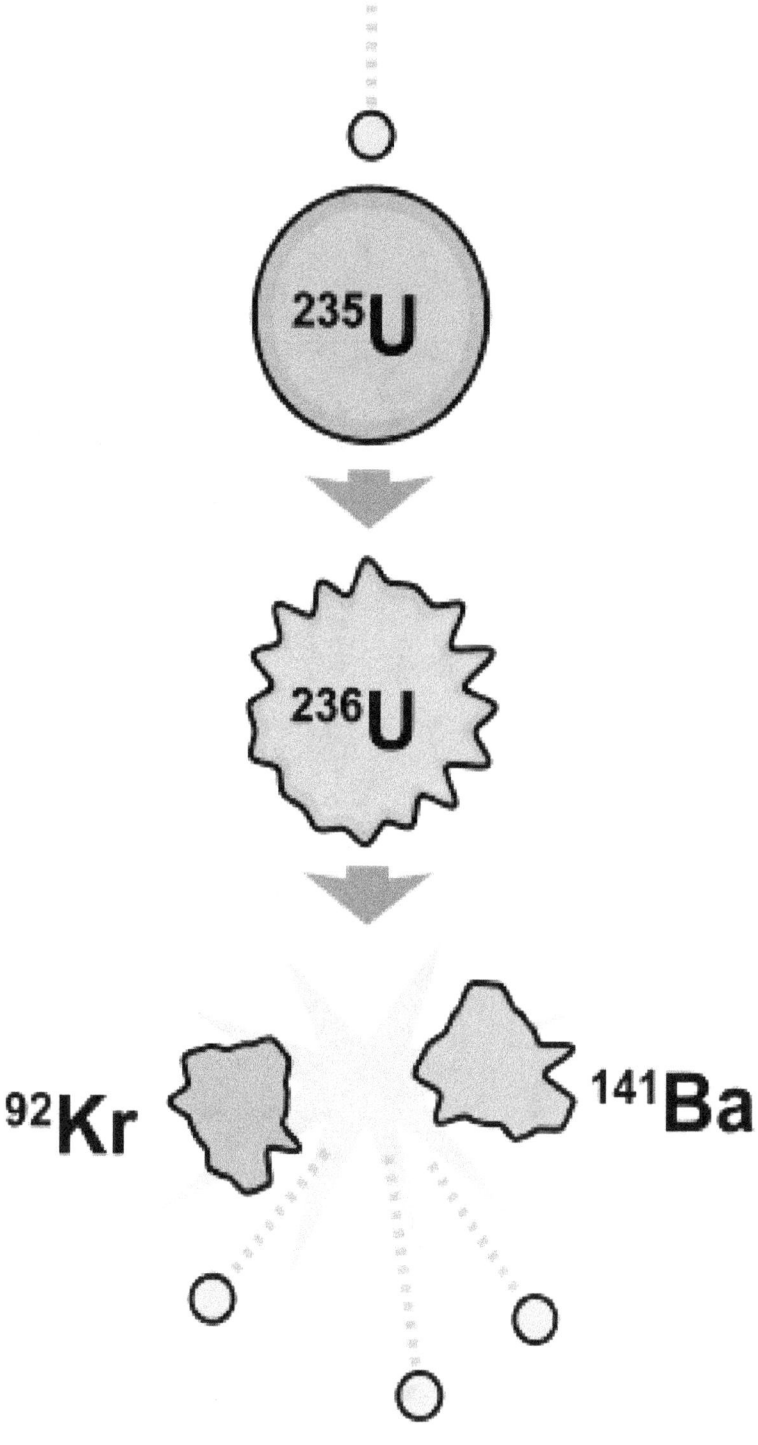

An induced fission reaction. A neutron is absorbed by a uranium-235 nucleus, turning it briefly into
an excited uranium-236 nucleus, with the excitation energy provided by the kinetic energy of the neutron
plus the forces that bind the neutron. The uranium-236, in turn, splits into fast-moving lighter elements
(fission products) and releases three free neutrons. At the same time, one or more "prompt gamma rays"
(not shown) are produced, as well.

The mushroom cloud produced by Tsar Bomba, currently the largest man-made nuclear device detonated in history, next to other mushroom clouds of various nuclear devices.

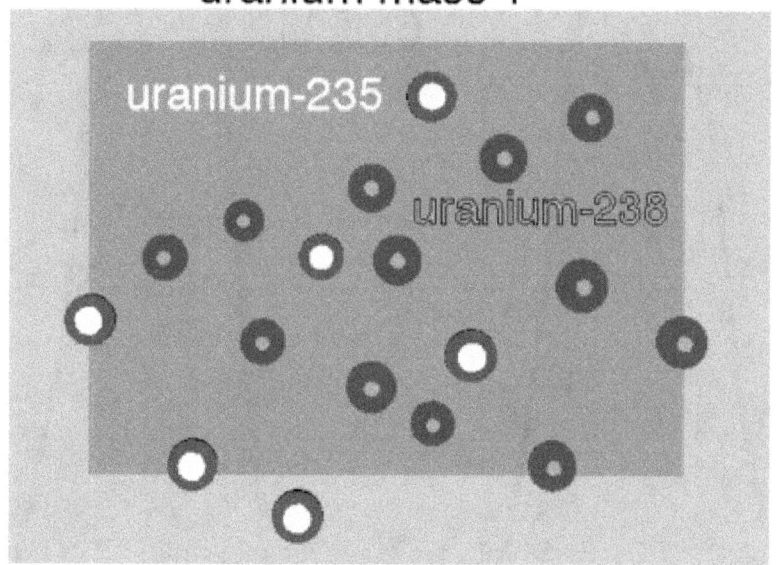

A visual representation of an induced nuclear fission event where a slow-moving neutron is absorbed by the nucleus of a uranium-235 atom, which fissions into two fast-moving lighter elements (fission products) and additional neutrons. Most of the energy released is in the form of the kinetic velocities of the fission products and the neutrons.

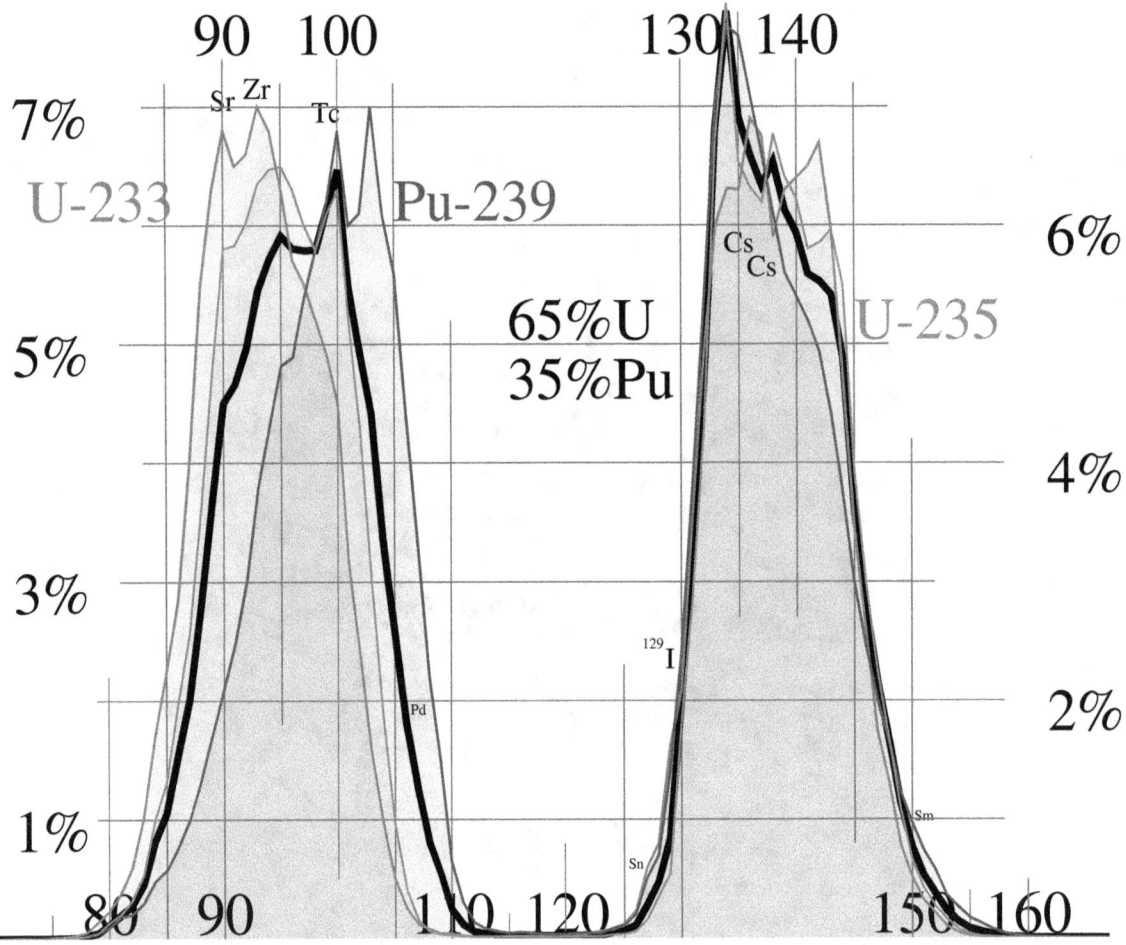

Fission product yields by mass for thermal neutron fission of U-235, Pu-239, a combination of the two typical of current nuclear power reactors, and U-233 used in the thorium cycle.

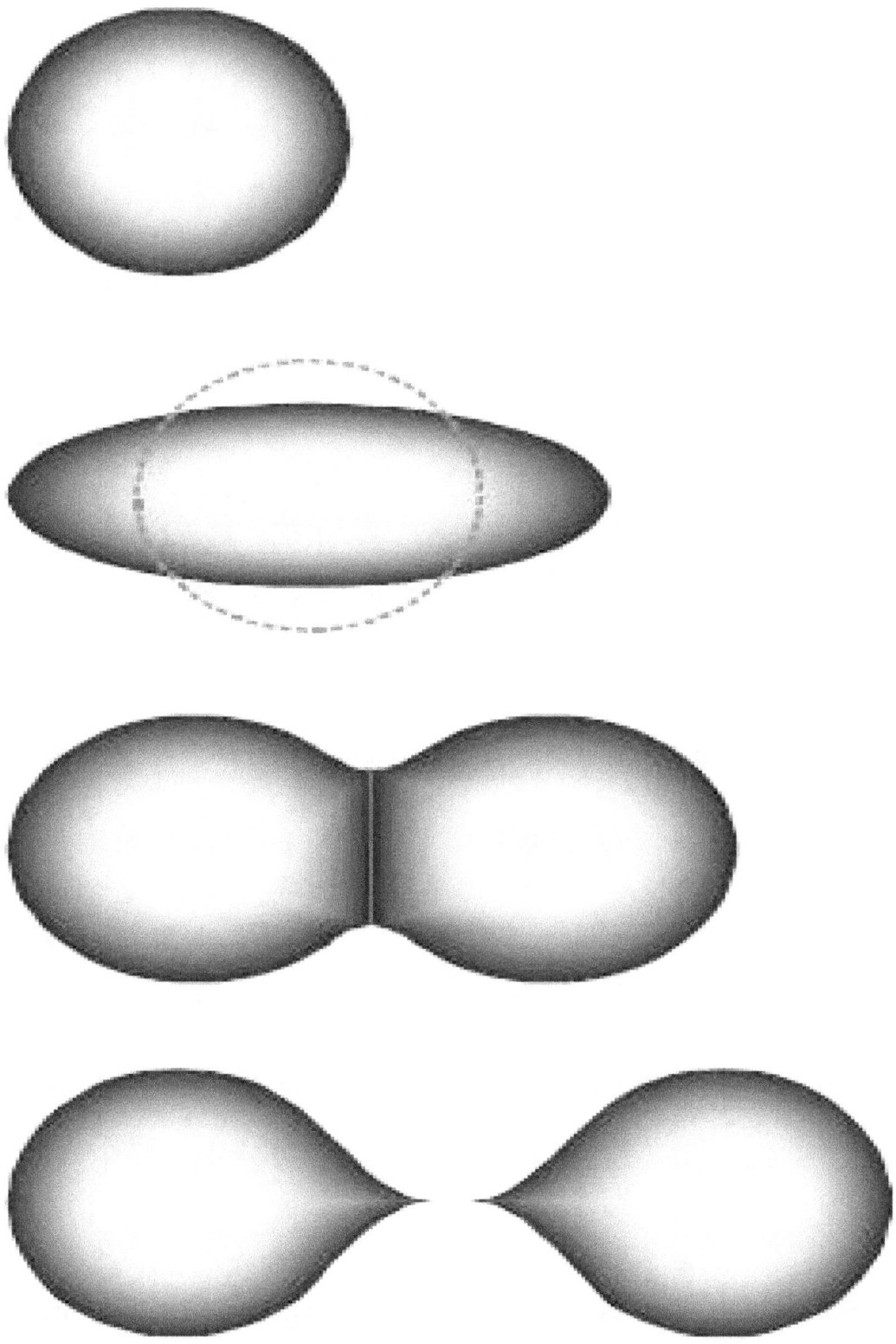

The stages of binary fission in a liquid drop model. Energy input deforms the nucleus into a fat "cigar" shape, then a "peanut" shape, followed by binary fission as the two lobes exceed the short-range nuclear force attraction distance, then are pushed apart and away by their electrical charge. In the liquid drop model, the two fission fragments are predicted to be the same size. The nuclear shell model allows for them to differ in size, as usually experimentally observed.

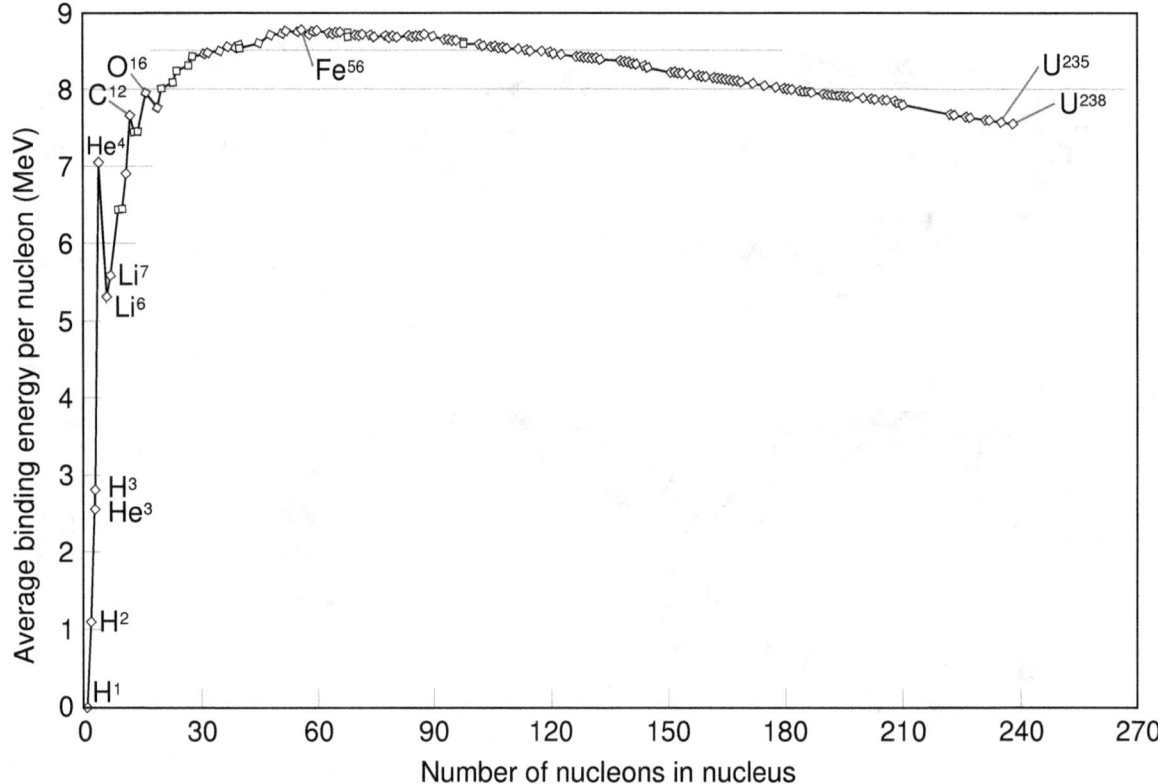

The "curve of binding energy": A graph of binding energy per nucleon of common isotopes.

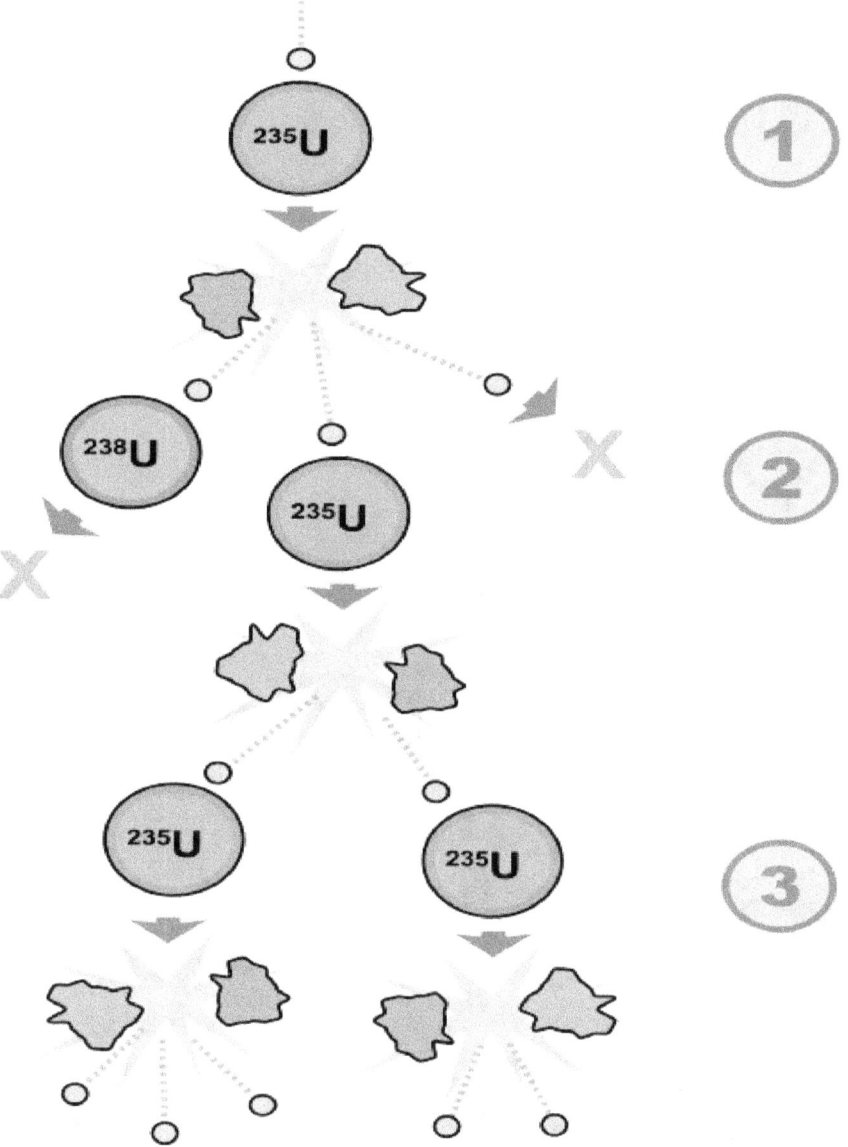

A schematic nuclear fission chain reaction. 1. Auranium-235 atom absorbs a neutron and fissions into two new atoms (fission fragments), releasing three new neutrons and some binding energy. 2.
One of those neutrons is absorbed by an atom of uranium-238 and does not continue the reaction. Another neutron is simply lost and does not collide with anything, also not continuing the reaction. However, the one neutron does collide with an atom of uranium-235, which then fissions and releases two neutrons and some binding energy. 3. Both of those neutrons collide with uranium-235 atoms, each of which fissions and releases between one and three neutrons, which can then continue the reaction.

The cooling towers of the Philippsburg Nuclear Power Plant, in Germany.

The mushroom cloud of the atom bomb dropped on Nagasaki, Japan in 1945 rose some 18 kilometres (11 mi) above the bomb's hypocenter. The bomb killed at least 60,000 people.[12]

The experimental apparatus with which Otto Hahn and Fritz Strassmann discovered nuclear fission in 1938

German stamp honoring Otto Hahn and his discovery of nuclear fission (1979)

Drawing of the first artificial reactor, Chicago Pile-1.

Chapter 30

Ultracold neutrons

Ultracold neutrons (UCN) are free neutrons which can be stored in traps made from certain materials. The storage is based on the reflection of UCN by such materials under any angle of incidence.

30.1 Properties

The reflection is caused by the coherent strong interaction of the neutron with atomic nuclei. It can be quantum-mechanically described by an effective potential which is commonly referred to as the *Fermi pseudo potential* or the *neutron optical potential*. The corresponding velocity is called the *critical velocity* of a material. Neutrons are reflected from a surface if the velocity component normal to the reflecting surface is less or equal the critical velocity.

As the neutron optical potential of most materials is below 300 neV, the kinetic energy of incident neutrons must not be higher than this value to be reflected under any angle of incidence, especially for normal incidence. The kinetic energy of 300 neV corresponds to a maximum velocity of 7.6 m/s or a minimum wavelength of 52 nm. As their density is usually very small, UCN can also be described as a very thin ideal gas with a temperature of 3.5 mK.

Due to the small kinetic energy of an UCN, the influence of gravitation is significant. Thus, the trajectories are parabolic. Kinetic energy of an UCN is transformed into potential (height) energy with ~102 neV/m.

The magnetic moment of the neutron, produced by its spin, interacts with magnetic fields. The total energy changes with ~60 neV/T.

30.2 History

It was Enrico Fermi who realized first that the coherent scattering of slow neutrons would result in an effective interaction potential for neutrons traveling through matter, which would be positive for most materials.[1] The consequence of such a potential would be the total reflection of neutrons slow enough and incident on a surface at a glancing angle. This effect was experimentally demonstrated by Fermi and Walter Henry Zinn [2] and Fermi and Leona Marshall.[3] The storage of neutrons with very low kinetic energies was predicted by Yakov Borisovich Zel'dovich[4] and experimentally realized simultaneously by groups at Dubna [5] and Munich.[6]

30.3 UCN production

In 1979, three methods for the production of UCN were described: 1. By F.L. Shapiro: The use of a horizontal evacuated tube from the reactor, curved so all but UCN would be absorbed by the walls of the tube before reaching the detector. 2. By Albert Steyerl: Neutrons transported from the reactor though a vertical evacuated guide about 11 meters long are slowed down by gravity, so only those that happened to have ultracold energies can reach the detector at the top of the tube. 3. A neutron turbine in which neutrons at 50 m/s are directed against a bladed turbine with receding tangential velocity 25 m/s, from which neutrons emerged after multiple reflections with a speed of about 5 m/s.[7]

30.4 Reflecting materials

Any material with a positive neutron optical potential can reflect UCN. The table on the right gives an (incomplete) list of UCN reflecting materials including the height of the neutron optical potential (*VF*) and the corresponding critical velocity (*vC*). The height of the neutron optical potential is isotope-specific. The highest known value of VF is measured for ^{58}Ni: 335 neV (vC=8.14 m/s). It defines the upper limit of the kinetic energy range of UCN.

The most widely used materials for UCN wall coatings are Beryllium, Beryllium oxide, Nickel (including ^{58}Ni) and more recently also diamond-like carbon (DLC).

Non-magnetic materials such as DLC are usually preferred for the use with polarized neutrons. Magnetic centers in e.g. Ni can lead to de-polarization of such neutrons upon reflection. If a material is magnetized, the neutron optical potential is different for the two polarizations, caused by

$$V_F(pol.) = V_F(unpol.) \pm \mu_N \cdot B$$

where μ_N is the magnetic moment of the neutron and $B = \mu_0 \cdot M$ the magnetic field created on the surface by the magnetization.

Each material has a specific **loss probability** per reflection,

$$\mu(E, \theta) = 2\eta \sqrt{\frac{E \cos^2 \theta}{V_F - E \cos^2 \theta}}$$

which depends on the kinetic energy of the incident UCN (*E*) and the angle of incidence (θ). It is caused by absorption and thermal upscattering. The **loss coefficient** η is energy-independent and typically of the order of 10^{-4} to 10^{-3}.

30.5 Experiments with UCN

The production, transportation and storage of UCN is currently motivated by their usefulness as a tool to determine properties of the neutron and to study fundamental physical interactions. Storage experiments have improved the accuracy or the upper limit of some neutron related physical values.

30.5.1 Measurement of the neutron lifetime

Today's world average value for the neutron lifetime is $885.7 \pm 0.8 \ s$,[10] to which the experiment of Arzumanov et al.[11] contributes strongest. Ref.[11] measured $\tau_n = 885.4 \pm 0.9_{stat} \pm 0.4_{syst} \ s$ by storage of UCN in a material bottle covered with Fomblin oil. Using traps with different surface to volume ratios allowed them to separate storage decay time and neutron lifetime from each other. There is another result, with even smaller uncertainty, but which is not included in the World average. It was obtained by Serebrov et al.,[12] who found $878.5 \pm 0.7_{stat} \pm 0.4_{syst} \ s$. Thus, the two most precisely measured values deviate by 5.6σ

30.5.2 Measurement of the neutron electric dipole moment

Main article: Neutron electric dipole moment

The neutron electric dipole moment (nEDM) is a measure for the distribution of positive and negative charge inside the neutron. No nEDM has been found until now (May 2008). Today's lowest value for the upper limit of the nEDM was measured with stored UCN (see main article).

30.5.3   Observation of the gravitational interactions of the neutron

Physicists have observed quantized states of matter under the influence of gravity for the first time. Valery Nesvizhevsky of the Institute Laue-Langevin and colleagues found that cold neutrons moving in a gravitational field do not move smoothly but jump from one height to another, as predicted by quantum theory. The finding could be used to probe fundamental physics such as the equivalence principle, which states that different masses accelerate at the same rate in a gravitational field (V Nesvizhevsky *et al.* 2001 Nature 415 297). UCN spectroscopy has been used to limit scenarios including dark energy, chameleon fields,[13] and new short range forces[14]

30.5.4   Measurement of the neutron-anti-neutron oscillation time

30.5.5   Measurement of the A-coefficient of the neutron beta decay correlation

The first reported measurement of the beta-asymmety using UCN is from a Los Alamos group in 2009.[15] The LANSCE group published precision measurements with polarized UCN the next year.[16] Further measuments by these groups and others have led to the current world average:[17]

$$A_0 = -0.1184 \pm 0.0010$$

30.6   References

[1] E. Fermi, Ricerca Scientifica **7** (1936) 13

[2] E. Fermi, W.H. Zinn, Phys. Rev. **70** (1946) 103

[3] E. Fermi, L. Marshall, Phys. Rev. **71** (1947) 666

[4] Ya.B. Zeldovich, Sov. Phys. JETP-**9** (1959) 1389

[5] V.I. Lushikov *et al.*, Sov. Phys. JETP Lett. **9** (1969) 23

[6] A. Steyerl, Phys. Lett. **B29** (1969) 33

[7] R. Golub, W. Mampe, J. M. Pendelbury & P. Ageron *Scientific American*, June 1979

[8] R. Golub, D. Richardson, S.K. Lamoreaux, *Ultra-Cold Neutrons*, Adam Hilger (1991), Bristol

[9] V.K. Ignatovich, *The Physics of Ultracold Neutrons*, Clarendon Press (1990), Oxford, UK

[10] W.-M. Yao *et al.* (Particle Data Group), J. Phys. G **33**, 1 (2006) and 2007 partial update for edition 2008 (URL: http://pdg.lbl.gov)

[11] S. Arzumanov, L. Bondarenko, S. Chernyavsky, W. Drexel *et al.*, Phys. Lett. B **483** (2000) 15

[12] A. Serebrov, V. Varlamov, A. Kharitonov, A. Fomin *et al.*, Phys. Lett. B **605** (2005) 72

[13] "Gravity Resonance Spectroscopy Constrains Dark Energy and Dark Matter Scenarios". *PRL.* 16 April 2014.

[14] "Constraints on New Gravitylike Forces in the Nanometer Range". *Phys.Rev.Lett.114,161101.*22April2015.doi1101.

[15] "First Measurement of the NeutronβAsymmetry with Ultracold Neutrons".*PRL102,012301.*5January2009.do012301.

[16] "http://arxiv.org/abs/1007.3790". *Phys.Rev.Lett. 105, 181803*. Jul 2010. doi:10.1103/PhysRevLett.105.181803.

[17] K.A. Olive et al. (Particle Data Group) (2014). "e− ASYMMETRY PARAMETER A".

Chapter 31

Neutron capture nucleosynthesis

Neutron capture nucleosynthesis describes two nucleosynthesis pathways: the r-process and the s-process, for *rapid* and *slow* neutron captures, respectively. R-process describes neutron capture in a region of high neutron flux, such as during supernova nucleosynthesis after core-collapse, and yields neutron-rich nuclides. S-process describes neutron capture that is slow relative to the rate of beta decay, as for stellar nucleosynthesis in some stars, and yields nuclei with stable nuclear shells. Each process is responsible for roughly half of the observed abundances of elements heavier than iron. The importance of neutron capture to the observed abundance of the chemical elements was first described in 1957 in the B^2FH paper.[1]

31.1 References

[1] Wallerstein, George; Icko Iben, Jr.; Peter Parker; Ann Merchant Boesgaard et al. (1997). "Synthesis of the elements in stars: forty years of progress". *Reviews of Modern Physics* **69** (4):995. Bibcode:1997RvMP...69..995W. doi:10.1103/RevModPhy.

31.2 External links

• E. M. Burbidge; G. R. Burbidge; W. A. Fowler; F. Hoyle (1957). "Synthesis of the Elements in Stars". *Reviews of Modern Physics* **29** (4): 547–650. Bibcode:1957RvMP...29..547B. doi:10.1103/RevModPhys.29.547.

Chapter 32

Neutron bomb

A **neutron bomb**, officially known as one type of **Enhanced Radiation Weapon** (ERW) , is a low yield fission-fusion thermonuclear weapon (hydrogen bomb) in which the burst of neutrons generated by a fusion reaction is intentionally allowed to escape the weapon, rather than being absorbed by its other components.[3] The weapon's radiation case, usually made from relatively thick uranium, lead or steel in a standard bomb, is, instead, made of as thin a material as possible, to facilitate the greatest escape of fusion-produced neutrons. The "usual" nuclear weapon yield—expressed as kilotons of TNT equivalent—is not a measure of a neutron weapon's destructive power. It refers only to the energy released (mostly heat and blast), and does not express the lethal effect of neutron radiation on living organisms.

Compared to a pure fission bomb with an identical explosive yield, a neutron bomb would emit about ten times[4] the amount of neutron radiation. In a fission bomb, at sea level, the total radiation pulse energy which is composed of both gamma rays and neutrons is approximately 5% of the entire energy released; in the neutron bomb it would be closer to 40%. Furthermore, the neutrons emitted by a neutron bomb have a much higher average energy level (close to 14 MeV) than those released during a fission reaction (1–2 MeV).[5] Technically speaking, all low yield nuclear weapons are radiation weapons, including non-enhanced variants. Up to about 10 kilotons in yield, all nuclear weapons have prompt neutron radiation[6] as their most far reaching lethal component, after which point the lethal blast and thermal effects radius begins to out-range the lethal ionizing radiation radius.[7][8][9] Enhanced radiation weapons also fall into this same yield range and simply enhance the intensity and range of the neutron dose for a given yield.

32.1   History and deployment to present

Conception of the neutron bomb is generally credited to Samuel T. Cohen of the Lawrence Livermore National Laboratory, who developed the concept in 1958.[10] Initial development was carried out as part of projects DOVE and STARLING, and an early device was tested underground in early 1962. Designs of a "weaponized" version were carried out in 1963.[11][12]

Development of two production designs for the Army's MGM-52 Lance short-range missile began in July 1964, the W63 at Livermore and the W64 at Los Alamos. Both entered Phase 3 testing in July 1964, and the W64 was cancelled in favor of the W63 in September 1964. The W63 was in turn cancelled in November 1965 in favor of the W70 (Mod 0), a conventional design.[11] By this time, the same concepts were being used to develop warheads for the Sprint missile, an anti-ballistic missile (ABM), with Livermore designing the W65 and Los Alamos the W66. Both entered Phase 3 testing in October 1965, but the W65 was cancelled in favor of the W66 in November 1968. Testing of the W66 was carried out in the late 1960s, and entered production in June 1974,[11] the first neutron bomb to do so. Approximately 120 were built, with about 70 of these being on active duty during 1975 and 1976 as part of the Safeguard Program. When that program was shut down they were placed in storage, and eventually decommissioned in the early 1980s.[11]

Development of ER warheads for Lance continued, but in the early 1970s attention had turned to using modified versions of the W70, the W70 Mod 3.[11] Development was subsequently postponed by President Jimmy Carter in 1978 following protests against his administration's plans to deploy neutron warheads to ground forces in Europe.[13] On November

17, 1978, in a test the USSR detonated its first similar-type bomb.[14] President Ronald Reagan restarted production in 1981.[13] The Soviet Union began a propaganda campaign against the US's neutron bomb in 1981 following Reagan's announcement. In 1983 Reagan then announced the Strategic Defense Initiative, which surpassed neutron bomb production in ambition and vision and with that the neutron bomb quickly faded from the center of the public's attention.[14]

Three types of enhanced radiation weapons (ERW) were built by the United States.[15] The W66 warhead, for the anti-ICBM Sprint missile system, was deployed in 1975 and retired the next year, along with the missile system. The W70 Mod 3 warhead was developed for the short-range, tactical Lance missile, and the W79 Mod 0 was developed for artillery shells. The latter two types were retired by President George H. W. Bush in 1992, following the end of the Cold War.[16][17] The last W70 Mod 3 warhead was dismantled in 1996,[18] and the last W79 Mod 0 was dismantled by 2003, when the dismantling of all W79 variants was completed.[19]

According to the Cox Report, as of 1999 the United States had never deployed a neutron weapon. The nature of this statement is not clear; it reads "The stolen information also includes classified design information for an enhanced radiation weapon (commonly known as the "neutron bomb"), which neither the United States, nor any other nation, has ever deployed."[20] However, the fact that neutron bombs had been produced by the US was well known at this time and part of the public record. Sam Cohen suggests the report is playing with the definitions; the US bombs were never deployed *to Europe*, they remained stockpiled in the US.[21]

In addition to the two superpowers, France and China are known to have tested neutron or enhanced radiation bombs. France conducted an early test of the technology in 1967[22] and tested an "actual" neutron bomb in 1980.[23] China conducted a successful test of neutron bomb principles in 1984 and a successful test of a neutron bomb in 1988. However, neither of those countries chose to deploy the neutron bomb. Chinese nuclear scientists stated prior to the 1988 test that China had no need for the neutron bomb, but it was developed to serve as a "technology reserve," in case the need arose in the future.[24]

Although no country is currently known to deploy them in an offensive manner, all thermonuclear dial-a-yield warheads that have about 10 kiloton and lower as one dial option, with a considerable fraction of that yield derived from fusion reactions, can be considered capable of being neutron bombs in actuality if not in name. The only country definitively known to deploy dedicated (that is, not dial-a-yield) neutron warheads for any length of time is Russia, which inherited the USSR's neutron warhead equipped ABM-3 Gazelle missile program. This anti-ballistic missile (ABM) system contains at least 68 neutron warheads with a 10 kiloton yield each and it has been in service since 1995, with inert missile testing approximately every other year since then (2014). The system is designed to destroy incoming "endo-atmospheric" level nuclear warheads aimed at Moscow and other targets and is the lower-tier/last umbrella of the A-135 anti-ballistic missile system (NATO reporting name: ABM-3).[25]

By 1984, according to Mordechai Vanunu, Israel was mass-producing neutron bombs.[26] A number of analysts believe that the Vela incident was an Israeli neutron bomb experiment.[27]

Considerable controversy arose in the U.S. and Western Europe following a June 1977 *Washington Post* exposé describing U.S. government plans to purchase the bomb. The article focused on the fact that it was the first weapon specifically intended to kill humans with radiation.[28][29] Lawrence Livermore National Laboratory director Harold Brown and Soviet General Secretary Leonid Brezhnev both described the neutron bomb as a "capitalist bomb", because it was designed to destroy people while preserving property.[30][31] Science fiction author Isaac Asimov also stated that "Such a neutron bomb or N bomb seems desirable to those who worry about property and hold life cheap."[32]

32.2 Use

Neutron bombs are purposely designed with explosive yields lower than other nuclear weapons. Since neutrons are absorbed by air,[6] neutron radiation effects drop off very rapidly with distance in air, there is a sharper distinction, as opposed to thermal effects, between areas of high lethality and areas with minimal radiation doses.[3] All high yield (more than ~10 kiloton) "neutron bombs", such as the extreme example of a device that derived 97% of its energy from fusion, the 50 megaton Tsar Bomba, are not able to radiate sufficient neutrons beyond their lethal blast range when detonated as a surface burst or low altitude air burst and so are no longer classified as neutron bombs, thus limiting the yield of neutron bombs to a maximum of about 10 kilotons. The intense pulse of high-energy neutrons generated by a neutron bomb is the principal killing mechanism, not the fallout, heat or blast.

The inventor of the neutron bomb, Sam Cohen, criticized the description of the W70 as a neutron bomb since it could be configured to yield 100 kilotons:

> the W-70 ... is not even remotely a "neutron bomb." Instead of being the type of weapon that, in the popular mind, "kills people and spares buildings" it is one that both kills and physically destroys on a massive scale. The W-70 is not a discriminate weapon, like the neutron bomb—which, incidentally, should be considered a weapon that "kills enemy personnel while sparing the physical fabric of the attacked populace, and even the populace too."[36]

Although neutron bombs are commonly believed to "leave the infrastructure intact", with current designs that have explosive yields in the low kiloton range,[37] detonation in a built up area would still cause considerable, although not total, destruction through blast and heat effects out to a considerable radius.[38]

Neutron bombs were to be used as tactical nuclear weapons, intended for use against armored forces. The neutron bomb was originally conceived by the U.S. military as a weapon that could stop massed Soviet armored divisions from overrunning allied nations without destroying the infrastructure of the allied nation.[41][42] As the Warsaw Pact tank strength was over twice that of NATO, and Soviet Deep Battle doctrine was likely to be to use this numerical advantage to rapidly sweep across continental Europe if the Cold War ever turned hot, any weapon that could break up their intended mass tank formation deployments and force them to deploy their tanks in a thinner, more easily dividable manner,[41] would aid ground forces in the task of hunting down solitary tanks and firing anti-tank missiles upon them,[43] such as the contemporary M47 Dragon and BGM-71 TOW missiles, which NATO had hundreds of thousands of.[44]

Rather than making extensive preparations for battlefield nuclear combat in Central Europe, "The Soviet military leadership believed that conventional superiority provided the Warsaw Pact with the means to approximate the effects of nuclear weapons and achieve victory in Europe without resort to those weapons."[45]

Neutron bombs, or more precisely, enhanced [neutron] radiation weapons were also to find use as strategic anti-ballistic missile weapons,[38] and in this role they are believed to remain in active service within Russia's Gazelle (missile).[46]

32.2.1 Effects

Upon detonation, a 1 kiloton neutron bomb near the ground, in an airburst would produce a large blast wave, and a powerful pulse of both thermal radiation and ionizing radiation, mostly in the form of fast (14.1 MeV) neutrons. The thermal pulse would cause third degree burns to unprotected skin out to approximately 500 meters. The blast would create at least 4.6 PSI out to a radius of 600 meters, which would severely damage all non-reinforced concrete structures, at the conventional effective combat range against modern main battle tanks and armored personnel carriers (<690–900 m) the blast from a 1 kt neutron bomb will destroy or damage to the point of non-usability almost all un-reinforced civilian buildings. Thus the use of neutron bombs to stop an enemy armored attack by rapidly incapacitating the crew with a dose of 8000+ rads of radiation,[48] which would require exploding large numbers of them to blanket the enemy forces, would also destroy all normal civilian buildings in the same immediate area ~600 meters,[48][49] and via neutron activation it would make many building materials in the city radioactive, such as zinc coated steel/galvanized steel (see area denial use below). Although at this ~600 meter distance the 4-5 PSI blast overpressure would cause very few direct casualties as the human body is resistant to sheer overpressure, the powerful winds produced by this overpressure are capable of throwing human bodies into objects or throwing objects—including window glass at high velocity—both with potentially lethal results, rendering casualties highly dependent on surroundings, including on if the building they are in collapses.[50] The pulse of neutron radiation would cause immediate and permanent incapacitation to unprotected outdoor humans in the open out to 900 meters,[4] with death occurring in one or two days. The lethal dose (LD50) of 600 rads would extend to about 1350–1400 meters for those unprotected and outdoors,[48] where approximately half of those exposed would die of radiation sickness after several weeks.

However a human residing within, or simply shielded by, at least one of the aforementioned concrete buildings with walls and ceilings 30 centimeters/12 inches thick, or alternatively of damp soil 24 inches thick, would receive a neutron radiation exposure reduced by a factor of 10.[51][52] Even near ground zero, Basement sheltering or buildings with similar radiation shielding characteristics, would drastically reduce the radiation dose.[53]

Furthermore, the neutron absorption spectrum of air is disputed by some authorities and depends in part on absorption by hydrogen from water vapor. It therefore might vary exponentially with humidity, making neutron bombs immensely more deadly in desert climates than in humid ones.[48]

32.2.2 Questionable effectiveness in modern anti-tank role

See also: Centurion Tank § Nuclear tests, Object 279 and Neutron transport
The questionable effectiveness of ER weapons against modern tanks is cited as one of the main reasons that these weapons are no longer fielded or stockpiled. With the increase in average tank armor thickness since the first ER weapons were fielded, tank armor protection approaches the level where tank crews are now almost completely protected from radiation effects. Therefore, for an ER weapon to incapacitate a modern tank crew through irradiation, the weapon must now be detonated at such a close proximity to the tank that the nuclear explosion's blast would now be equally effective at incapacitating it and its crew.[54] However this assertion was regarded as dubious in a reply in 1986[55] by a member of the Royal Military College of Science as neutron radiation from a 1 kiloton neutron bomb would incapacitate the crew of a tank with a protection factor of 35 out to a range of 280 meters, but the incapacitating blast range, depending on the exact weight of the tank, is much less, from 70 to 130 meters. However although the author did note that effective neutron absorbers and neutron poisons such as boron carbide can be incorporated into conventional armor and strap on neutron moderating hydrogenous material (hydrogen atom containing substances), such as explosive reactive armor, can both increase the protection factor, the author holds that in practice combined with neutron scattering, the actual average total tank area protection factor is rarely higher than 15.5 to 35.[56] According to the Federation of American Scientists, the neutron protection factor of a "tank" can be as low as 2,[2] without qualifying whether the statement implies a light tank, medium tank, or main battle tank.

A composite high density concrete, or alternatively, a laminated Graded Z shield, 24 units thick of which 16 units are iron and 8 units are polyethylene containing boron (BPE), and additional mass behind it to attenuate neutron capture gamma rays is more effective than just 24 units of pure iron or BPE alone, due to the advantages of both iron and BPE in combination. Iron is effective in slowing down/scattering high-energy neutrons in the 14-MeV energy range and attenuating gamma rays, while the hydrogen in polyethylene is effective in slowing down these now slower fast neutrons in the few MeV range, and boron 10 has a high absorption cross section for thermal neutrons and a low production yield of gamma rays when it absorbs a neutron.[57][58][59][60] The Soviet T72 tank, in response to the neutron bomb threat, is cited as having fitted a boronated,[61] polyethylene liner, which has had its neutron shielding properties simulated.[52][62]

However some tank armor material contains depleted uranium (DU), common in the US's M1A1 Abrams tank, which "incorporates steel-encased depleted uranium armour",[63] a substance that will fast fission when it captures a fast, fusion generated neutron, and therefore upon fissioning it will produce fission neutrons and fission products embedded within the armor, products which emit amongst other things, penetrating gamma rays. Although the neutrons emitted by the neutron bomb may not penetrate to the tank crew in lethal quantities, the fast fission of DU within the armor could still ensure a lethal environment for the crew and maintenance personnel by fission neutron and gamma ray exposure,[64] largely depending on the exact thickness and elemental composition of the armor—information usually hard to attain. Despite this, DUCRETE—which has an elemental composition similar to, but not identical to the ceramic 2nd generation heavy metal Chobham armor of the Abrams tank—is an effective radiation shield, to both *fission* neutrons and gamma rays due to it being a graded Z material.[65][66] Uranium being about twice as dense as lead is thus nearly twice as effective at shielding gamma ray radiation per unit thickness.[67]

32.2.3 Use against ballistic missiles

As an anti-ballistic missile weapon, the first fielded ER warhead, the W66, was developed for the Sprint missile system as part of the Safeguard Program to protect United States cities and missile silos from incoming Soviet warheads by damaging their electronic components with the intense neutron flux.[38] Ionization greater than 5,000 rads in silicon chips delivered over seconds to minutes will degrade the function of semiconductors for long periods.[68] Due to the rarefied atmosphere encountered high above the earth at the most likely intercept point of an incoming warhead by a neutron bomb/warhead, whether it be the retired Sprint missile's W66 neutron warhead or the still in service Russian counterpart, the ABM-3 Gazelle, at the Terminal phase point (10–30 km) of the incoming warheads flight, the neutrons generated by a mid- to high-altitude nuclear explosion (HANE) have an even greater range than that encountered after a low altitude air burst, as in the high altitude case, there is a lower density of air molecules that produces, by comparison, an appreciable reduction in the air shielding effect/half-value thickness.

However, although this neutron transparency advantage attained only increases at increased altitudes, neutron effects lose importance in the exoatmospheric environment, being overtaken by the range of another effect of a nuclear detonation, at approximately the same altitude as the end of the incoming missile's boost phase (~150 km), ablation producing soft X-rays are the chief nuclear effects threat to the survival of incoming missiles and warheads rather than neutrons.[69] A factor exploited by the other warhead of the Safeguard Program, the enhanced (X-ray) radiation W71 and its USSR/Russian counterpart, the warhead on the A-135 Gorgon missile.

Another method by which neutron radiation can be used to destroy incoming nuclear warheads is by serving as an intense neutron generator and to thus initiate fission in the incoming warhead's fissionable components by fast fission, potentially causing the incoming warhead to prematurely detonate in a fizzle if within sufficient proximity, but in most likely inter-ception ranges, requiring only that enough fissionable material in the warhead fissions to interfere with the functioning of the incoming warhead when it is later fuzed to explode (see related physics: Subcritical reactor).

Lithium-6 hydride (Li6H) is cited as being used as a countermeasure to reduce the vulnerability/"harden" nuclear warheads from the effects of externally generated neutrons.[70][71] Radiation hardening of the warhead's electronic components as a countermeasure to high altitude neutron warheads somewhat reduces the range that a neutron warhead could successfully cause an unrecoverable glitch by the *TREE* (transient radiation effects on electronics) mechanism.[72][73]

32.2.4   Use as an area denial weapon

In November 2012, during the planning stages of Operation Hammer of God, it was suggested by a British parliamentar-ian that multiple enhanced radiation reduced blast (ERRB) warheads could be detonated in the mountain region of the Afghanistan/Pakistan border to prevent infiltration.[74] He proposed to warn the inhabitants to evacuate, then irradiate the area, making it unusable and impassable.[75] Used in this manner, the neutron bomb(s), regardless of burst height, would release neutron activated casing materials used in the bomb, and depending on burst height, create radioactive soil activation products.

In much the same fashion as the area denial effect resulting from fission product (the substances that make up the majority of fallout) contamination in an area following a conventional surface burst nuclear explosion, as considered in the Korean War by Douglas MacArthur, it would thus be a form of radiological warfare - with the difference that neutron bombs produce half, or less, of the quantity of fission products when compared to the same-yield pure fission bomb. Radiological warfare with neutron bombs that rely on fission primaries would therefore still produce fission fallout, albeit a compara-tively "cleaner" and shorter lasting version of it in the area if air bursts were utilized, as little to no fission products would be deposited on the direct immediate area, instead becoming diluted global fallout.

However the most effective use of a neutron bomb with respect to area denial would be to encase it in a thick shell of material that could be neutron activated, and use a surface burst. In this manner the neutron bomb would be turned into a "salted bomb"; a case of zinc-64, produced as a byproduct of depleted zinc oxide enrichment, would for example probably be the most attractive from a military point of view, as when activated the zinc-65 that is created is a gamma emitter, with a half life of 244 days.[76]

32.2.5   Maintenance

Neutron bombs/warheads require considerable maintenance for their capabilities, requiring some tritium for fusion boost-ing and tritium in the secondary stage (yielding more neutrons), in amounts on the order of a few tens of grams[77] (10–30 grams[78] estimated). Because tritium has a relatively short half-life of 12.32 years (after that time, half the tritium has decayed), it is necessary to replenish it periodically in order to keep the bomb effective. (For instance: to maintain a constant level of 24 grams of tritium in a warhead, about 1 gram per bomb per year[79] must be supplied.) Moreover, tritium decays into helium-3, which absorbs neutrons[80] and will thus further reduce the bomb's neutron yield.

32.3 See also

- Atomic demolition munitions - similar strategic use, low yield nuclear weapons.

- Neutron activation

- Neutron transport

- Nuclear fallout

- Nuclear strategy

- Nuclear warfare

- Nuclear weapon design

- W54

32.4 References

[1] "Sci/Tech Neutron bomb: Why 'clean' is deadly".

[2] "CHAPTER 2 CONVENTIONAL AND NUCLEAR WEAPONS - ENERGY PRODUCTION AND ATOMIC PHYSICS SECTION I - GENERAL. Figure 2-IX".

[3] "The Neutron Bomb".

[4] Kistiakovsky, George (Sep 1978). "The folly of the neutron bomb". *Bulletin of the Atomic Scientists* **34**: 27. Retrieved 11 February 2011.

[5] Hafemeister, David W. (2007). *Physics of societal issues: calculations on national security, environment, and energy.* Springer. p. 18. ISBN 978-0-387-95560-5.

[6] "CHAPTER 2 CONVENTIONAL AND NUCLEAR WEAPONS - ENERGY PRODUCTION AND ATOMIC PHYSICS SECTION I - GENERAL. Table 2-III".

[7] "Mock up". Remm.nlm.gov. Retrieved 2013-11-30.

[8] "Range of weapons effects". Johnstonsarchive.net. Retrieved 2013-11-30.

[9] "Weapon designer Robert Christy discussing scaling laws, that is, how injuries from ionizing radiation do not linearly scale in lock step with the range of thermal flash injuries, especially as higher and higher yield nuclear weapons are used". Webofstories.com. Retrieved 2013-11-30.

[10] Robert D. McFadden (December 1, 2010). "Samuel T. Cohen, Neutron Bomb Inventor, Dies at 89". *The New York Times.* Retrieved 2010-12-02. After the war, he joined the RAND Corporation and in 1958 designed the neutron bomb as a way to strike a cluster of enemy forces while sparing infrastructure and distant civilian populations.

[11] Cochran, Thomas; Arkin, William; Hoenig, Milton (1987). *Nuclear Weapons Databook: U.S. nuclear warhead production. Volume 2.* Ballinger Publishing. p. 23.

[12] "About: Chemistry article", by Anne Marie Helmenstine, Ph. D

[13] "On this Day: 7 April". *BBC.* 1978-04-07. Retrieved 2010-07-02. Jimmy Carter's successor, Ronald Reagan, changed US policy and gave the order for the production of neutron warheads to start in 1981. ...

[14] "The Soviet neutron bomb at 30. March 07 2010. RT".

[15] "Nuclear Weapon News and Background". Archived from the original on 2007-09-29. Retrieved 2012-10-11.

[16] Christopher Ruddy (June 15, 1997). "Bomb inventor says U.S. defenses suffer because of politics". *Tribune-Review*. Retrieved 2010-07-03. With the fall of the Berlin Wall and the end of communism as we knew it, the Bush administration moved to dismantle all of our tactical nuclear weapons, including the Reagan stockpile of neutron bombs. In Cohen's mind, America was brought back to Square One. Without tactical weapons like the neutron bomb, America would be left with two choices if an enemy was winning a conventional war: surrender, or unleash the holocaust of strategic nuclear weapons.

[17] "Types of Nuclear Weapons". Nuclearweaponarchive.org. Retrieved 2012-10-12.

[18] John Pike. "March 13, 1996". Globalsecurity.org. Retrieved 2012-10-12.

[19] (12-03).pdf "National Nuclear Security Administration - Homepage" (PDF). Nnsa.doe.gov. Retrieved 2012-10-12.

[20] "Report Of The Select Committee On U.S. National Security And Military/Commercial Concerns With The People's Republic Of China: Chapter 2 - PRC Theft Of U.S. Thermonuclear Warhead Design Information".

[21] Cohen, Samuel (9 August 1999). *Insight on the News* http://www.highbeam.com/doc/1G1-55426724.html. Missing or empty |title= (help)

[22] "Neutron bomb: Why 'clean' is deadly". BBC News. 1999-07-15. Retrieved 2012-10-12.

[23] UK parliamentary question on whether condemnation was considered by Thatcher government

[24] Ray, Jonathan (January 2015). "Red China's "Capitalist Bomb": Inside the Chinese Neutron Bomb Program" (PDF). *China Strategic Perspectives* (Washington, DC: National Defense University Press) **8**.

[25] http://www.globalsecurity.org/wmd/world/russia/gazelle.htm

[26] *The Nuclear Express: A Political History of the Bomb and Its Proliferation*, By Thomas C. Reed, Danny B. Stillman (2010), page 181

[27] *The Nuclear Express: A Political History of the Bomb and Its Proliferation*, By Thomas C. Reed, Danny B. Stillman (2010), page 177

[28] Wittner, Lawrence S. (2009). *Confronting the bomb: a short history of the world nuclear disarmament movement*. Stanford University Press. pp. 132–133. ISBN 978-0-8047-5632-7.

[29] Auten, Brian J. (2008). *Carter's conversion: the hardening of American defense policy*. University of Missouri Press. p. 134. ISBN 978-0-8262-1816-2.

[30] National security for a new era: globalization and geopolitics after Iraq, Donald Snow

[31] Herken, Greff (2003). *Brotherhood of the Bomb: The Tangled Lives and Loyalties of Robert Oppenheimer, Ernest Lawrence, and Edward Teller*. Macmillan. p. 332. ISBN 978-0-8050-6589-3.

[32] Asimov, Isaac. The New Intelligent Man's Guide to Science. Basic Books, New York, 1965. Page 410.

[33] "Neutron bomb an explosive issue, 1981".

[34] "Neutron bomb an explosive issue, 1981".

[35] Healy, Melissa (October 3, 1987). "Senate Permits Study for New Tactical Nuclear Missile". *Los Angeles Times*. Retrieved 2012-08-08.

[36] "Check Your Facts: Cox Report Bombs". Insight on the News. 9 August 1999. Retrieved 5 June 2015. – via Questia (subscription required)

[37] "List of All U.S. Nuclear Weapons". Nuclearweaponarchive.org. 2006-10-14. Retrieved 2012-10-12.

[38] "What Is a Neutron Bomb? By Anne Marie Helmenstine, Ph.D.".

[39] Netherlands dual capable artillery, 1985

[40] LLNL achievements in the 1970s

[41] "Neutron bomb an explosive issue, 1981".

[42] Muller, Richard A. (2009). *Physics for Future Presidents: The Science Behind the Headlines.* W.W. Norton & Company. p. 148. ISBN 978-0-393-33711-2.

[43] "what is a neutron bomb "In strategic terms, the neutron bomb has a theoretical deterrent effect: discouraging an armoured ground assault by arousing the fear of neutron bomb counterattack"".

[44] "Neutron bomb an explosive issue, 1981".

[45] http://www.gwu.edu/~{}nsarchiv//nukevault/ebb285/

[46] Soviet Ballistic Missile Defense and the Western Alliance, By David Scott Yost, pg 67-68

[47] "Neutron bomb an explosive issue, 1981".

[48] "Fact-index, neutron bomb".

[49] Calculated from http://nuclearweaponarchive.org/Nwfaq/Nfaq5.html assuming 0.5 kt combined blast and thermal

[50] "1) Effects of blast pressure on the human body" (PDF). Retrieved 2012-10-12.

[51] "Field manual 3-4 chapter 4".

[52] "Applications of the Monte Carlo Adjoint Shielding Methodology - MIT".

[53] "Neutron bomb an explosive issue, 1981".

[54] *New Scientist March 13, 1986 pg 45*. 1986-03-13. Retrieved 2012-10-12.

[55]

[56] *New Scientist June 12, 1986 pg 62.*

[57] "Monte Carlo Calculations Using MCNP4B for an Optimal Shielding Design of a 14-MeV Neutron Source, Submitted to the Journal of Radiation Protection Dosimetry 1998" (PDF).

[58] "Neutron Interactions – Part 2 George Starkschall, Ph.D. Department of Radiation Physics." (PDF).

[59] "22.55 "Principles of Radiation Interactions"" (PDF).

[60] "THE PREPARATION OF POLYETHYLENE AND MINERAL MATERIAL COMPOSITES, AND EXPERIMENTAL AND THEORETICAL (USING MCNP CODE) VERIFICATION OF THEIR CHARACTERISTICS FOR NEUTRON BEAM ATTENUATION" (PDF).

[61] "What is a neutron bomb".

[62] *Terror Reigns Again By Ronan Strobing. pg 418.*

[63] "M1A1/2 Abrams Main Battle Tank, United States of America".

[64] ""For example, M-1 tank armor includes depleted uranium, which can undergo fast fission and can be made to be radioactive when bombarded with neutrons".".

[65] http://web.ead.anl.gov/uranium/pdf/ducretecosteffec.pdf Paper Summary Submitted to Spectrum 2000, Sept 24-28, 2000, Chattanooga, TN DUCRETE: A Cost Effective Radiation Shielding Material. Quote- "The Ducrete/DUAGG replaces the conventional aggregate in concrete producing concrete with a density of 5.6 to 6.4 g/cm3 (compared to 2.3 g/cm3 for conventional concrete). This shielding material has the unique feature of having both high Z and low Z elements in a single matrix. Consequently, it is very effective for the attenuation of gamma and neutron radiation..."

[66] M. J. Haire and S. Y. Lobach, "Cask size and weight reduction through the use of depleted uranium dioxide (DUO$_2$)-concrete material", Waste Management 2006 Conference,Tucson, Arizona, February 26–March 2, 2006.

[67] "Half-Value Layer (Shielding)".

[68] "FAS Nuclear Weapon Radiation Effects".

[69] "Nuclear Matters Handbook". Nuclear weapon-generated X-rays are the chief threat to the survival of strategic missiles in-flight above the atmosphere and to satellites...The Neutron and gamma ray effects dominate at lower altitudes where the air absorbs most of the X-rays.

[70] "Section 12.0 Useful Tables Nuclear Weapons Frequently Asked Questions". Due to moderating ability and light weight, used to harden weapons against outside neutron fluxes (especially in combination with Li-6)...The very high cross section of this reaction for thermalized neutrons, combined with the light weight of the Li-6 atom, make it useful in the form of lithium hydride for hardening of nuclear weapons against external neutron fluxes.

[71] "*Restricted Data Declassification Policy, 1946 to the Present RDD-1*". The fact that Li6H is used in unspecified weapons for hardening

[72] "The Nuclear Matters Handbook, F.13".

[73] "Transient Radiation Effects on Electronics (TREE) Handbook Formerly Design Handbook for TREE, Chapters 1-6".

[74] "Huffington Post". Retrieved 2012-11-27.

[75] "Lord Gilbert obituary, by Andrew Roth, 3 June 2013. "Nobody lives up in the mountains on the border between Afghanistan and Pakistan except for a few goats and a handful of people herding them," he observed. "If you told them that some ... warheads were going to be dropped there and that it would be a very unpleasant place to go, they would not go there."".

[76] "1.6 Cobalt Bombs and other Salted Bombs, Nuclear Weapons Archive, Carey Sublette.".

[77] Kalinowski, Martin (2004). *International control of tritium for nuclear nonproliferation and disarmament*. CRC Press. p. 10. ISBN 978-0-415-31615-6.

[78] Zerriffi, Hisham (January 1996). "Tritium: The environmental, health, budgetary, and strategic effects of the Department of Energy's decision to produce tritium". Institute for Energy and Environmental Research.

[79] After 12.32 years, half the 24g has decayed and thus about 12g is missing: to replenish these 12g during the 12 years they decayed, adding about 1g per year is needed.

[80] When absorbing neutrons, helium-3 produces back some tritium, but it comes too late in the reaction for fusion boosting and doesn't compensate for the decayed tritium missing at the start of the reaction.

32.5 Further reading

- Cohen, Sam, *The Truth About the Neutron Bomb: The Inventor of the Bomb Speaks Out*, William Morrow & Co., 1983, ISBN 0-688-01646-4

- Cohen, Sam, *F*** You! Mr. President: Confessions of the Father of the Neutron Bomb*, Xlibris Corporation, 2000

32.6 External links

- Strategic Implications of Enhanced Radiation Weapons

- Nuclear Files.org Definition and history of the neutron bomb

- Creator of Neutron Bomb Leaves an Explosive Legacy

- The Woodrow Wilson Center's Nuclear Proliferation International History Project or NPIHP is a global network of individuals and institutions engaged in the study of international nuclear history through archival documents, oral history interviews and other empirical sources.

The Soviet/Warsaw pact invasion plan, "Seven Days to the River Rhine" to seize West Germany. According to proponents, neutron bombs would blunt an invasion by Soviet tanks and armored vehicles without causing as much damage as other nuclear weapons would.[33] Neutron bombs would have been used if the REFORGER conventional response of NATO to the invasion, was too slow or ineffective.[34][35]

U.S. Army M110 howitzers in a 1984 REFORGER staging area prior to transport. Variants of this "dual capable",[39] howitzer would launch the W79 neutron bomb.[40]

Wood frame house in 1953 nuclear test, 5 psi overpressure, complete collapse. Although neutron bombs, such as that fitted on the MGM-52 Lance missile would cause similar levels of destruction as depicted here within the zone were ~1970s tank crews would also be incapacitation by neutron radiation. When compared to the range of destruction that would be caused by the comparatively higher yield conventional nuclear weapons that it supplanted, which had been needed to deliver the same range and intensity of neutron dose to neutralize tank crews, the range of civilian destruction and amount of fission product fallout generated by a neutron bomb is far more constrained.[47] Sparing the destruction of West Germany more than would otherwise be the case.

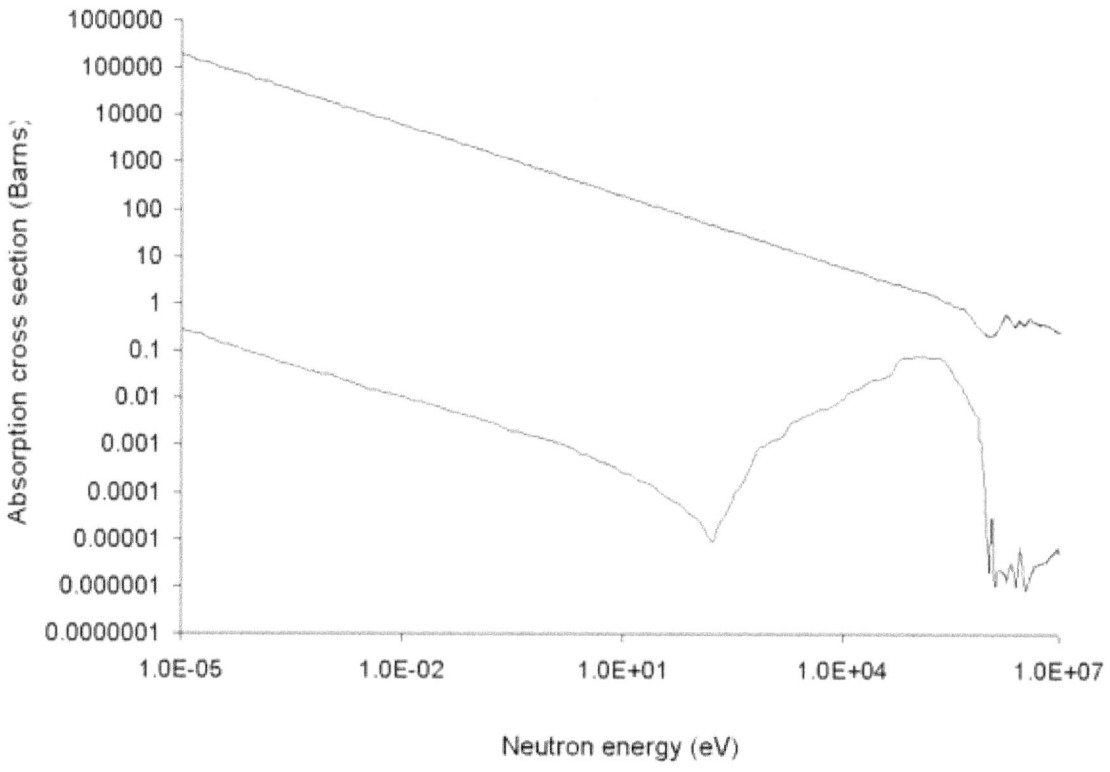

The Neutron cross section/ absorption probability in barns of the two natural Boron isotopes found in nature (top curve is for 10B and bottom curve for 11B. As neutron energy increases to 14 MeV, the absorption effectiveness, in general, decreases. Therefore, for boron containing armor to be effective, fast neutrons must first be slowed by another element by neutron scattering.

Radiation Weighting Factors for Neutrons

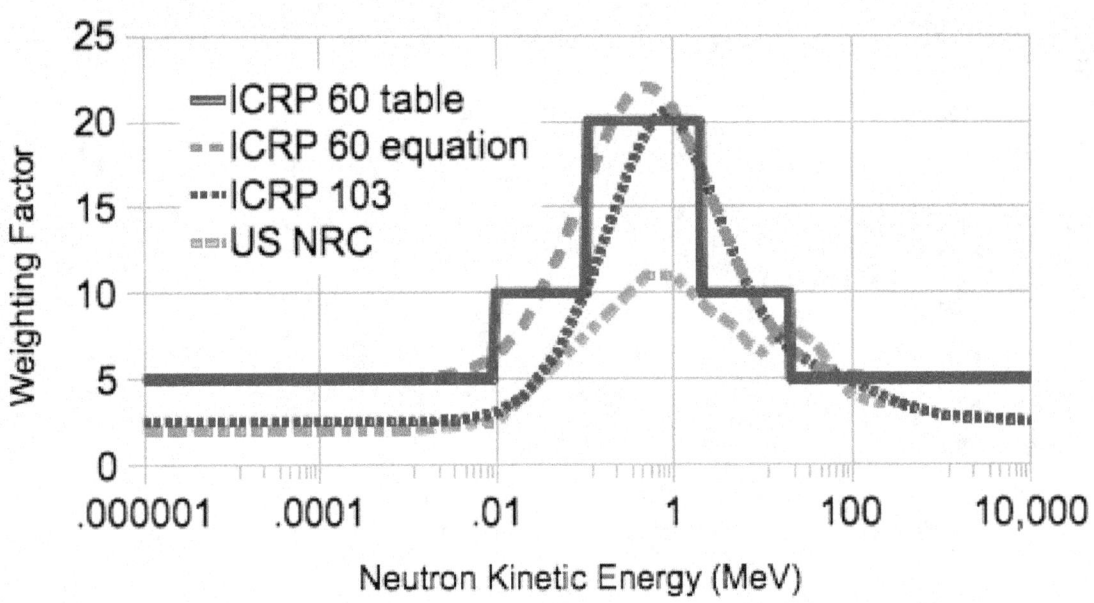

The radiation weighting factor for neutrons of various energy has been revised over time and certain agencies have different weighting factors, however despite the variation amongst the agencies, from the graph, for a given energy, A Fusion neutron (14 MeV) although more energetic, is less biologically deleterious than a Fission generated neutron or a Fusion neutron slowed to that energy, ~0.8 MeV .

Chapter 33

Neutron flux

The **neutron flux** is a quantity used in nuclear reactor physics corresponding to the total length travelled by all neutrons per unit time and volume,[1] or nearly equivalently number of neutrons travelling through a unit area in unit time.[2] The **neutron fluence** is defined as the neutron flux integrated over a certain time period.

33.1 Natural neutron flux

Neutron flux in asymptotic giant branch stars and in supernova is responsible for most of the natural nucleosynthesis producing elements heavier than iron. In stars there is a relatively low neutron flux on the order of 10^5 to 10^{11} neutrons per cm^2 per second, resulting in nucleosynthesis by the s-process (slow-neutron-capture-process). By contrast, after a core-collapse supernova, there is an extremely high neutron flux, on the order of 10^{22} neutrons per cm^2 per second, resulting in nucleosynthesis by the r-process (rapid-neutron-capture-process).

Atmospheric neutron flux, apparently from thunderstorms, can reach levels of $3 \cdot 10^2$ to $5 \cdot 10^2$ neutrons per cm^2 per sec.[3] However, recent results[4] obtained with unshielded scintillation neutron detectors actually show a decrease in the neutron flux during thunderstorms.

33.2 Artificial neutron flux

Further information: Neutron radiation

Artificial neutron flux refers to neutron flux which is man-made, either as byproducts from weapons or nuclear energy production or for specific application such as from a research reactor or by spallation. A flow of neutrons is often used to initiate the fission of unstable large nuclei. The additional neutron(s) may cause the nucleus to become unstable, causing it to decay (split) to form more stable products. This effect is essential in fission reactors and nuclear weapons.

Within a nuclear fission reactor the neutron flux is primarily the form of measurement used to control the reaction inside. The flux shape is the term applied to the density or relative strength of the flux as it moves around the reactor. Typically the strongest neutron flux occurs in the middle of the reactor core, becoming lower toward the edges. The higher the neutron flux the greater the chance of a nuclear reaction occurring as there are more neutrons going through an area.

A reactor vessel of a typical nuclear power plant (PWR) endures in 40 years (32 full reactor years) of operation approximately 3.5×10^{19} n/cm^2 (E>1MeV).[5] Neutron flux causes reactor vessels to suffer from embrittlement and the steel gets activated.

33.3 See also

- Neutron radiation

- Neutron transport

33.4 References

[1] Rudi J. J. Stamm'ler, Máximo Julio Abbate, Methods of steady-state reactor physics in nuclear design. ISBN 978-0126633207

[2] Neutron flux from the United States Nuclear Regulatory Commission, retrieved 30 May 2008

[3] Gurevich, A. V.; Antonova, V. P. (2012). "Strong Flux of Low-Energy Neutrons Produced by Thunderstorms". *Physical Review Letters* (Americal Physical Society) **108** (12). Bibcode:2012PhRvL.108l5001G. doi:10.1103/PhysRevLett.108.125001.

[4] Alekseenko, V.; Arneodo, F.; Bruno, G.; Di Giovanni, A.; Fulgion, W.; Gromushkin, D.; Shchegolev, O.; Stenkin, Yu.; Stepanov, V.; Sulakov, V.; Yashin, I. (2015). "Decrease of Atmospheric Neutron Counts Observed during Thunderstorms". *Physical Review Letters*(Americal Physical Society)**114**(12).Bibcode:2015PhRvL.114l5003A.doi:10.1103/PhysRevLett.114.1253.

[5] Nuclear Power Plant Borssele Reactor Pressure Vessel Safety Assessment, p. 29, 5.6 Neutron Fluence Calculation

Chapter 34

Neutron transport

Neutron transport is the study of the motions and interactions of neutrons with materials. Nuclear scientists and engineers often need to know where neutrons are in an apparatus, what direction they are going, and how quickly they are moving. It is commonly used to determine the behavior of nuclear reactor cores and experimental or industrial neutron beams. Neutron transport is a type of radiative transport.

34.1 Background

Neutron transport has roots in the Boltzmann equation, which was used in the 1800s to study the kinetic theory of gases. It did not receive large-scale development until the invention of chain-reacting nuclear reactors in the 1940s. As neutron distributions came under detailed scrutiny, elegant approximations and analytic solutions were found in simple geometries. However, as computational power has increased, numerical approaches to neutron transport have become prevalent. Today, with massively parallel computers, neutron transport is still under very active development in academia and research institutions throughout the world. It remains one of the most computationally challenging problems in the world since it depends on 3-dimensions of space, time, and the variables of energy span several decades (from fractions of meV to several MeV). Modern solutions use either discrete-ordinates or monte-carlo methods, or even a hybrid of both.

34.2 Neutron Transport Equation

The neutron transport equation is a balance statement that conserves neutrons. Each term represents a gain or a loss of a neutron, and the balance, in essence, claims that neutrons gained equals neutrons lost. It is formulated as follows:[1]

$$\left(\frac{1}{v(E)} \frac{\partial}{\partial t} + \hat{\mathbf{\Omega}} \cdot \nabla + \Sigma_t(\mathbf{r}, E, t) \right) \psi(\mathbf{r}, E, \hat{\mathbf{\Omega}}, t) =$$

$$\frac{\chi_p(E)}{4\pi} \int_0^\infty dE' \nu_p(E') \Sigma_f(\mathbf{r}, E', t) \phi(\mathbf{r}, E', t) + \sum_{i=1}^N \frac{\chi_{di}(E)}{4\pi} \lambda_i C_i(\mathbf{r}, t) +$$

$$\int_{4\pi} d\Omega' \int_0^\infty dE' \Sigma_s(\mathbf{r}, E' \to E, \hat{\mathbf{\Omega}}' \to \hat{\mathbf{\Omega}}, t) \psi(\mathbf{r}, E', \hat{\mathbf{\Omega}}', t) + s(\mathbf{r}, E, \hat{\mathbf{\Omega}}, t)$$

Where:

Symbol	Meaning	Comments
$\mathbf{r}$	Position vector (i.e. x,y,z)	
E	Energy	
$\hat{\Omega} = \dfrac{\mathbf{v}(E)}{|\mathbf{v}(E)|} = \dfrac{\mathbf{v}(E)}{v(E)}$	Unit vector (solid angle) in direction of motion	
t	Time	
$\mathbf{v}(E)$	Neutron velocity vector	
$\psi(\mathbf{r}, E, \hat{\Omega}, t)dr\, dE\, d\Omega$	Angular neutron flux Amount of neutron track length in a differential volume dr about r, associated with particles of a differential energy in dE about E, moving in a differential solid angle in $d\Omega$ about $\hat{\Omega}$, at time t.	Note integrating over all angles yields scalar neutron flux $\phi = \displaystyle\int_{4\pi} d\Omega \psi$
$\phi(\mathbf{r}, E, t)dr\, dE$	Scalar neutron flux Amount of neutron track length in a differential volume dr about r, associated with particles of a differential energy in dE about E, at time t.	
ν_p	Average number of neutrons produced per fission (e.g., 2.43 for U-235).[2]	
$\chi_p(E)$	Probability density function for neutrons of exit energy E from all neutrons produced by fission	
$\chi_{di}(E)$	Probability density function for neutrons of exit energy E from all neutrons produced by delayed neutron precursors	
$\Sigma_t(\mathbf{r}, E, t)$	Macroscopic total cross section, which includes all possible interactions	
$\Sigma_f(\mathbf{r}, E', t)$	Macroscopic fission cross section, which includes all fission interactions in dE' about E'	
$\Sigma_s(\mathbf{r}, E' \to E, \hat{\Omega}' \to \hat{\Omega}, t)dE'd\Omega'$	Double differential scattering cross section Characterizes scattering of a neutron from an incident energy E' in dE' and direction $\hat{\Omega}'$ in $d\Omega'$ to a final energy E and direction $\hat{\Omega}$	
N	Number of delayed neutron precursors	
λ_i	Decay constant for precursor i	
$C_i(\mathbf{r}, t)$	Total number of precursor i in $\mathbf{r}$ at time t	
$s(\mathbf{r}, E, \hat{\Omega}, t)$	Source term	

The transport equation can be applied to a given part of phase space (time t, energy E, location r, and direction of travel).

The first term represents the time rate of change of neutrons in the system. the second terms describes the movement of neutrons into or out of the volume of space of interest.
The third term accounts for all neutrons that have a collision

in that phase space. The first term on the right hand side is the production of neutrons in this phase space due to fission, while the second term on the right hand side is the production of neutrons in this phase space due to delayed neutron precursors (i.e., unstable nuclei which undergo neutron decay). The third term on the right hand side is in-scattering, these are neutrons that enter this area of phase space as a result of scattering interactions in another. The fourth term on the right is a generic source. The equation is usually solved to find $\phi(\mathbf{r}, E)$, since that will allow for the calculation of reaction rates, which are of primary interest in shielding and dosimetry studies.

34.3 Types of neutron transport calculations

Several basic types of neutron transport problems exist, depending on the type of problem being solved.

34.3.1 Fixed Source

A fixed source calculation involves imposing a known neutron source on a medium and determining the resulting neutron distribution throughout the problem. This type of problem is particularly useful for shielding calculations, where a designer would like to minimize the neutron dose outside of a shield while using the least amount of shielding material. For instance, a spent nuclear fuel cask requires shielding calculations to determine how much concrete and steel is needed to safely protect the truck driver who is shipping it.

34.3.2 Criticality

Fission is the process through which a nucleus splits into (typically two) smaller atoms. If fission is occurring, it is often of interest to know the asymptotic behavior of the system. A reactor is called "critical" if the chain reaction is self-sustaining and time-independent. If the system is not in equilibrium the asymptotic neutron distribution, or the fundamental mode, will grow or decay exponentially over time.

Criticality calculations are used to analyze steady-state multiplying media (multiplying media can undergo fission), such as a critical nuclear reactor. The loss terms (absorption, out-scattering, and leakage) and the source terms (in-scatter and fission) are proportional to the neutron flux, contrasting with fixed-source problems where the source is independent of the flux. In these calculations, the presumption of time invariance requires that neutron production exactly equals neutron loss.

Since this criticality can only be achieved by very fine manipulations of the geometry (typically via control rods in a reactor), it is unlikely that the modeled geometry will be truly critical. To allow some flexibility in the way models are set up, these problems are formulated as eigenvalue problems, where one parameter is artificially modified until criticality is reached. The most common formulations are the time-absorption and the multiplication eigenvalues, also known as the alpha and k eigenvalues. The alpha and k are the tunable quanitites.

K-eigenvalue problems are the most common in nuclear reactor analysis. The number of neutrons produced per fission is multiplicatively modified by the dominant eigenvalue. The resulting value of this eigenvalue reflects the time dependence of the neutron density in a multiplying medium.

- *keff* < 1, subcritical: the neutron density is decreasing as time passes;

- *keff* = 1, critical: the neutron density remains unchanged; and

- *keff* > 1, supercritical: the neutron density is increasing with time.

In the case of a nuclear reactor, neutron flux and power density are proportional, hence during reactor start-up *keff* > 1, during reactor operation *keff* = 1 and *keff* < 1 at reactor shutdown.

34.4   Computational Methods

Both fixed-source and criticality calculations can be solved using deterministic methods or stochastic methods. In deterministic methods the transport equation (or an approximation of it, such as diffusion theory) is solved as a differential equation. In stochastic methods such as Monte Carlo discrete particle histories are tracked and averaged in a random walk directed by measured interaction probabilities. Deterministic methods usually involve multi-group approaches while Monte Carlo can work with multi-group and continuous energy cross-section libraries. Multi-group calculations are usually iterative, because the group constants are calculated using flux-energy profiles, which are determined as the result of the neutron transport calculation.

34.4.1   Discretization in Deterministic Methods

To numerically solve the transport equation using algebraic equations on a computer, the spatial, angular, energy, and time variables must be discretized.

- Spatial variables are typically discretized by simply breaking the geometry into many small regions on a mesh. The balance can then be solved at each mesh point using finite difference or by nodal methods.

- Angular variables can be discretized by discrete ordinates and weighting quadrature sets (giving rise to the SN methods), or by functional expansion methods with the spherical harmonics (leading to the PN methods).

- Energy variables are typically discretized by the multi-group method, where each energy group represents one constant energy. As few as 2 groups can be sufficient for some thermal reactor problems, but fast reactor calculations may require many more.

- The time variable is broken into discrete time steps, with time derivatives replaced with difference formulas.

34.4.2   Computer Codes Used In Neutron Transport

Probabilistic codes

- **OpenMC** - An MIT developed open source Monte Carlo code [3]

- **MCNP** - A LANL developed Monte Carlo code for general radiation transport

- **KENO** - An ORNL developed Monte Carlo code for criticality analysis

- **MCBEND** - An ANSWERS Software Service developed Monte Carlo code for general radiation transport

- **Serpent** - A Finnish developed Monte Carlo neutron transport code [4]

- **TRIPOLI** - 3D general purpose continuous energy Monte Carlo Transport code developed at CEA, France [5]

Deterministic codes

- **Attila** - A commercial transport code

- **DRAGON** - An open-source lattice physics code

- **PHOENIX/ANC** - A proprietary lattice-physics and global diffusion code suite from Westinghouse Electric

- **PARTISN** - A LANL developed transport code based on the discrete ordinates method

- **NEWT** - An ORNL developed 2-D SN code

- **DIF3D/VARIANT** - An Argonne National Laboratory developed 3-D code originally developed for fast reactors

- **DENOVO** - A massively parallel transport code under development by ORNL

- **DANTSYS**

- **RAMA** - A proprietary 3D method of characteristics code with arbitrary geometry modeling, developed for EPRI by TransWare Enterprises Inc. [6]

- **RAPTOR-M3G** - A proprietary parallel radiation transport code developed by Westinghouse Electric Company

- **OpenMOC** - An MIT developed open source parallel method of characteristics code [7]

- **MPACT** - A parallel 3D method of characteristics code under development by the University of Michigan

- **DORT** - Discrete Ordinates Transport

- **APOLLO** - A lattice physics code [8]

34.5 See also

- Nuclear Reactor

- Boltzmann equation

- TINTE

- Neutron scattering

- Monte Carlo N-Particle Transport Code

34.6 References

[1] Adams, Marvin L. (2009). *Introduction to Nuclear Reactor Theory*. Texas A&M University.

[2] "ENDF Libraries".

[3] "OpenMC".

[4] "PSG2 Serpent".

[5] "TRIPOLI-4".

[6] "RAMA".

[7] "OpenMOC".

[8] "APOLLO3" (PDF).

- Lewis, E., & Miller, W. (1993). Computational Methods of Neutron Transport. American Nuclear Society. ISBN 0-89448-452-4.

- Duderstadt, J., & Hamilton, L. (1976). Nuclear Reactor Analysis. New York: Wiley. ISBN 0-471-22363-8.

- Marchuk, G. I., & V. I. Lebedev (1986). Numerical Methods in the Theory of Neutron Transport. Taylor & Francis. p. 123. ISBN 978-3-7186-0182-0.

34.7   External links

- LANL MCNP6 website
- LANL MCNPX website
- VTT Serpent website
- MIT CRPG OpenMOC website

34.8 Text and image sources, contributors, and licenses

34.8.1 Text

- **Neutron** *Source:* https://en.wikipedia.org/wiki/Neutron?oldid=682674206 *Contributors:* AxelBoldt, Tobias Hoevekamp, Chenyu, Trelvis, Calypso, Mav, Bryan Derksen, The Anome, AstroNomer~enwiki, Malcolm Farmer, Andre Engels, Xaonon, Danny, XJaM, Roadrunner, Jaknouse, Olivier, Patrick, Michael Hardy, Valery Beaud, Ixfd64, TakuyaMurata, NuclearWinner, Looxix~enwiki, ArnoLagrange, Mkweise, Ellywa, Ahoerstemeier, Cyp, Andrewa, Aarchiba, Julesd, Glenn, Nikai, Andres, Stone, Denni, Kbk, Tarosan~enwiki, Maximus Rex, Donarreiskoffer, Gentgeen, Robbot, Fredrik, Romanm, Merovingian, Rursus, Wikibot, Alan Liefting, Dave6, Giftlite, Mikez, Art Carlson, Herbee, Xerxes314, Everyking, Dratman, NeoJustin, Bensaccount, Poupoune5, Jorge Stolfi, Christofurio, Knutux, Karol Langner, Aecarol, Icairns, Zfr, Cglassey, Peter bertok, Frau Holle, M1ss1ontomars2k4, Sparky2002b, Mike Rosoft, Guanabot, Vsmith, Dbachmann, Bender235, Kjoonlee, AlDragon, Geoking66, Neko-chan, RJHall, CanisRufus, El C, Susvolans, Femto, CDN99, Bobo192, O18, Smalljim, SpeedyGonsales, Kjkolb, Obradovic Goran, Sam Korn, Nsaa, Jakew, Eddideigel, Jumbuck, Patsw, Alansohn, Interiot, Riana, Wtmitchell, BRW, NickMartin, Vuo, DV8 2XL, HenryLi, Tchaika, Forteblast, Falcorian, Richard Arthur Norton (1958-), JarlaxleArtemis, WadeSimMiser, Sega381, SDC, Jon Harald Søby, Prashanthns, Abd, LexCorp, Graham87, Magister Mathematicae, Doughboy, Ketiltrout, Rjwilmsi, Nightscream, Zbxgscqf, Strait, AySz88, Oo64eva, Rangek, FlaBot, Nihiltres, Goudzovski, Srleffler, Ronebofh, King of Hearts, Chobot, DVdm, YurikBot, RobotE, Bambaiah, JWB, TSO1D, Jimp, Phantomsteve, KyleDantarin, Stephenb, Gaius Cornelius, Yyy, Salsb, NawlinWiki, Tupungato, Wiki alf, Complainer, Grafen, Długosz, Voidxor, Scottfisher, Kkmurray, Spute, Dna-webmaster, Wknight94, Stefan Udrea, Mike Serfas, Closedmouth, Reyk, Modify, Alchie1, CWenger, RG2, Paul Erik, Triple333, Attilios, SmackBot, Caiyern, Melchoir, Wiki Tiki God, Unyoyega, Jrockley, Dr.Science, Edgar181, Yamaguchi⬚⬚, Kdliss, Wigren, Chris the speller, Rajeevmass~enwiki, ·Persian Poet Gal, SchfiftyThree, Complexica, DHN-bot~enwiki, Sbharris, Colonies Chris, Brainblaster52, Can't sleep, clown will eat me, DéRahier, Juancnuno, SundarBot, DFriend, Aldaron, KunalKathuria, Nakon, Mwtoews, DMacks, Soarhead77, Bdushaw, Pilotguy, Renafaye77, SashatoBot, Demicx, Tim bates, Mgiganteus1, Slakr, Citicat, Asyndeton, BranStark, Shoeofdeath, Newone, Tawkerbot2, Atomobot, Mosaffa, CmdrObot, Wafulz, Dycedarg, Rwflammang, Joelholdsworth, Lokal Profil, Karenjc, Myasuda, Safalra, Icek~enwiki, Badseed, Nick Y., Gogo Dodo, Chasingsol, Phydend, Gimmetrow, Thijs!bot, Epbr123, Montazmeahii, Goods21, Tsogo3, N5iln, Oerjan, Headbomb, Marek69, SouthernMan, RoboServien, Escarbot, Aadal, WikiSlasher, AntiVandalBot, Seaphoto, Naturalnumber, Spencer, Astavats, Husond, CosineKitty, Medconn, TheEditrix2, Bongwarrior, VoABot II, Kuyabribri, JamesBWatson, WODUP, Mother.earth, Animum, BatteryIncluded, Dirac66, LorenzoB, DerHexer, JaGa, Hans Moravec, Hyray, Patstuart, MartinBot, Church of emacs, Gnuarm, Mennoblaauw, Andre.holzner, Rettetast, J.delanoy, Dbiel, Extransit, Acalamari, Ncmvocalist, TomasBat, MetsFan76, Joshmt, Heavens is the world, Scott Illini, TraceyR, Idioma-bot, Mviduka4197, VolkovBot, Tourbillon, Thedjatclubrock, Jeff G., Mocirne, Seattle Skier, TXiKiBoT, DoctorPiouk, Dev 176, Martin451, ABigGreenHippo, Abdullais4u, FreeFull, Wikiisawesome, Scarymaryfwfc, RadiantRay, Roomyt, W1k13rh3nry, Antixt, Deanlsinclair, Enviroboy, Burntsauce, Brianga, AlleborgoBot, EmxBot, Neparis, D. Recorder, Ponyo, YohanN7, SieBot, Cwkmail, Yintan, Agesworth, JerrySteal, Keilana, RadicalOne, Toddst1, Tiptoety, JetLover, Arjen Dijksman, Sbowers3, Aruton, Oxymoron83, AnonGuy, Beej175560, Techman224, Anyeverybody, Nergaal, Denisarona, Lord Shivan, Naturespace, ClueBot, RudolfSchmidt, PipepBot, Fasettle, Fyyer, The Thing That Should Not Be, Starkiller88, Industrieman, Mild Bill Hiccup, Polyamorph, Shjacks45, ChandlerMapBot, DragonBot, Gnome de plume, Jusdafax, Ju7kik8ol568r, Cenarium, Jotterbot, Vboo-belarus, Subash.chandran007, Plasmic Physics, Versus22, XLinkBot, Dark Mage, PL290, SkyLined, Addbot, Taschna, DOI bot, Ronhjones, Mr. Wheely Guy, LaaknorBot, CarsracBot, JBukon, Favonian, LinkFA-Bot, 5 albert square, AgadaUrbanit, Morgrimm, Numbo3-bot, Ehrenkater, LarryFrank, Tide rolls, Lightbot, Teles, Legobot, Luckas-bot, Yobot, Велетень, 2D, Tohd8BohaithuGh1, Cabb99, AnakngAraw, AnomieBOT, Bsimmons666, Jim1138, AdjustShift, Bluerasberry, Materialscientist, Hdehuer, The High Fin Sperm Whale, Citation bot, Satan's Kitchen, Maxis ftw, Raven1977, ArthurBot, Marshallsumter, Xqbot, Gopal81, Capricorn42, Drilnoth, DSisyphBot, Gilo1969, Paula Pilcher, Faatoafe90, Goostyyy, WaveEtherSniffer, GrouchoBot, Abce2, Amaury, Doulos Christos, Gordonrox24, Shadowjams, A. di M., Samwb123, R8R Gtrs, FrescoBot, LucienBOT, Paine Ellsworth, Cannolis, Citation bot 1, Ecko15, Biker Biker, Pinethicket, HRoestBot, Jonesey95, Nicklcms, Seattle Jörg, Abhinav paulite, Double sharp, Darrell cosare, کاشف عقیل, Mr.98, Diannaa, Ironnickel, Andrea105, Onel5969, TjBot, MagnInd, Jackehammond, Jimmy be, Robert Johnson 10, EmausBot, Green Day143, WikitanvirBot, Unkenruf, GoingBatty, Illdz, Psturm~enwiki, Pcorty, Wikipelli, Hhhippo, ZéroBot, John Cline, Brazmyth, Quondum, GianniG46, Copper.nanotube, Brandmeister, L Kensington, Epicstonemason, Sjkimminau, Chris857, VictorianMutant, DASHBotAV, Whoop whoop pull up, ClueBot NG, Nebulosus, CocuBot, Satellizer, Letoya123, TruPepitoM, OverQuantum, Heyheyheyhohoho, Rezabot, Android1188, Widr, Diyar se, Ieditpagesincorrectly, Bibcode Bot, Neutronscattering, Wiki13, Metricopolus, Contact '97, Universuminkeisari, Nathanrohler, Zedshort, Hamish59, Nitrobutane, Hobos-r-us12, Oznitecki, BattyBot, MeowMeowArf, Dansalmo, ChrisGualtieri, GeorgEhlers, Ducknish, Gladiator222, Dexbot, Mogism, 331dot, TwoTwoHello, Lugia2453, Graphium, FaerieChilde, Fossilsnout, Morg00, Cldorian, Xuanmingzi, DihllonJessie, Jesse.johns, The Herald, Zenibus, Darkch2, Jwratner1, Javierha, My name is not dave, Cytokinetics, EtymAesthete, DudeWithAFeud, Abitslow, Aspaas-Bekkelund, Bballbro62, Mahusha, Light on the wall, Monkbot, Profesionalpretzels, Jayakumar RG, Haftswinch532, Selmatoed50, Istillcant, HMSLavender, Petahr, Orduin, Kethrus, Pulkit 4325, DiscantX, TSchonfeldt, Matan Kovac, KasparBot, Kafishabbir, Lord Wingus The Third and Anonymous: 554

- **Discovery of the neutron** *Source:* https://en.wikipedia.org/wiki/Discovery_of_the_neutron?oldid=678521647 *Contributors:* Gparker, Bgwhite, Bdushaw, Oerjan, Headbomb, Dirac66, Lamro, Citation bot, Jbhunley, Bibcode Bot, Dexbot, Knife-in-the-drawer and Anonymous: 1

- **Standard Model** *Source:* https://en.wikipedia.org/wiki/Standard_Model?oldid=679087631 *Contributors:* AxelBoldt, Derek Ross, CYD, Bryan Derksen, The Anome, Ed Poor, Andre Engels, Roadrunner, David spector, Isis~enwiki, Youandme, Ram-Man, Stevertigo, Edward, Patrick, Boud, Michael Hardy, SebastianHelm, Looxix~enwiki, Julesd, Glenn, AugPi, Mxn, Raven in Orbit, Reddi, Phr, Tpbradbury, Populus, Haoherb428, Phys, Floydian, Bevo, Pierre Boreal, AnonMoos, BenRG, Jeffq, Dmytro, Drxenocide, Robbot, Nurg, Securiger, Texture, Roscoe x, Fuelbottle, Mattflaschen, Tobias Bergemann, Alan Liefting, Ancheta Wis, Giftlite, Dbenbenn, Harp, Herbee, Monedula, LeYaYa, Xerxes314, Dratman, Alison, JeffBobFrank, Dmmaus, Pharotic, Brockert, Bodhitha, Andycjp, Sonjaaa, HorsePunchKid, APH, Icairns, AmarChandra, Gscshoyru, Kate, Arivero, FT2, Rama, Vsmith, David Schaich, Xezbeth, D-Notice, Dfan, Bender235, Pt, El C, Laurascudder, Shanes, Drhex, Fogger~enwiki, Brim, Rbj, Jeodesic, Jumbuck, Alansohn, Gary, ChristopherWillis, Guy Harris, Axl, Sligocki, Kocio, Stillnotelf, Alinor, Wtmitchell, Egg, TenOfAllTrades, H2g2bob, Killing Vector, Linas, Mindmatrix, Benbest, Dodiad, Mpatel, Faethon, TPickup, Faethon34, Palica, Dysepsion, Faethon36, Qwertyca, Drbogdan, Rjwilmsi, Zbxgscqf, Macumba, Strangethingintheland, Dstudent, R.e.b., Bubba73, Drrngrvy, Agasicles, FlaBot, Naraht, Agasides, DannyWilde, Dave1g, Itinerant1, Gparker, Jrtayloriv, Goudzovski, Chobot, Bgwhite, FrankTobia, YurikBot, Bambaiah, Ohwilleke, VoxMoose, Bhny, JabberWok, Bovineone, Krbabu, SCZenz, JulesH, Davemck, Lomn, E2mb0t~enwiki, Dna-webmaster, Jrf, Dv82matt, Tetracube, Hirak 99, Arthur Rubin, Netrapt, JLaTondre, Caco de vidro, RG2, GrinBot~enwiki, That Guy, From That Show!, Hal peridol, SmackBot, YellowMonkey, Tom Lougheed, Melchoir, Bazza 7, KocjoBot~enwiki, Jagged 85, Thunderboltz, Setanta747 (locked), Skizzik, Dauto, Chris the speller, Bluebot, TimBentley, Sirex98, Silly rabbit, Complexica, Metacomet, DHN-bot~enwiki, MovGP0, QFT, Kittybrewster, Addshore, Jmnbatista, Cybercobra, Jgwacker, BullRangifer, Soarhead77, Daniel.Cardenas, Yevgeny Kats, Byelf2007, TriTertButoxy, Craig Bolon, Ajnosek, Ekjon Lok, Bjankuloski06, Tarcieri, Waggers, JarahE, Michaelbusch, Lottamiata, Newone, Twas Now, IanOfNorwich, Srain, Patrickwooldridge, J Milburn, Mosaffa, Gatortpk, Vessels42, Geremia, Van helsing, Harrigan, Phatom87, Cydebot, David edwards, Verdy p, Michael C Price, Xantharius, Crum375, JamesAM, Thijs!bot, Epbr123, Headbomb, Phy1729, Stannered, Tariqhada, Seaphoto, Orionus, Voyaging, Gnixon, Jbaranao, Jrw@pobox.com, Len Raymond, Narssarssuaq, Bakken, CattleGirl, Davidoaf, Vanished user ty12kl89jq10, Lvwarren, Taborgate, Leyo, HEL, J.delanoy, Hans Dunkelberg, Stephanwehner, Wbellido, Aoosten, Jacksonwalters, The Transliterator, DadaNeem, Student7, Joshmt, WJBscribe, Jozwolf, Hexane2000, BernardZ, Awren, Sheliak, Physicist brazuca, Schucker, Goop Goop, Fences and windows, Dextrose, Mcewan, Swamy g, TXiKiBoT, Sharikkamur, Thrawn562, Voorlandt, Escalona, Setreset, PDFbot, Pleroma, UnitedStatesian, Piyush Sriva, Kacser, Billinghurst, Francis Flinch, Moose-32, Ptrslv72, David Barnard, SieBot, ShiftFn, Robdunst, Jim E. Black, SheepNotGoats, Gerakibot, Nozzer42, Mr swordfish, Wing gundam, Bamkin, Likebox, Arthur Smart, HungarianBarbarian, Commutator, KathrynLybarger, Iomesus, C0nanPayne, Crazz bug 5, ClueBot, Superwj5, Wwheaton, Garyzx, SuperHamster, Elsweyn, Maldmac, DragonBot, Djr32, Diagramma Della Verita, Nymf, Eeekster, Brews ohare, NuclearWarfare, PhySusie, Ordovico, Mastertek, DumZiBoT, BodhisattvaBot, Guarracino, Mitch Ames, Truthnlove, Stephen Poppitt, Tayste, Addbot, Deepmath, Eric Drexler, DWHalliday, Mjamja, Leszek Jańczuk, NjardarBot, Mwoldin, Bassbonerocks, Barak Sh, AgadaUrbanit, Lightbot, Smeagol 17, Abjiklam, Ve744, Luckas-bot, Yobot, Orion11M87, AnomieBOT, JackieBot, Icalanise, Citation bot, ArthurBot, Northryde, LilHelpa, Xqbot, Sionus, Professor J Lawrence, Tomwsulcer, Edsegal, GrouchoBot, Trongphu, QMarion II, Ernsts, A. di M., Bytbox, FrescoBot, Paine Ellsworth, Aliotra, Steve Quinn, Citation bot 1, Rameshngbot, MJ94, RedBot, MastiBot, Aknochel, Sijothankam, Puzl bustr, Beta Orionis, Physics therapist, Bj norge, Innotata, Jesse V., RjwilmsiBot, Mathewsyriac, Afteread, EmausBot, Bookalign, WikitanvirBot, Wilhelm-physiker, Bdijkstra, DerNeedle, Kenmint, Dbraize, Tanner Swett, HeptishHotik, بهار منشیون مهمن, Suslindisambiguator, Quondum, Webbeh, UniversumExNihilo, Vanished user fijw983kjaslkekfhj45, Maschen, RockMagnetist, Stormymountain, Ζeτα ζ, Whoop whoop pull up, Isocliff, ClueBot NG, Smtchahal, Snotbot, Tonypak, O.Koslowski, CharleyQuinton, Dsperlich, Theopolisme, ZakMarksbury, Helpful Pixie Bot, Bibcode Bot, BG19bot, Tirebiter78, AvocatoBot, Lukys~enwiki, Stapletongrey, Ownedroad9, Chip123456, ChrisGualtieri, Khazar2, Billyfesh399, Rhlozier, JYBot, Dexbot, Doom636, Rongended, Cerabot~enwiki, CuriousMind01, Cjean42, Jayanta mallick, Joeinwiki, Kowtje, JPaestpreornJeolhlna, Eyesnore, Euan Richard, Nigstomper, Particle physicist, Prokaryotes, Jernahthern, Ginsuloft, Dimension10, JNrgbKLM, Krabaey, 1codesterS, FelixRosch, Monkbot, Delbert7, BradNorton1979, Lathamboyle, Tetra quark, KasparBot, Buckbill10 and Anonymous: 357

- **Nuclear force** *Source:* https://en.wikipedia.org/wiki/Nuclear_force?oldid=682319479 *Contributors:* Agtx, Jakohn, Owain, GreatWhiteNortherner, Xerxes314, D6, Hidaspal, ESkog, Rbj, Nk, Riana, Kocio, Kusma, Forteblast, Linas, Tabletop, Scroipt, RE, Yamamoto Ichiro, Eubot, Nihiltres, Chobot, Cshay, WriterHound, YurikBot, Wavelength, Borgx, Mushin, Bambaiah, Hairy Dude, Phmer, Jimp, Mconst, SCZenz, Voidxor, Deane@gooroos.com, FyzixFighter, SmackBot, Incnis Mrsi, Chris the speller, DHN-bot~enwiki, Croquant, Sbharris, Voyajer, Jgwacker, Bdushaw, Safalra, Kanags, Cricketgirl, Headbomb, Bobblehead, Mhaitham.shammaa, Magioladitis, Bongwarrior, Cgingold, JJ Harrison, Robin S, J.delanoy, VolkovBot, JohnBlackburne, Clarince63, Tamorlan, Biasoli, Hoopssheaffer, Abortz, SieBot, Sanya3, Proton666, Dolphin51, Bschaeffer~enwiki, ClueBot, Binksternet, Chris Illert, Manishearth, Djr32, RexxS, Ladsgroup, Karthik bala2009, SilvonenBot, Addbot, Tcncv, Mjamja, WFPM, CarsracBot, Tide rolls, Luckas-bot, Yobot, VanishedUser sdu9aya9fasdsopa, Orange Knight of Passion, Kingpin13, Bci2, Δζ, Pinethicket, RedBot, Ripchip Bot, Mema mema12, Gfoley4, XinaNicole, Wikipelli, ZéroBot, Quondum, Brandmeister, ClueBot NG, Helpful Pixie Bot, Bibcode Bot, Krenair, Vkpd11, Mdahir, Kylegodbey, Ginsuloft, Marj2117, JellyPatotie, Tetra quark and Anonymous: 70

- **Binding energy** *Source:* https://en.wikipedia.org/wiki/Binding_energy?oldid=667349242 *Contributors:* Bryan Derksen, Patrick, Tim Starling, Kku, SebastianHelm, The Anomebot, Taxman, Robbot, Fredrik, Securiger, Auric, Harp, BenFrantzDale, Fastfission, LeYaYa, Jason Quinn, ConradPino, Creidieki, Pinnerup, Zowie, RJHall, Cap'n Refsmmat, Neilrieck, CDN99, Mike Schwartz, La goutte de pluie, Kjkolb, Larry V, Nsaa, Jjron, Jhd, Riana, EagleFalconn, H2g2bob, Zereshk, Capecodeph, Blaze Labs Research, Firsfron, Reinoutr, Woohookitty, Cruccone, Robert K S, Rjwilmsi, Wikibofh, Kazrak, FlaBot, Margosbot~enwiki, Fragglet, Gurch, Fresheneesz, DVdm, Dstrozzi, YurikBot, Wavelength, Spacepotato, JWB, Van der Hoorn, Wikimachine, Howcheng, Goffrie, Kkmurray, Covington, Petri Krohn, Benonemusic, CWenger, RG2, Phr en, Bibliomaniac15, SmackBot, Jrockley, RobotJcb, Fetofs, Croquant, Sbharris, Colonies Chris, Javalenok, OrphanBot, Voyajer, ArglebargleIV, Sophia, Euchiasmus, Herr apa, Zarniwoot, Dan Gluck, Igoldste, JRSpriggs, Cydebot, Nick Y., Bvcrist, DumbBOT, Thijs!bot, Barticus88, Headbomb, PJtP, EdJohnston, WhaleyTim, Canarris, Seaphoto, CPWinter, Jayron32, Larrybaxter, Kiliman, Bongwarrior, VoABot II, Luctuosa, Stratford15, Mollwollfumble, Su-no-G, Robin S, Natsirtguy, T.vanschaik, Panguanwen, Gombang, Rominandreu, Brian Pearson, BigHairRef, Potatoswatter, MetsFan76, BrianScanlan, Kamparius, Signalhead, VolkovBot, TXiKiBoT, A4bot, BotKung, YohanN7, Caltas, Jimmycleveland, Proton666, Dabears36, ClueBot, Boing! said Zebedee, Erebus Morgaine, Jamespitt, Iohannes Animosus, SoxBot III, Gnowor, Addbot, DOI bot, WFPM, NjardarBot, Kyle1278, Tide rolls, Legobot, Luckas-bot, Yobot, AnomieBOT, Xqbot, GrouchoBot, Pepemonbu, Sumankumardutta, A. di M., Dave3457, Darkskynet, I dream of horses, Abductive, Tom.Reding, Achim1999, Gistmass, V.V.93, EngineerFromVega, Chriss.2, Techhead7890, Ibbn, John Cline, Mast3rj3di, Quondum, Xrayburst1, ClueBot NG, CocuBot, Wikihelperhelper, Silvrous, Lmfaoolmfaa, Radoslaw.c, Makecat-bot, DSCrowned, Khuzema Ali, Jocelyndurrey and Anonymous: 141

- **Neutron emission** *Source:* https://en.wikipedia.org/wiki/Neutron_emission?oldid=678062193 *Contributors:* Eric119, Angela, Nikai, Pstudier, Donarreiskoffer, Wikibot, Mike40033, Karol Langner, Icairns, Creidieki, Grm wnr, Randwicked, Vsmith, Roo72, Joanjoc~enwiki, RxS, Crazynas, Chobot, Tone, YurikBot, JWB, SmackBot, Man with two legs, Sbharris, Tsca.bot, DerHerrMigo, Lavateraguy, A876, Joeyfox10, Headbomb, Dirac66, CommonsDelinker, Sheliak, Wenli, Polyamorph, Nagika, WikHead, Addbot, Cesiumfrog, Spacy73, TaBOT-zerem, Thehelpfulbot, Minivip, Double sharp, K6ka, ZéroBot, Quondum, Voomoo, Mikhail Ryazanov, Latifahphysics, BG19bot, Glevum, Teddyktchan, MaddieeMariee, LinVee, LianaFoxx, InnaPhy and Anonymous: 20

- **Pauli exclusion principle** *Source:* https://en.wikipedia.org/wiki/Pauli_exclusion_principle?oldid=679741150 *Contributors:* Chenyu, CYD, Andre Engels, Graham Chapman, XJaM, PierreAbbat, Roadrunner, Maury Markowitz, Montrealais, Stevertigo, Michael Hardy, Tim Starling, EddEdmondson, Alan Peakall, Dominus, Shellreef, Graue, Ahoerstemeier, Stevenj, Glenn, Andres, Emperorbma, Charles Matthews, Wikiborg,

ElusiveByte, Fibonacci, Robbot, Owain, Modeha, Jheise, Tobias Bergemann, Enochlau, Giftlite, Harp, BenFrantzDale, HorsePunchKid, Gunnar Larsson, Karol Langner, Tsemii, Iwilcox, FT2, Bender235, RJHall, Kaszeta, Liberatus, Army1987, Smalljim, John Vandenberg, Cherlin, Obradovic Goran, Voltagedrop, RJFJR, Pol098, Isnow, Crucis, Graham87, ElCharismo, Enzo Aquarius, Eyu100, Jehochman, FlaBot, RexNL, Fresheneesz, JohnMarkStrain, Srleffler, Chobot, Sharkface217, DVdm, YurikBot, Wavelength, Wolfmankurd, RJC, JabberWok, Okedem, Ravindraia, Rsrikanth05, Wiki alf, Grafen, Ripper234, Enormousdude, RG2, Phr en, Mejor Los Indios, That Guy, From That Show!, SmackBot, Thorseth, David G Brault, Shai-kun, Georgelulu, Complexica, DHN-bot~enwiki, Canice, Gurevichar, Otis182, BTDenyer, Maximum bobby, Michalchik, Sadi Carnot, Vina-iwbot~enwiki, Andrei Stroe, DJIndica, WhiteHatLurker, Meco, Ambuj.Saxena, D Hill, T.O. Rainy Day, Wfructose, Blehfu, Zipz0p, JRSpriggs, Pseudospin, Xcentaur, Mattbr, Sohum, Vyznev Xnebara, Icek~enwiki, Cydebot, UncleBubba, Michael C Price, Gromonger-17, Headbomb, FourBlades, CharlotteWebb, Greg L, Escarbot, Gioto, Seaphoto, Orionus, Strami, Glennwells, CosineKitty, Wasell, Trugster, Mrfunkyostrich, Aryabhatta, Tanvirzaman, Animum, Dirac66, TristramBrelstaff, Custos0, Melamed katz, Maurice Carbonaro, Mrhsj, Cpiral, Katalaveno, P.wormer, Gombang, Zoedill, Ontarioboy, Keenman76, Sheliak, VolkovBot, John Darrow, Kriak, Nxavar, Anonymous Dissident, LeaveSleaves, UnitedStatesian, Pishogue, Spiral5800, Riick, AlleborgoBot, SieBot, Soler97, Adabow, Likebox, Henry Delforn (old), KoshVorlon, Iain99, Jakeng, Jonlandrum, Anchor Link Bot, ClueBot, Trojancowboy, Bryangv, Dvorsky~enwiki, Dlabtot, Chief buffalo chip, DragonBot, IEROslippersBRYAR, Brews ohare, Thingg, RQG, BodhisattvaBot, SilvonenBot, Quaint and curious, Addbot, Narayansg, CanadianLinuxUser, Chamal N, AgadaUrbanit, Lightbot, Zorrobot, Luckas-bot, Yobot, Heisenbergthechemist, IW.HG, AnomieBOT, Rubinbot, Materialscientist, Citation bot, ArthurBot, Xqbot, Nickkid5, Beelize23, GrouchoBot, Omnipaedista, Nathanielvirgo, ▢▢, Craig Pemberton, Cognitivelydissonant, Relke, RedBot, Tjlafave, Hickorybark, Halteres, Korepin, EmausBot, Ethereal-Blade, RA0808, H3llBot, Quondum, ChuispastonBot, Mikhail Ryazanov, ClueBot NG, Asalrifai, Helpful Pixie Bot, Bibcode Bot, Krishnaprasaths, BG19bot, GKFX, Alexander1102, BattyBot, JYBot, Tony Mach, Frosty, Serten, Septate, Yakamashi, Kfitzell29 and Anonymous: 197

- **Inverse beta decay** *Source:* https://en.wikipedia.org/wiki/Inverse_beta_decay?oldid=621637968 *Contributors:* Pcb21, Bearcat, Xerxes314, RussBot, Malcolma, 2over0, Jgwacker and Zedshort

- **Beta decay***Source:*https://en.wikipedia.org/wiki/Beta_decay?oldid=682280135*Contributors:*AxelBoldt, Chenyu, Trelvis, Mav, Peterlin~enwiki,Ellywa, Andrewa, Cyan, Mxn, Robertb-dc, Shizhao, Pstudier, Jusjih, Twang, Donarreiskoffer, Robbot, Enochlau, Giftlite, Donvinzk, Harp,Herbee, Xerxes314, Radius, Mdob, Antandrus, Icairns, Ukexpat, Jørgen Friis Bak, Discospinster, Guanabot, Vsmith, Roo72, Gianluigi,Joanjoc~enwiki, Neilrieck, Bobo192, Army1987, Drw25, Nk, Haham hanuka, AjAldous, Wtmitchell, Saga City, Falcorian, Flying fish,Eleassar777, Gimboid13, Graham87, Rjwilmsi, FlaBot, Ground Zero, Itinerant1, Goudzovski, Chobot, AllyD, Bgwhite, Roboto de Ajvol,YurikBot, JWB, Jimp, JabberWok, Romanc19s, Spike Wilbury, Johantheghost, Reyk, Geoffrey.landis, JLaTondre, MacsBug, SmackBot, In-cnis Mrsi, Ixtli, Gilliam, Skizzik, Dauto, Chris the speller, Octahedron80, Yurigerhard, Sbharris, Audriusa, Tsca.bot, Vladis1 av, Decltype,Akulkis, Hgilbert, Polonium, Adj08, JorisvS, Steipe, Mets501, Tuttt, Happy-melon, Civil Engineer III, Timrem, CRGreathouse, JohnCD,Rwflammang, Joelholdsworth, WeggeBot, Kanags, HPaul, Neil9999, Barticus88, Headbomb, Escarbot, AntiVandalBot, Edokter, LibLord,Salgueiro~enwiki, MSBOT, .anacondabot, VoABot II, SHCarter, Pixel ;-), Geekmansworld, Kevinmon, Johnbibby, Dirac66, Edward321,Geboy, Andre.holzner, Catmoongirl, Happyfacesrock, It Is Me Here, Rominandreu, Mcat2, KylieTastic, DorganBot, Y2H, Sheliak, Clubhouse, Milesisgreat, VolkovBot, Hqb, JhsBot, BotKung, Pishogue, FMasic, Cnilep, Tresiden, PlanetStar, Jasondet, Paolo.dL, Arjen Dijks-man, Sean.hoyland, Rjc34, Muhends, Sidhu ghanta, Loren.wilton, ClueBot, R000t, Maxtitan, EhJJ, Mikaey, SoxBot III, Directormq, RP459,SkyLined, Addbot, Yakiv Gluck, DOI bot, Man utd suger, Ayrenz, Mdnahas, Zorrobot, Spacy73, Skippy le Grand Gourou, आशीष भटनागर,Luckas-bot, Yobot, Fraggle81, TaBOT-zerem, THEN WHO WAS PHONE?, Azylber, Kulmalukko , AnomieBOT, SamuraiBot, Citation bot,Xqbot, FrescoBot, Tekmeme, Jonesey95, Achim1999, Minivip, ApusChin, Double sharp, Bj norge, Andrea105, John of Reading, Acather96,Dewritech, GoingBatty, JSquish, Δ, Coasterlover1994, Illinikiwi, ClueBot NG, Movses-bot, Widr, Meea, MerlIwBot, Bibcode Bot, BG19bot,ElphiBot, Onewhohelps, Cadiomals, Ragnarstroberg, Glevum, Currb, CeraBot, Idenshi, Goyala1, Stigmatella aurantiaca, ChrisGualtieri, Pvoy-tas, Monkbot, Raytuzio, Hyperclassic, Scipsycho, KasparBot and Anonymous: 174

- **Electron capture** *Source:* https://en.wikipedia.org/wiki/Electron_capture?oldid=681029045 *Contributors:* Mav, Andre Engels, Imran, GaryW, Pstudier, Twang, Donarreiskoffer, Gentgeen, Romanm, SpellBott, Mikez, Art Carlson, Dratman, Icairns, B.d.mills, Hax0rw4ng, Newhoggy, Discospinster, Vsmith, Sunborn, Joanjoc~enwiki, Brim, Foobaz, Riana, DV8 2XL, Forteblast, Richard Arthur Norton (1958-), Benbest, Rjwilmsi, Chobot, DVdm, Tone, YurikBot, Spacepotato, Hairy Dude, Shawn81, Shaddack, Anomalocaris, Dna-webmaster, LeonardoRob0t, Incnis Mrsi, Jagged 85, Betacommand, Sbharris, Vladis1av, BIL, Drphilharmonic, Daniel.Cardenas, Untitleduser, C.jeynes, Diverman, Magere Hein, Icek~enwiki, Michael C Price, Headbomb, Hcobb, Roches, Dirac66, LorenzoB, Vinograd19, AstroHurricane001, Howa0082, Yonidebot, Jutiphan, Vatic7, Sheliak, VolkovBot, TXiKiBoT, Pamputt, SieBot, YonaBot, Flyer22, ClueBot, Cmj91uk, SchreiberBike, Oldnoah, NellieBly, SkyLined, Debzer, Addbot, LaaknorBot, Zorrobot, Skippy le Grand Gourou, Luckas-bot, AnomieBOT, ArthurBot, Xqbot, GrouchoBot, FrescoBot, PigFlu Oink, Minivip, Miracle Pen, AndyHe829, MartinThoma, Bibcode Bot, Snow Rise, Eio, Zedshort, Paní Slepičková, JPBrod, Maysens, BsGTeo, Spyglasses, Meteor sandwich yum, Monkbot, Haveasweater, Jsaur, Wqwt, Alma.f.r, KasparBot and Anonymous: 53

- **Neutron electric dipole moment** *Source:* https://en.wikipedia.org/wiki/Neutron_electric_dipole_moment?oldid=682683284 *Contributors:* Michael Hardy, RJFJR, Rjwilmsi, Jimp, Reyk, Elonka, Alaibot, Headbomb, Cyclonenim, Leyo, Tadpole9, Useight, Brewcrewer, HumphreyW, Stefan Ritt, Addbot, SpellingBot, Morgrimm, TheTsax, Materialscientist, Citation bot, PVMNT, ▢▢, Rush8799, 1414rwbt, Citation bot 1, Tom.Reding, WikitanvirBot, Quondum, JohnnyLurg, ClueBot NG, Bibcode Bot, SoledadKabocha, RookJameson, Thenameis24, Monkbot, GZsigmond and Anonymous: 17

- **Neutron magnetic moment***Source:*https://en.wikipedia.org/wiki/Neutron_magnetic_moment?oldid=678984039*Contributors:*Herbee, Pol098 ,Rjwilmsi, BradBeattie, Kkmurray, SmackBot, Chris the speller, Bdushaw, Jonathan A Jones, Rotiro, Cydebot, DumbBOT, Headbomb, MarshBot, Desertsky85451, R'n'B, Rogermong2, Hersfold, Kropotkine 113, Azo bob, Mild Bill Hiccup, Excirial, Addbot, Poco a poco, AnomieBOT, Zad68, Rothbrad, Mnmngb, GliderMaven, RedBot, John of Reading, Quondum, QEDK, Barzo Naqishbandy, Bibcode Bot, ChrisGualtieri, Mark viking, Monkbot, ScrapIronIV and Anonymous: 16

- **Antineutron** *Source:* https://en.wikipedia.org/wiki/Antineutron?oldid=661667229 *Contributors:* Grendelkhan, Millosh, DocWatson42, Herbee, Icairns, Jh51681, Mike Rosoft, Pavel Vozenilek, MBisanz, Jag123, TenOfAllTrades, Redvers, Kbdank71, Benanhalt, Wrightbus, Chobot, YurikBot, Bambaiah, Jimp, Conscious, Dobromila, Salsb, Malcolma, SmackBot, Hmains, Bdushaw, Candamir, Aeluwas, Stikonas, Newone, Doug Weller, Zalgo, Thijs!bot, Headbomb, Altamel, STBot, Numbo3, Rod57, TXiKiBoT, Escalona, Spinningspark, SieBot, JerrySteal, Antonio Lopez, ClueBot, Djr32, BOTarate, Coopman86, MystBot, SkyLined, Addbot, LaaknorBot, 16dan44, Lightbot, Marimarina, Luckas-bot,

monk, Zonafan39, Mono, Michael9422, Extra999, HawkE65, Ioan Wynne-Jones, Zucchini7, Рулин, TjBot, Jpatros, EmausBot, Vanished user zq46pw21, Tommy2010, K6ka, Hhhippo, ZéroBot, Josve05a, MithrandirAgain, StringTheory11, Druzhnik, Medeis, Cline92, Lone St4lk3r, David J Johnson, Coolbob2422, KKPie, Inka 888, Zueignung, Mainy1996, Carmichael, JavinComi, ChuispastonBot, Robertschulze, Ζeta ζ, Czeror, Mhvk, Whoop whoop pull up, Madibootay, ClueBot NG, Gilderien, Frietjes, Jlattimer, Doctree, Helpful Pixie Bot, Calabe1992, Bibcode Bot, IzackN, BG19bot, Albert instine, Vicky.singh092, MusikAnimal, Exobiologist, FiveColourMap, Robert the Devil, Cadiomals, Shulse123, GregorDS, Thisdick1300794318, MrJohnnyMorales, Chowder98100, Zedshort, Doctorwhofan2013, Danijm, BattyBot, Cyberbot II, EuroCarGT, Wtrebla, G.Kiruthikan, Zacattack147, Reatlas, Randompoopcake, Rfassbind, Chris90nz, Sageattorney, Tentinator, Brobof, Nestrs, Raphael.concorde, Jwratner1, Johndric Valdez, Mfb, Fuckyourmomma2, Signoredexter, Monkbot, Raichu234352, Christometh, Sebgod, Steampunk09, AntHerder, SkyFlubbler, Poiuytrewqvtaatv123321, Tetra quark, Dcs2020, Rmh2020, Corvus-TAU, Znbn, Pulkitmidha, KasparBot, Ilikchese, Mosovon64 and Anonymous: 543

- **Neutron detection** *Source:* https://en.wikipedia.org/wiki/Neutron_detection?oldid=678984024 *Contributors:* Charles Matthews, Taxman, Omegatron, Topbanana, Sho Uemura, Karol Langner, H Padleckas, Forteblast, Jeff3000, Rjwilmsi, Williamborg, Kolbasz, Goudzovski, Anomalocaris, Syrthiss, Kkmurray, SmackBot, Fantasizer, Pfaff9, Bluebot, Sergio.ballestrero, KunalKathuria, Soarhead77, Jaeger5432, Vyznev Xnebara, Cydebot, Thijs!bot, Dougsim, Headbomb, Roggg, Liveste, Squids and Chips, Amog, AlleborgoBot, Malcolmxl5, Hoplon, Atif.t2, 718 Bot, Mcwescott, DumZiBoT, Naksatro, XLinkBot, Hiraku.n, Addbot, DOI bot, Luckas-bot, Yobot, Citation bot, George2001hi, Citation bot 1, DrilBot, Philémon Cyclone, Trappist the monk, RjwilmsiBot, Dewritech, ZéroBot, Storesund.hetland, ClueBot NG, Bibcode Bot, A2-33, BattyBot, ChrisGualtieri, KennethWeston, Nicholishiell, Monkbot, 2theta, Fahrenheit08 and Anonymous: 23

- **Neutron capture** *Source:* https://en.wikipedia.org/wiki/Neutron_capture?oldid=673059446 *Contributors:* Docu, Stone, Alexf, Karol Langner, Humblefool, D6, Vsmith, Nk, ACW, MattWade, RJFJR, Benbest, Pol098, Mandarax, Chobot, Tone, JWB, Limulus, Shaddack, GraemeL, Nekura, ThePromenader, A5b, HPaul, DumbBOT, M mattera, Uruiamme, Yellowdesk, Sophie means wisdom, Belg4mit, Pagw, Nono64, Leyo, Sheliak, Amikake3, Hqb, Sakkura, Meisterkoch, R000t, Saddhiyama, Gwguffey, Shinkolobwe, Dj manton, Chaosdruid, XLinkBot, Zinger0, Addbot, Cuaxdon, Fluffernutter, Favonian, Kein Einstein, Yobot, Ptbotgourou, AnomieBOT, Materialscientist, LilHelpa, MauritsBot, Biem, Ironboy11, Cs32en, 10metreh, RedBot, MastiBot, Minivip, Gryllida, Double sharp, Miracle Pen, RjwilmsiBot, 1947enkidu, AndyHe829, EmausBot, Wikipelli, ZéroBot, Aschwole, LikeLakers2, ClueBot NG, Ggabriel, Lanthanum-138, Helpful Pixie Bot, Neutronscattering, Kay Uwe Böhm, Kay Uwe Böhm 4, Kay Uwe Böhm 5, Kay Uwe Böhm 6, Daveturnr, Monkbot and Anonymous: 48

- **Neutron source** *Source:* https://en.wikipedia.org/wiki/Neutron_source?oldid=672869464 *Contributors:* Obok, Twanvl, Anonymous56789, Mkweise, Aarchiba, Kbk, Omegatron, Pstudier, Pakaran, Sanders muc, Wwoods, Just Another Dan, Karol Langner, H Padleckas, Iantresman, Sparky2002b, Moulding, CDN99, Bobo192, Vuo, DV8 2XL, Christopher Thomas, Josh Parris, Krash, Fivemack, Gurch, Kebes, Goudzovski, Phmer, Bhny, Shaddack, Welsh, Petri Krohn, CWenger, SmackBot, Dr.Science, Sbharris, FRocchi, Rwflammang, Stormwyrm, A876, Thijs!bot, Dtgriscom, Robina Fox, Rtcoles, STBot, Rod57, Sintaku, JsePrometheus, DumZiBoT, Addbot, Edoe, Ciphers, Marshallsumter, Obersachsebot, Thehelpfulbot, DrilBot, EmausBot, John of Reading, GoingBatty, Hhhippo, ZéroBot, Ὁ οἶστρος, MajorVariola, Circuitboardsushi, BG19bot, Neutronscattering, Morg00, Objuan Kanini and Anonymous: 30

- **Neutron generator** *Source:* https://en.wikipedia.org/wiki/Neutron_generator?oldid=674789119 *Contributors:* Gbleem, Kbk, Tempshill, Oaktree b, Art Carlson, Jonabbey, Niteowlneils, Rich Farmbrough, CDN99, Eric Kvaalen, Wtshymanski, RJFJR, Cfrjlr, Eleassar777, Christopher Thomas, Miken32, Rjwilmsi, Shaddack, CWenger, SmackBot, Dr.Science, Sbharris, Teveten, DabMachine, Cydebot, Thijs!bot, Nyq, Email4mobile, CardinalDan, Cyfal, ClueBot, Mild Bill Hiccup, Badgernet, Yobot, Materialscientist, Tollsjo, EmausBot, Ὁ οἶστρος, Wayne Slam, ClueBot NG, Millermk, Bibcode Bot, Comfr, ChrisGualtieri, Mogism, Izy ze Frog, Objuan Kanini, Skjain anita, FourViolas, Sfi llc and Anonymous: 45

- **Radioactive decay** *Source:* https://en.wikipedia.org/wiki/Radioactive_decay?oldid=682565072 *Contributors:* The Anome, Danny, Roadrunner, Mrwojo, Spiff~enwiki, Patrick, Ahoerstemeier, Andrewa, LittleDan, Kricke, Samw, Mxn, Smack, Hike395, Hollgor, Chuljin, Jitse Niesen, Audin, Furrykef, Populus, Omegatron, Topbanana, Pstudier, Finlay McWalter, PuzzletChung, Robbot, Romanm, Chancemill, Securiger, Merovingian, Pengo, Giftlite, Fudoreaper, Netoholic, Herbee, Everyking, Snowdog, Curps, Eequor, Jackol, Mmm~enwiki, Manuel Anastácio, Utcursch, Andycjp, LiDaobing, Antandrus, Beland, DragonflySixtyseven, Icairns, GeoGreg, Urhixidur, Syvanen, Olivier Debre, Deglr6328, Kate, Running, Mike Rosoft, Mormegil, Freakofnurture, Discospinster, Rydel, Rama, Vsmith, Mjpieters, Mani1, Night Gyr, Bender235, ESkog, Sunborn, Tompw, El C, J-Star, Lankiveil, Joanjoc~enwiki, Hayabusa future, RoyBoy, Orestes~enwiki, Grick, Bobo192, Stesmo, Smalljim, Indio~enwiki, Cohesion, Kjkolb, Nsaa, Storm Rider, Alansohn, Mr Adequate, AjAldous, Seans Potato Business, Ynhockey, Velella, Harej, RainbowOfLight, Dirac1933, Sciurinæ, Mikeo, DV8 2XL, Paraphelion, Zntrip, Ocollard, StradivariusTV, Duncan.france, Miss Madeline, CharlesC, Wdanwatts, Jacj, Qwertyus, Jclemens, Scuzzman, Martinevos~enwiki, Rjwilmsi, Jmcc150, Nneonneo, Bubba73, Watcharakorn, Lionelbrits, Ground Zero, Old Moonraker, RexNL, Kolbasz, Dalef, Fresheneesz, Guliolopez, Gwernol, Roboto de Ajvol, Wavelength, Phmer, Kymacpherson, RussBot, Jengelh, Shawn81, Kerowren, David Woodward, Gaius Cornelius, CambridgeBayWeather, Rsrikanth05, Bovineone, Tungsten, Grafen, Jaxl, Welsh, ONEder Boy, Ino5hiro, DJ John, Lomn, Scottfisher, DeadEyeArrow, Jeremy Visser, Ignitus, Wknight94, FF2010, Light current, Sefarkas, Closedmouth, Јованвб, Reyk, CharlesHBennett, CWenger, Fourohfour, Caco de vidro, Moomoomoo, Sbyrnes321, DVD R W, CIreland, Xtraeme, Eog1916, Itub, MacsBug, SmackBot, FocalPoint, Jclerman, Lcarsdata, Incnis Mrsi, KnowledgeOfSelf, Joonhon, Hydrogen Iodide, NoahWolfe, Jmulvey, Blue520, CMD Beaker, Jrockley, Yamaguchi🔲🔲, Gilliam, Carl.bunderson, TRosenbaum, Ati3414, Chris the speller, Bluebot, Kurykh, Agateller, Cadmium, MK8, Metacomet, Uthbrian, Reko, Sbharris, Rogermw, NYKevin, Can't sleep, clown will eat me, Ajaxkroon, Shalom Yechiel, Abyssal, Vladislav, Ioscius, KaiserbBot, Rrburke, VMS Mosaic, Rsm99833, Addshore, Mrdempsey, Megamix, Flyguy649, Smooth O, Xyzzy n, Dreadstar, -Ozone-, Lcarscad, Cockneyite, Drphilharmonic, DMacks, Where, Bidabadi~enwiki, Cyberevil, Lambiam, SuperTycoon, Sanya, JoshuaZ, Accurizer, Minna Sora no Shita, IronGargoyle, 16@r, Ryulong, Peyre, Squirepants101, Dan Gluck, BranStark, Pegasus1138, CP\M, Freelance Intellectual, Fdp, Tawkerbot2, Chetvorno, Bstepp99, Conrad.Irwin, INkubusse, Xcentaur, RSido, Vyznev Xnebara, Nunquam Dormio, Solargenerator9.5, MarsRover, Leujohn, Smoove Z, Myasuda, J. Tyler, Island Dave, Quinnculver, Kanags, Gogo Dodo, HPaul, Mad-rick, Rracecarr, Skittleys, Christian75, FastLizard4, Gmoney650, The real avenger, Mikewax, Thijs!bot, Epbr123, Plmoknijb, Dougsim, Headbomb, Marek69, Deschreiber, Davidhorman, Meteoritekid, FourBlades, Stannered, Mentifisto, AntiVandalBot, Quintote, Jj137, Panu Petteri Höglund, Hanzoro5, Myanw, JAnDbot, Arch dude, Andonic, Xact, Snowynight, Acroterion, Geniac, Freedomlinux, Bongwarrior, VoABot II, AuburnPilot, Hillgentleman, JNW, Estonofunciona~enwiki, DMcanada, Klausok, Pixel ;-), Colinsweet, SparrowsWing, Indon, Animum, Dirac66, 28421u2232nfenfcenc, LorenzoB, Tswsl1989, JoergenB, Squidonius, Lewismatson, Chuckwatson, NatureA16, MartinBot, Mermaid from the Baltic Sea, Bus stop, R'n'B,

Leyo, J.delanoy, Trusilver, Bogey97, Maurice Carbonaro, Cpiral, Gzkn, Stan J Klimas, DarkFalls, Dynetrekk~enwiki, Tarotcards, Pyrospirit, Sara0202, Chikinsawsage, Fountains of Bryn Mawr, Ohms law, Treisijs, Jim Swenson, Useight, Xiahou, RJASE1, Idioma-bot, ACSE, Cuzkatzimhut, Malik Shabazz, Deor, Matt1191, VolkovBot, ABF, VasilievVV, Philip Trueman, TXiKiBoT, Oshwah, Xenophrenic, Technopat, Hqb, Jcherbak, Someguy1221, Kirkpthompson, LeaveSleaves, Bearian, 0x539, Spiral5800, MichaelMorrill, Enigmaman, Yk Yk Yk, Bryan26, Synthebot, Falcon8765, Jluo, Sylent, Xxxlilbritxxx, Insanity Incarnate, Kehrbykid, Alytkin, Borne nocker, Brettdog, Deconstructhis, Starkrm, D. Recorder, Drawde22, SieBot, Tiddly Tom, Scarian, Viskonsas, Caltas, Soler97, Keilana, Nic92, TJHarrison, Oxymoron83, Faradayplank, Lightmouse, RW Marloe, Arnobarnard, Rj39pooch2, Nergaal, Babakathy, Martarius, ClueBot, HujiBot, Avenged Eightfold, GorillaWarfare, Fasettle, Bobathon71, Pvineet131, The Thing That Should Not Be, Plastikspork, VsBot, Wysprgr2005, Denna Haldane, Skäpperöd, CounterVandalismBot, Akash1209, Dougdp, MindstormsKid, Jersey emt, Opaltehjerkzors, Robert Skyhawk, Jusdafax, Erebus Morgaine, Huzzy92, 06multan, Arjayay, Radiogenic, PhySusie, Iohannes Animosus, Francisco Albani, IXella007, Dekisugi, La Pianista, Thingg, Aitias, Jonverve, Plasmic Physics, Megachad, Party, OpusAtrum, Johnson-gray, MystBot, Angerfist~enwiki, Thatguyflint, Hobbema, CalumH93, Amezcackle, Addbot, Proofreader77, Chorro22, Magus732, Smb6009, Laurinavicius, CanadianLinuxUser, Leszek Jańczuk, WFPM, Cst17, LaaknorBot, PranksterTurtle, Exor674, Lordlosss2, Tide rolls, Jarble, Legobot, Luckas-bot, Yobot, TaBOT-zerem, Legobot II, Theropod, Amble, Ayrton Prost, Hurricaneguy, AnomieBOT, DemocraticLuntz, Killiondude, Jim1138, Piano non troppo, AdjustShift, Scuzzer, Law, Materialscientist, The High Fin Sperm Whale, Citation bot, E2eamon, Bob Burkhardt, LilHelpa, Xqbot, Transity, Capricorn42, Richarddgill, Webkinzgirl101, Omnipaedista, RibotBOT, Amaury, Doulos Christos, Eugene-elgato, Pumpmaster60, FrescoBot, Surv1v4l1st, Wusel007, LucienBOT, Wvilhellm, Tobby72, Pepper, Oldlaptop321, MagnaGraecia, Footyfanatic3000, HJ Mitchell, Cannolis, Citation bot 1, Arthree, Pinethicket, Edderso, 10metreh, Odyssey xg, A8UDI, Minivip, Meaghan, Double sharp, TobeBot, Trappist the monk, Lotje, Ndkartik, TheBFG, Mozi17, Comet Tuttle, Math.geek3.1415926, Dinamik-bot, Vrenator, Tobias1984, Bluefist, Specs112, SilverbladeGR, Cfsgfds, Fastilysock, Cutelyaware, Sampathsris, Minimac, TjBot, TomBeasley, KuanRyan, Androstachys, Alison22, DASHBot, TGCP, BotdeSki, John of Reading, WikitanvirBot, Lunaibis, RedHab, ScottyBerg, Yt95, RenamedUser01302013, Kulmeetster, Wikipelli, K6ka, Sydneyanders, JSquish, ZéroBot, John Cline, PBS-AWB, Mkevinjnr, Suslindisambiguator, Elio96, Gz33, QEDK, Aschwole, L Kensington, MonoAV, Maschen, Donner60, Scientific29, ChuispastonBot, RockMagnetist, Ryan Pianesi, Newtrend19, Petrb, ClueBot NG, Crazyman121, Littleal38, Verpies, Satellizer, Baseball Watcher, Slartibartfastibast, Widr, Dasetwundabal, Oddbodz, Helpful Pixie Bot, Ciro612, Strike Eagle, Calabe1992, Bibcode Bot, Jeraphine Gryphon, BG19bot, Teiu88, Northamerica1000, Wiki13, ElphiBot, MusikAnimal, Cynaide, Shampa1, Flying hippo705, Glevum, DynamicDino, Adebish, Zedshort, Hamish59, Mgoelzer, SfHuIcTk, Thegreatgrabber, Achowat, Imawesome12345678910, ArrakisFrance, 555snowy, Kisokj, Ezekiel25q, Wolf11235, Cyprien 1997, BrightStarSky, Apples122, Ultimatewikimaster12345, Reatlas, Joeinwiki, Cavisson, Tentinator, Awesome boss 69 69, Bond064, Jyotmankad, CloudStrifeNBHM, Jwratner1, Applezpi3, Genome0514, StevenD99, Bkilli1, Ilikethemchickenwing$, Andthewinneris...Cole, Zane7777, Shbew, Monkbot, UDDM, Vieque, Thenapster1426, TheFireRises, Micbattle064, Paul2lyfe, Amortias, Pacifist peeta, Radioactiveisreallyawesome, KasparBot, Cerberus123, Never gonna See me, Lexi sioz, Subhajit07, Soumik Pattanayak, Bigdaddyyyyy69 and Anonymous: 804

- **Neutron activation** *Source:* https://en.wikipedia.org/wiki/Neutron_activation?oldid=677950403 *Contributors:* Andre Engels, Maury Markowitz, Aarchiba, DocWatson42, Poupoune5, Chad.netzer, Karol Langner, H Padleckas, Brim, Lectonar, Piotrr~enwiki, Mushin, JWB, Shaddack, David R. Ingham, CWenger, MaeseLeon, SmackBot, Jrockley, Bluebot, Ottawakismet, Sbharris, Giancarlo Rossi, Tiger99, Magere Hein, Novangelis, Rifleman 82, Thijs!bot, Headbomb, Uruiamme, SkoreKeep, CosineKitty, Penubag, Aleksander.adamowski, Pgrouse, Jcwf, Adam37, J 496, Dodger67, Martarius, Rdevany, PixelBot, ChrisHodgesUK, Addbot, Luckas-bot, AnomieBOT, LilHelpa, Biem, MastiBot, MagnInd, John of Reading, ClueBot NG, BG19bot, ChrisGualtieri, Cephas Atheos, JParker555, Graemem56 and Anonymous: 23

- **Neutron temperature** *Source:* https://en.wikipedia.org/wiki/Neutron_temperature?oldid=676379411 *Contributors:* Patrick, Julesd, Xanzzibar, Sho Uemura, Achurch, Smalljim, Kjkolb, Tpikonen, Forteblast, Pol098, Athelek, Bgwhite, Roboto de Ajvol, YurikBot, JWB, Limulus, Chichui, PotatoSamurai, Arthur Rubin, Petri Krohn, SmackBot, Dr.Science, Gaff, Gilliam, Kdliss, Chris the speller, Sbharris, Colonies Chris, Tsca.bot, Jmnbatista, Rwflammang, Phatom87, Saintrain, Thijs!bot, Montazmeahii, Alphachimpbot, Lklundin, Magioladitis, JaGa, NReitzel, Maurice Carbonaro, Rominandreu, Kamparius, Pamputt, MajorHazard, Southtown, Polyamorph, Nicker000, Lwnf360, Megiddo1013, Avm1, Allsvartr, Addbot, Hst1977, Lightbot, זהוריט55, Yobot, AnomieBOT, Abce2, Armando-Martin, Unkenruf, GoingBatty, RenamedUser01302013, Markmassie, Frietjes, Qbgeekjtw, Iankhou, Morethananumber, KennethWeston, Tony Mach, I am One of Many, Morg00, YiFeiBot, ARUNEEK and Anonymous: 45

- **Neutron diffraction** *Source:* https://en.wikipedia.org/wiki/Neutron_diffraction?oldid=666441199 *Contributors:* Andre Engels, Ahoerstemeier, Darkwind, Doradus, Zoicon5, Wikibot, Jorge Stolfi, Karol Langner, Sword~enwiki, Brim, Jag123, Benjah-bmm27, Vuo, Gene Nygaard, Rjwilmsi, Jaraalbe, Hellbus, PhilBentley, Eno-ja, Kipmaster, Kdliss, Bluebot, Smokefoot, Soarhead77, John, Uuhuụ, SimonD, Majora4, Gonzo fan2007, YK Times, Andyfaff, Appraiser, Lawrie.skinner, R'n'B, Andrewcwalters, Rod57, Jcwf, VolkovBot, Judge Nutmeg, Saber girl08, V81, Eco76, Nafradi, Polyamorph, Puppy8800, Addbot, Tomer shalev, Tassedethe, Lightbot, Romaioi, Pietrow, PV=nRT, Yobot, GoOhm, Xqbot, RibotBOT, Ludvig14, Sophus Bie, Mnmngb, Albris, FrescoBot, Royal Wulff, Tom.Reding, TobeBot, Marie Poise, RjwilmsiBot, John of Reading, Sp33dyphil, Hhhippo, ZéroBot, ClueBot NG, Metabolizer, Snotbot, X-men2011, Bibcode Bot, Neutronscattering, A2-33, Chrisanion and Anonymous: 42

- **Inelastic neutron scattering** *Source:* https://en.wikipedia.org/wiki/Inelastic_neutron_scattering?oldid=602278150 *Contributors:* Joachim Wuttke, Kkmurray, Chris the speller, Bluebot, Kukini, Paradoxsociety, Mkresch, Alaibot, D-rew, Squids and Chips, Polyamorph, Addbot, Paula Pilcher, Marie Poise, MagnInd, Neutronscattering, Jamesx12345, Pseudocubic and Anonymous: 11

- **Neutron tomography** *Source:* https://en.wikipedia.org/wiki/Neutron_tomography?oldid=679428827 *Contributors:* H Padleckas, Xezbeth, Arcadian, Mgiganteus1, Smith609, PamD, PDFbot, Yobot, Citation bot, Erik9bot, AManWithNoPlan, Dexbot and Anonymous: 1

- **Fast neutron therapy** *Source:* https://en.wikipedia.org/wiki/Fast_neutron_therapy?oldid=640537391 *Contributors:* Kku, Art LaPella, Arcadian, Wouterstomp, Avenue, Kolbasz, BorgQueen, SmackBot, Melchoir, Myrryam, Fvasconcellos, KNM, Gholson, Cydebot, Natalie Erin, Magioladitis, WolfmanSF, Rod57, Nigelloring, Steven3045, Hinder54321, Jamison 33, Lightmouse, Bgordski, Afernand74, ImageRemovalBot, Polyamorph, DumZiBoT, Heeero60, Addbot, Yobot, Wantdouble, Slightsmile, Snyderm, BG19bot, ChrisGualtieri, Polambda and Anonymous: 15

- **Neutron capture therapy of cancer** *Source:* https://en.wikipedia.org/wiki/Neutron_capture_therapy_of_cancer?oldid=678984010 *Contributors:* H Padleckas, Wouterstomp, Vuo, Kolbasz, Chris Capoccia, Gaius Cornelius, Tony1, Chris the speller, Sbharris, Hgrosser, DMacks,

- **Neutron capture nucleosynthesis** *Source:* https://en.wikipedia.org/wiki/Neutron_capture_nucleosynthesis?oldid=670143660 *Contributors:* BD2412, Rjwilmsi, 2over0, Headbomb, Disabling, Citation bot 1, Tom.Reding, BattyBot and Monkbot

- **Neutron bomb** *Source:* https://en.wikipedia.org/wiki/Neutron_bomb?oldid=682496609 *Contributors:* Trelvis, Mav, Bryan Derksen, Rmhermen, Ray Van De Walker, Maury Markowitz, Stevertigo, Patrick, Michael Hardy, Cyde, Theanthrope, Delirium, Minesweeper, Tregoweth, Ellywa, William M. Connolley, Mark Foskey, Julesd, Nikai, Evercat, Lommer, Rami Neudorfer, Arteitle, Vroman, Hashar, Guaka, Adam Bishop, Malcohol, Zoicon5, Shizhao, Pakaran, Denelson83, Riddley, Lowellian, Diderot, LGagnon, Wikibot, Victor, DocWatson42, Nunh-huh, Tom harrison, Everyking, Mboverload, Jorge1000xl, Gadfium, Jamougha, VoX, LiDaobing, LucasVB, Beland, Eroica, Karol Langner, Oneiros, Clarknova, Æ, MementoVivere, Flex, SYSS Mouse, N328KF, Supercoop, Rupertslander, Nharmon, Sunborn, Ben Webber, El C, RoyBoy, PatrikR, R. S. Shaw, Davidsmind, Schnolle, Still, Njaard, Joshbaumgartner, Axl, Rwendland, Hohum, Velella, Isaac, Max rspct, Pauli133, DV8 2XL, Axeman89, Crosbiesmith, Richard Arthur Norton (1958-), Woohookitty, GrouchyDan, Madmardigan53, CyrilleDunant, BillC, Jacobolus, Bbatsell, Chris Buckey, Kralizec!, Tmrobertson, RckmRobot, Ashmoo, Tovias, MC MasterChef, KaiMartin, Yuletide, Gordon Stangler, Amhaun01, Tajgenie, Gold Stur, Ttsalo, Ewlyahoocom, Gurch, Kolbasz, Diza, Ourboldhero, Chobot, Bgwhite, Cactus.man, Gwernol, YurikBot, Hairy Dude, Ineedbettername, Xihr, Fuzzy901, Raquel Baranow, Stephenb, C777, Gaius Cornelius, Daveswagon, NawlinWiki, Slarson, Flup, MakeChooChooGoNow, MrBark, Eyal0, Daniel C, Ninly, Arthur Rubin, Aurax, JoanneB, ThunderBird, Mikkow, Allens, Drcwright, SmackBot, Od Mishehu, Blue520, Clpo13, Jrockley, WayneConrad, Man with two legs, Gaff, Bluebot, Weeniemann, Persian Poet Gal, Thumperward, Hibernian, Mje, DevSolar, Gimeral, D97rolph, Ian01, EVula, Derek R Bullamore, Luís Felipe Braga, Giancarlo Rossi, Parrot of Doom, Will Beback, Jonnty, Esrever, Aaron Lawrence, Kuru, Alexcollins, CenozoicEra, Colin002, Fig wright, Mark Lungo, Tls, CompIsMyRx, Muadd, Therealhazel, Flaphead, QuilaBird, Hu12, Iridescent, Lord Anubis, Jookypipe, The Letter J, Chetvorno, SkyWalker, Wpmccray, CRGreathouse, CmdrObot, Carpenoctem, Kylu, Airport 1975, Abeg92, Besieged, UncleBubba, Gogo Dodo, Hanfuzzy, Otto4711, Give Peace A Chance, Photocopier, Tkynerd, Nabokov, Optimist on the run, Bob Stein - VisiBone, Omicronpersei8, Cancun771, Smiteri, Thijs!bot, Kubanczyk, LooseArrow, Keraunos, Moonshadow Rogue, Headbomb, Kinglink, Catsmoke, Sean William, Navdar, Slive~enwiki, Jj137, Zachwoo, ARTEST4ECHO, Mccollou, Ingolfson, JAnDbot, Fil-mex91, Avaya1, Magioladitis, Bg007, SHCarter, CodeCat, Gundato, Vreemdst, LittleOldMe old, Drewwiki, Thaurisil, Slow Riot, Andareed, Some Sort Of Anarchist Nutter, Matisia, Adamdaley, Cadwaladr, Пётр Петров, Blood Oath Bot, Jomomm, A.Ou, Sjosa2, Emeraldcrown, Philip Trueman, JayEsJay, TXiKiBoT, Chris-marsh-usa, Sintaku, IllaZilla, Tricky Wiki44, Foshowmo, LeaveSleaves, NKEISK, Doug, Wasted Sapience, NPguy, Tresiden, Scarian, WereSpielChequers, Phebot, Jon joy 1999, Digwuren, Mandsford, Anyeverybody, PlantTrees, Sfan00 IMG, ClueBot, Qsaw, Polyamorph, Niceguyedc, Justtryn2help, DragonBot, Nymf, Crywalt, Vivio Testarossa, Sun Creator, Casi233, Levent, John Paul Parks, DumZiBoT, Faulcon DeLacy, PL290, Wgwells, Thatguyflint, Scabbed Angel, Addbot, DavidNotDave, TutterMouse, Vanished user oerjio4kdm3, Shrogen, Fluffernutter, Innocent Byproduct, Yobot, AnomieBOT, Mintrick, T34CH, ArthurBot, .45Colt, KrisBogdanov, GrouchoBot, 图图图图, Mattg82, Stratocracy, Cekli829, FrescoBot, HAHAHAHATHATS4HAS, Siddharth 1999, Slastic, IVAN3MAN, Neurotip, Diannaa, Reach Out to the Truth, RjwilmsiBot, EmausBot, John of Reading, Boundarylayer, Peaceray, Wikipelli, K6ka, Illegitimate Barrister, A2soup, Canine virtuoso, SporkBot, Sailsbystars, ClueBot NG, Mesoderm, Zakblade2000, Rezabot, Oddbodz, Helpful Pixie Bot, Mbarland, The Mark of the Beast, Robert the Devil, TROPtastic, Nuke1st, Zedshort, Casimirck, Fixing the lie, Cutoffyourjib, Webclient101, Al Bestose, Jdc843, Jakec, Tototo30, Ajaythomas0007, Biblioworm, WC Jay, Guy Cox, Hairykrishna, KasparBot and Anonymous: 377

- **Neutron flux** *Source:* https://en.wikipedia.org/wiki/Neutron_flux?oldid=665564077 *Contributors:* Dmd3e, Robbot, Lowellian, Wereon, Mattflaschen, Bobblewik, Karol Langner, H Padleckas, Rich Farmbrough, TaintedMustard, Gene Nygaard, Linas, Benbest, Bluemoose, Wdanwatts, Radiant!, Eric Burnett, Physchim62, Bgwhite, Phmer, RussBot, Spike Wilbury, Mejor Los Indios, SmackBot, Jrockley, Basalisk, Will Beback, Robofish, Van helsing, Cancun771, Headbomb, CharlotteWebb, Roman à clef, Sophie means wisdom, Potatoswatter, Weavkd, Anyeverybody, Polyamorph, MystBot, Addbot, Yobot, Ptbotgourou, Paula Pilcher, FrescoBot, Mild fuzz, WikitanvirBot, ZéroBot, Sjjamsa, Bibcode Bot, Belromain, BG19bot, Stratoprutser and Anonymous: 15

- **Neutron transport** *Source:* https://en.wikipedia.org/wiki/Neutron_transport?oldid=678485972 *Contributors:* The Anome, Wapcaplet, Andycjp, Karol Langner, H Padleckas, Dryazan, Ottosix, Lectonar, Gene Nygaard, Linas, Oo64eva, Tone, Ntouran, SmackBot, Nkrupans, Theanphibian, Tomhubbard, Twitchax, Xwingchewie, Aeons, Dlohcierekim, WinBot, Dougher, R'n'B, BigrTex, Liveste, Fences and windows, Phebot, ClueBot, Hiraku.n, Addbot, Henriksjostrand, Yobot, Omnipaedista, Tmitchellloyd, Thehelpfulbot, FrescoBot, DexDor, Helwr, EmausBot, 2andrewknyazev, Nuceraccoon, Helpful Pixie Bot, BG19bot, Wbinventor, Jesse.johns, Jones.nuke and Anonymous: 28

34.8.2 Images

- **File:15-137-CircinusX1-XRayLightRings-NeutronStar-Chandra-20150624.jpg***Source:*https://upload.wikimedia.org/wikipedia/commons/6/60/15-137-CircinusX1-XRayLightRings-NeutronStar-Chandra-20150624.jpg*License:*Public domain*Contributors:*http://chandra.si.edu/photo/2015/cirx1/(image link)*Original artist:*X-ray: NASA/CXC/Univ. of Wisconsin-Madison/S. Heinz, et al.; Optical: DSS

- **File:2004_stellar_quake_full.jpg** *Source:* https://upload.wikimedia.org/wikipedia/commons/e/ed/2004_stellar_quake_full.jpg *License:* Public domain *Contributors:* http://www.nasaimages.org/luna/servlet/detail/nasaNAS~{}20~{}20~{}120777~{}227479:Stellar-Quakes *Original artist:* NASA

- **File:Alfa_beta_gamma_radiation.svg** *Source:* https://upload.wikimedia.org/wikipedia/commons/d/d6/Alfa_beta_gamma_radiation.svg *License:* CC BY 2.5 *Contributors:* Traced from this PNG image. *Original artist:* User:Stannered

- **File:Alpha_Decay.svg** *Source:* https://upload.wikimedia.org/wikipedia/commons/7/79/Alpha_Decay.svg *License:* Public domain *Contributors:* This vector image was created with Inkscape. *Original artist:* Inductiveload

- **File:Ambox_important.svg** *Source:* https://upload.wikimedia.org/wikipedia/commons/b/b4/Ambox_important.svg *License:* Public domain *Contributors:* Own work, based off of Image:Ambox scales.svg *Original artist:* Dsmurat (talk · contribs)

- **File:Ambox_wikify.svg** *Source:* https://upload.wikimedia.org/wikipedia/commons/e/e1/Ambox_wikify.svg *License:* Public domain *Contributors:* Own work *Original artist:* penubag

- **File:Atomic_rearrangement_following_an_electron_capture.svg** *Source:* https://upload.wikimedia.org/wikipedia/commons/b/b1/Atomic_rearrangement_following_an_electron_capture.svg *License:* CC BY-SA 4.0 *Contributors:* Own work *Original artist:* Pamputt

- **File:Neutristor_in_its_simplest_form.JPG**_Source:_https://upload.wikimedia.org/wikipedia/commons/b/ba/Neutristor_in_its_simplest. JPG _License:_ CC BY-SA 3.0 _Contributors:_ Own work _Original artist:_ Objuan Kanini

- **File:Neutristor_test_sample.jpg** _Source:_ https://upload.wikimedia.org/wikipedia/commons/3/3f/Neutristor_test_sample.jpg _License:_ CC BY-SA 3.0 _Contributors:_ Own work _Original artist:_ Objuan Kanini

- **File:Neutron.svg** _Source:_ https://upload.wikimedia.org/wikipedia/commons/d/dd/Neutron.svg _License:_ Public domain _Contributors:_ ? _Original artist:_ ?

- **File:NeutronCaptureTherapyImage.jpg** _Source:_ https://upload.wikimedia.org/wikipedia/commons/4/4c/NeutronCaptureTherapyImage.jpg _License:_ Public domain _Contributors:_ Scientific American _Original artist:_ Rolf F. Barth, Albert H. Soloway and Ralph G. Fairchild

- **File:NeutronPort.gif** _Source:_ https://upload.wikimedia.org/wikipedia/en/3/34/NeutronPort.gif _License:_ PD _Contributors:_ ? _Original artist:_ ?

- **File:Neutron_radiation_weighting_factor_as_a_function_of_kinetic_energy.gif** _Source:_ https://upload.wikimedia.org/wikipedia/ 5/5f/Neutron_radiation_weighting_factor_as_a_function_of_kinetic_energy.gif_License:_CC BY-SA3.0_Contributors:_Made this graph using OpenOffice based on data from reference documents_Original artist:_Ytrottier

- **File:Neutron_spin_dipole_field.jpg** _Source:_ https://upload.wikimedia.org/wikipedia/commons/1/15/Neutron_spin_dipole_field.jpg _License:_ CC BY-SA 4.0 _Contributors:_ Own work _Original artist:_ Bdushaw

- **File:Neutron_star_cross_section.svg** _Source:_ https://upload.wikimedia.org/wikipedia/commons/9/9e/Neutron_star_cross_section.svg _License:_ CC BY-SA 3.0 _Contributors:_ Own work _Original artist:_ Robert Schulze

- **File:Neutroncrosssectionboron.png** _Source:_ https://upload.wikimedia.org/wikipedia/commons/5/5c/Neutroncrosssectionboron.png _License:_ Public domain _Contributors:_ w:Image:Neutroncrosssectionboron.jpg _Original artist:_ wikipedia:en:user:Cadmium

- **File:Neutronstar_2Rs.svg** _Source:_ https://upload.wikimedia.org/wikipedia/commons/a/a3/Neutronstar_2Rs.svg _License:_ CC BY-SA 2.0 de _Contributors:_ File:Neutronstar 2Rs.png _Original artist:_ Derivative work: Mouagip

- **File:NuclearReaction.png** _Source:_ https://upload.wikimedia.org/wikipedia/commons/7/7d/NuclearReaction.png _License:_ CC BY-SA 3.0 _Contributors:_ Own work _Original artist:_ Michalsmid

- **File:Nuclear_Fission_Experimental_Apparatus_1938_-_Deutsches_Museum_-_Munich.jpg** _Source:_ https://upload.wikimedia.org/ /commons/2/23/Nuclear_Fission_Experimental_Apparatus_1938_-_Deutsches_Museum_-_Munich.jpg_License:_ CC BY-SA2.0_Contribu-tors:_ originally posted to_Flickr_as_Nuclear Fission Deutsches Museum_Original artist:_J Brew

- **File:Nuclear_Force_anim_smaller.gif** _Source:_ https://upload.wikimedia.org/wikipedia/commons/3/35/Nuclear_Force_anim_smaller.gif _License:_ CC BY-SA 3.0 _Contributors:_ Own work _Original artist:_ Manishearth

- **File:Nuclear_fission.svg** _Source:_ https://upload.wikimedia.org/wikipedia/commons/1/15/Nuclear_fission.svg _License:_ Public domain _Contributors:_ ? _Original artist:_ ?

- **File:Nuvola_apps_edu_mathematics_blue-p.svg**_Source:_https://upload.wikimedia.org/wikipedia/commons/3/3e/Nuvola_apps_edu_ blue-p.svg _License:_ GPL _Contributors:_ Derivative work from Image:Nuvola apps edu mathematics.png and Image:Nuvola apps edu mathematics-p.svg _Original artist:_ David Vignoni (original icon); Flamurai (SVG convertion); bayo (color)

- **File:Nuvola_apps_katomic.png** _Source:_ https://upload.wikimedia.org/wikipedia/commons/7/73/Nuvola_apps_katomic.png _License:_ LGPL _Contributors:_ http://icon-king.com _Original artist:_ David Vignoni / ICON KING

- **File:Office-book.svg** _Source:_ https://upload.wikimedia.org/wikipedia/commons/a/a8/Office-book.svg _License:_ Public domain _Contributors:_ This and myself. _Original artist:_ Chris Down/Tango project

- **File:Otto_Hahn_und_Lise_Meitner.jpg** _Source:_ https://upload.wikimedia.org/wikipedia/commons/2/2d/Otto_Hahn_und_Lise_Meitner.jpg _License:_ Public domain _Contributors:_ Search + 558596 • <a data-x-rel='nofollow' class='external text' href='http://research.archives.gov/ description/558596'>_Pioneering atomic physicist Ernest Rutherford in his laboratory, ca. 1925._ OPA at National Archives</a> <a data-x-rel='nofollow' class='external text' href='http://www.osti.gov/manhattan-project-history/images/meitnerhahnimage.htm'>_Lise Meit-ner and Otto Hahn, Kaiser-Wilhelm Institute, Berlin_</a> • "...the National Archives identifies the man as Ernest Rutherford, but other sources agree in labeling this a picture of Meitner and Hahn...". The U.S. DOE Office of History _Original artist:_ Unknown

- **File:PIA18848-PSRB1509-58-ChandraXRay-WiseIR-20141023.jpg**_Source:_https://upload.wiki58-ChandraXRay-WiseIR-20141023. jpg _License:_ Public domain _Contributors:_ http://www.nasa.gov/sites/default/files/pia18848-wisefacepalm.jpg _Original artist:_ NASA/CXC/SAO (X-Ray); NASA/JPL-Caltech (Infrared)

- **File:Periodic_Table_Stability_&_Radioactivity.png** _Source:_ https://upload.wikimedia.org/wikipedia/commons/c/c4/Periodic_Table_ %26_Radioactivity.png_License:_CC BY-SA2.5_Contributors:_https://commons.wikimedia.org/wiki/File:Periodic_Table_Radioactivity.svg_Original artist:_Alessio Rolleri(et al),Lexi sioz

- **File:Philippsburg2.jpg** _Source:_ https://upload.wikimedia.org/wikipedia/commons/e/e8/Philippsburg2.jpg _License:_ CC BY-SA 2.5 _Contrib-utors:_ Karlsruhe:Bild:Philippsburg2.jpg _Original artist:_ Lothar Neumann, Gernsbach [1]

- **File:Pierre_and_Marie_Curie.jpg** _Source:_ https://upload.wikimedia.org/wikipedia/commons/6/6c/Pierre_and_Marie_Curie.jpg _License:_ Pub-lic domain _Contributors:_ hp.ujf.cas.cz (uploader=--Kuebi 18:28, 10 April 2007 (UTC)) _Original artist:_ Unknown

- **File:Pn_Scatter_Quarks.svg** _Source:_ https://upload.wikimedia.org/wikipedia/commons/2/2e/Pn_Scatter_Quarks.svg _License:_ CC BY-SA 4.0_Contributors:_<a href='//commons.wikimedia.org/wiki/File:Pn_scatter_quarks.png' class='image'><img alt='Pn scatter quarks.png' src='https: //upload.wikimedia.org/wikipedia/commons/thumb/3/31/Pn_scatter_quarks.png/100px-Pn_scatter_quarks.png' width='100' height='64' srcset= 'https://upload.wikimedia.org/wikipedia/commons/thumb/3/31/Pn_scatter_quarks.png/150px-Pn_scatter_quarks.png 1.5x, https://upload. wikimedia.org/wikipedia/commons/thumb/3/31/Pn_scatter_quarks.png/200px-Pn_scatter_quarks.png 2x' data-file-width='810' data-file-height ='516' /></a>_Original artist:_Fred the Oyster